BLACKBERRIES
AND THEIR HYBRIDS

CROP PRODUCTION SCIENCE IN HORTICULTURE SERIES

Series Editor: Jeff Atherton, Professor of Tropical Horticulture, University of the West Indies, Barbados

This series examines economically important horticultural crops selected from the major production systems in temperate, subtropical and tropical climatic areas. Systems represented range from open field and plantation sites to protected plastic and glass houses, growing rooms and laboratories. Emphasis is placed on the scientific principles underlying crop production practices rather than on providing empirical recipes for uncritical acceptance. Scientific understanding provides the key to both reasoned choice of practice and the solution of future problems.

Students and staff at universities and colleges throughout the world involved in courses in horticulture, as well as in agriculture, plant science, food science and applied biology at degree, diploma or certificate level, will welcome this series as a succinct and readable source of information. The books will also be invaluable to progressive growers, advisers and end-product users requiring an authoritative, but brief, scientific introduction to particular crops or systems. Keen gardeners wishing to understand the scientific basis of recommended practices will also find the series very useful.

The authors are all internationally renowned experts with extensive experience of their subjects. Each volume follows a common format covering all aspects of production, from background physiology and breeding, to propagation and planting, through husbandry and crop protection, to harvesting, handling and storage. Selective references are included to direct the reader to further information on specific topics.

BLACKBERRIES AND THEIR HYBRIDS

Edited by

Harvey K. Hall, M.S.

Shekinah Berries Ltd, New Zealand

and

Richard C. Funt, PhD

The Ohio State University (Emeritus), USA

CABI

CABI is a trading name of CAB International

CABI	CABI
Nosworthy Way	745 Atlantic Avenue
Wallingford	8th Floor
Oxfordshire OX10 8DE	Boston, MA 02111
UK	USA

Tel: +44 (0)1491 832111	T: +1 (617)682-9015
Fax: +44 (0)1491 833508	E-mail: cabi-nao@cabi.org
E-mail: info@cabi.org	
Website: www.cabi.org	

British Library Cataloguing-in-Publication Data

A catalogue record for this book is available from the British Library.

Library of Congress Cataloging-in-Publication Data

Names: Hall, H. K. (Harvey K.), editor. | Funt, Richard C., editor.
Title: Blackberries and their hybrids / edited by Harvey K. Hall (M.S.), Richard C. Funt (Ph.D.).
Other titles: Crop production science in horticulture ; 26.
Description: Boston, MA : CABI, [2017] | Series: Crop production science in horticulture series ; 26 | Includes bibliographical references and index.
Identifiers: LCCN 2017017944 | ISBN 9781780646688 (pbk. : alk. paper) | ISBN 9781780646701 (epub)
Subjects: LCSH: Blackberries.
Classification: LCC SB386.B6 B53 2017 | DDC 634/.713--dc23 LC record available at https://lccn.loc.gov/2017017944

ISBN: 978 1 78064 668 8 (paperback)
     978 1 78064 669 5 (e-book)
     978 1 78064 670 1 (e-pub)

Commissioning editor: Rachael Russell
Editorial assistant: Alexandra Lainsbury
Production editor: Alan Worth

Typeset by AMA DataSet Ltd, Preston, UK

CONTENTS

DEDICATION

This book is dedicated to the amazing collection of researchers, growers, and marketers, who during the 19th and 20th centuries brought blackberries and their hybrids to millions of people around the world; and to those present and future generations, who continue to improve the blackberry, we present this manuscript. Specifically, we are indebted to plant breeders, plant pathologists, virologists, entomologists, plant scientists, agricultural economists, agricultural engineers, researchers, outreach specialists, agricultural editors, and to many others who have spent a lifetime working to improve the blackberry. Passionate people have produced nursery plants, conducted field trials, tested and sought chemical registration for use on this minor fruit crop. Many graduate students have contributed field and medical research projects and published their results in scientific publications and bulletins.

Agricultural engineers improved drip irrigation through the implementation of ultraviolet-resistant material, created uniform distribution of water (pressure compensating emitters) over long distances and hilly terrain, and developed daily watering schedules to optimize yields. They improved the effectiveness and efficiency of pesticide application equipment, as well as blackberry mechanical harvesters that can reduce the need for hand harvesting; and they have designed and built polythene tunnels for year-round production and sustained fruit quality, in spite of the vagaries of weather and climate variations.

Land Grant small fruit specialists tested insecticides, fungicides, and herbicides that could be safely used on blackberry plants and have worked closely with the USEPA (United States Environmental Protection Agency) for crop clearance of new complex chemicals and to provide safe working conditions for growers and their employees. The Land Grant System worked hand-in-hand with producers of quality fruit to supply a safe and affordable product in the market place for consumers to enjoy. And just as important are the scientists, who wrote vast amounts of new information in an understandable format for all to understand. Further, the communications staff at the Land Grant

System and USDA (United States Department of Agriculture) provided staff to edit and format the information for growers and the agricultural industry.

The authors of this book are especially indebted to the collaborative research and outreach from the laboratory testing and human clinical trials being conducted by medical biochemists, particularly since 2000. In this collaboration, there are horticulturists, food scientists, and human nutritionists, who have discovered that a diet containing blackberries can impact the human immune system and prevent certain cancers. This will have a great impact on human health in the 21st century.

We would be remiss if we did not include the many growers, nursery plant producers, and equipment company representatives who have walked the fields to solve problems, organized grower groups, winter grower meetings and university conferences, those who sought government funding to support virus-free and disease-free programs, and the medical investigations and trials that are benefiting the general public. Over the past 60 years, we have had the distinct pleasure of meeting and working alongside many of these people mentioned in this dedication.

It is now time to collectively bring together the vast amount of information from the laboratory, applied field studies, the medical clinical research, and the postharvest handling and marketing of a perishable commodity for the enjoyment and health benefits of the millions of people living on this planet. Therefore, the purpose of this book is to cover all aspects of the blackberry for students, growers, berry industry suppliers, and breeders of the 21st century, who will continue the effort to learn, produce quality fruit, provide agricultural supplies, and create a better, larger berry with improved flavor and increased levels of medicinal components.

Harvey K. Hall
Richard C. Funt
January 2017

Contributors

John R. Clark, PhD, Distinguished Professor, Department of Horticulture, 316 Plant Sciences Building, University of Arkansas, Fayetteville, Arkansas 72701, USA, e-mail: jrclark@uark.edu

Michael Duffy, PhD, Professor Emeritus, Department of Economics, 478E Heady Hall, Iowa State University, Ames, Iowa 50011, USA, e-mail: mduffy@iastate.edu

Chad E. Finn, PhD, Research Geneticist, USDA-ARS, Horticultural Crops Research Lab., 3420 NW Orchard Avenue, Corvallis, Oregon 97330, USA, e-mail: Chad.Finn@ars.usda.gov

Richard C. Funt, PhD, Professor Emeritus, Department of Horticulture and Crop Science, The Ohio State University, 2001 Fyffe Court, Columbus, Ohio 43210, USA, e-mail: richardfunt@sbcglobal.net

Gary Gao (Yu Igao Gao), PhD, Extension Horticulturist and Associate Professor, OSU South Centers, 1864 Shyville Road, Piketon, Ohio 45661, USA, e-mail: Gao.2@osu.edu

Harvey K. Hall, MS, Shekinah Berries, Ltd, 11 Ellesmere Close, Pyes Pa, Tauranga 3112, New Zealand, e-mail: shekinahberries@icloud.com

Eric Hanson, PhD, Professor, Department of Horticulture, 1066 Bogue St., Room A338, Michigan State University, East Lansing, Michigan 48824, USA, e-mail: hansone@msu.edu

Kim E. Hummer, PhD, Research Leader, USDA-ARS, National Clonal Germplasm Repository, 33447 Peoria Road, Corvallis, Oregon 97333-2521, USA, e-mail: Kim.Hummer@ars.usda.gov

Jungmin Lee, PhD, Research Food Technologist, USDA-ARS, Horticultural Crops Research Unit, Worksite, 29603 U of I Lane, Parma, Idaho 83660, USA, e-mail: Jungmin.Lee@ars.usda.gov

Robert R. Martin, PhD, Research Leader, Plant Pathology, USDA-ARS, Horticultural Crops Research Lab., 3420 NW Orchard Avenue, Corvallis, Oregon 97330, USA, e-mail: Bob.Martin@ars.usda.gov

Gail Nonnecke, PhD, University Professor, Morrill Professor, Department of Horticulture, Iowa State University, 105 Horticulture Hall, Ames, Iowa 50011, USA, e-mail: nonnecke@iastate.edu

Penelope Perkins-Veazie, PhD, Professor, Plants for Human Health Institute, Department of Horticulture Science, North Carolina State University, 600 Laureate Way, Suite 1321, Kannapolis, North Carolina 28081, USA, e-mail: Penelope_perkins@ncsu.edu

Sukalya Poothong, PhD, Lecturer, School of Agriculture and Natural Resources, University of Phayao, 19 Moo 2, Tambon Maeka, Muang District, Phayao 56000, Thailand, e-mail: sukalya_p@hotmail.com

Marvin Pritts, PhD, Professor, Horticulture Section, School of Integrative Plant Science, Cornell University, Ithaca, New York 14853, USA, e-mail: mpp3@cornell.edu

Barbara Reed, PhD, USDA-ARS, Research Plant Physiologist, Retired, National Clonal Germplasm Repository, 33447 Peoria Road, Corvallis, Oregon 97333-2521, USA, e-mail: reedba@onid.oregonstate.edu

David S. Ross, PhD, Professor Emeritus, Department of Environmental Science and Technology, University of Maryland, College Park, Maryland 20742, USA, e-mail: dsross@umd.edu

Annemiek C. Schilder, PhD, Associate Professor, Department of Plant Pathology, 578 Wilson, Rm. 105, CIPS, Michigan State University, East Lansing, Michigan 48824, USA, e-mail: Schilder@msu.edu

Michele Stanton, MS, Kenton County Extension Agent for Horticulture, 10990 Marshall Road, Covington, Kentucky 41015-9326, e-mail: michele.stanton@uky.edu

Bernadine C. Strik, PhD, Professor of Horticulture, Extension Berry Crops Specialist, Berry Crops Research Leader, NWREC, 4017 ALS, Department of Horticulture, Oregon State University, Corvallis, Oregon 97331, USA, e-mail: Bernadine.Strik@oregonstate.edu

Fumiomi Takeda, PhD, Research Horticulturist, USDA-ARS, Appalachian Fruit Research Station, 2217 Wiltshire Road, Kearneysville, West Virginia 25430, USA, e-mail: Fumi.Takeda@ars.usda.gov

Ellen Thompson, MS, Research and Breeding Director, Pacific Berry Breeding, LLC, 1611 Bunker Hill Way, Suite 200, Salinas, CA 93906, USA, e-mail: Ellen@pacificberrybreeding.com

Ioannis E. Tzanetakis, PhD, Professor, Department of Plant Pathology, Plant Sciences Building 213, University of Arkansas, 495 N. Campus Drive, Fayetteville, Arkansas 72701, USA, e-mail: itzaneta@uark.edu

Reviewer: **Shirley M. Funt, MEd**
Assistant Editor: **Alexandra Lainsbury**

PREFACE

Cultivated blackberries are a relatively new crop in the world as compared to other fruit crops. Historically, blackberries were harvested from the wild in Europe and North America. Now with new cultivars and cultural practices, they are grown and available around the world. Production regions have expanded internationally where innovated methods have shown much promise. With the evidence of human health benefits, blackberries are being consumed year round in North and South America, New Zealand, Asia, and Europe.

Breeders have accumulated germplasm that provides a wealth of traits for creating new cultivars and opportunities for production of blackberries and their hybrids. Fruit size, flavor, and longer storage capabilities have vastly improved the blackberry. Essentially, blackberry fruit can be available for year-round marketing. Further, the 'thornless' trait has enabled easier pruning and harvesting of berries by hand or machine and removed the risk of legal liability for having thorns in the final consumed product. Horticultural researchers have demonstrated profitable methods of growing blackberries in protected systems for extended seasonal harvest and for protection from winter cold. These systems have allowed the expansion of blackberries to regions where climate change has a significant effect on temperature, wind, rainfall, and pests.

Blackberries require climatic conditions that are conducive for production of quality fruit. Along with being produced on a good site, soils need to be well drained, fertile, and effectively prepared before planting. Clean water sources are required. New technologies, such as drip irrigation, fertigation, and machine harvest, are explained. The economics and business management of the blackberry firm is an important segment of production.

The authors in this book have used their talents to explain the many complex steps involved with producing a conventional or organic crop of blackberries for the fresh and processed market. This book is a complete source of information for students, growers, breeders, and nurserymen. Much information has been gathered from global sources and is appropriate for areas that can produce blackberries for the local, domestic, and/or export markets. Additional information for an in-depth understanding of all topics can be found in the references at the end of each chapter.

Turn the pages and increase your knowledge about blackberries and their hybrids.

Harvey K. Hall
Richard C. Funt
January 2017

Acknowledgments

The editors are indebted to the many people who have helped organize and develop this document into a body of information and knowledge to be available to all people worldwide. This includes the early development of the contents that created a flow from chapter to chapter through the appendices, which provide an in-depth level of information on new technologies, as well as the glossaries, which define scientific and economic terms to those who are not familiar with them. The authors wish to recognize Bernadine C. Strik, Chad E. Finn, John R. Clark, and Fumiomi Takeda for their assistance in writing, reviewing, and consultation on several chapters. Without the dedication of all of the authors this exemplary publication would not have been completed.

Also, we want to recognize the diligence and 'eagle eye' of Shirley M. Funt who edited, formatted, and captured many issues, such as converting SI and English units, consistency in the use of scientific terms, and the standardization of the references.

DISCLAIMER

The publishers, editors, and contributors have attempted to provide details of the latest production and postharvest methods, growing techniques, pest and disease control measures, and fertilizer management for blackberries, and will not be held liable for use or misuse of information provided here. The reader or user is urged to consult a local professional advisor on the suitability of application of products or management techniques in his or her region. Where the use of named (trade names) products or chemicals is described, no discrimination toward other products is implied or intended.

CONVERSION FACTORS FOR WEIGHTS AND MEASURES: EQUIVALENTS

	S.I. Units	U.S. Units
Length	1 millimeter (mm)	0.039 inch (in)
	1 centimeter (cm), 10 mm	0.39 inch (in)
	1 meter (m), 100 cm	39.4 inches (in)
	1 kilometer (km), 1000 m	0.62 mile (mi)
Area	1 square centimeter m^2	0.155 square inch (in^2)
	1 square meter m^2	1.2 square yards (yd^2)
	1 hectare (ha), 10,000 m^2	2.471 acres
	1 square kilometer (km^2), 100 ha	247.1 acres (0.386102 mi^2)
Weight	1 gram (g)	0.035274 ounces (oz)
	1 kilogram (kg), 1000 g	2.204623 pounds (lb)
	1 tonne (metric ton), 1000 kg	1.10231 tons (US)
Volume	1 milliliter (ml)	0.033814 fluid ounces (fl oz)
	1 liter (l), 1000 ml	1.056688 quarts (qt)
	1 cubic meter (m^3), 1000 l	264.172052 gallons (gal) (US)
	1 megaliter (ML)	0.810708 acre feet (ac ft)
Ratios	1 gram per m^2 ($g.m^{-2}$)	0.003277 ounce/ft^2
	1 gram per m^3 ($g.m^{-3}$)	0.0009988 ounce/ft^3
	1 tonne per hectare ($T.ha^{-1}$)	0.446090 tons/acre (t/ac)
	1 tonne per hectare ($T.ha^{-1}$)	892.179122 pounds/acre (lb/ac)
	1 gram per liter ($g.l^{-1}$)	0.008345 pounds/gallon (lb/gal)
	1 gram per liter ($g.l^{-1}$)	1 part per million (ppm)
Energy	1 joule (J)	0.0009478 btu
	1 kilojoule (kJ)	0.9478 btu
Temperature	°C (Celsius)	(°F − 32) ÷ 1.8

	U.S. Units	S.I. Units
Length	1 inch (in)	2.54 centimeters (cm)
	1 foot (ft), 12 in	30.48 cm
	1 yard (yd), 3 feet	0.9144 meters (m)
	1 mile (mi), 5280 feet	1.609344 kilometers (km)
Area	1 square inch, (in^2)	6.5 square centimeters (cm^2)
	1 square foot (ft^2), 144 in^2	930 (cm^2)
	1 square yard (yd^2), 9 ft^2	0.84 square meters (m^2)
	1 acre, 43,560 ft^2	0.405 hectares (ha)
	1 square miles (mi^2), 640 acres	259 ha (2.589988 km^2)
Weight	1 ounce (oz)	28.3 grams (g)
	1 pound (lb), 16 oz	0.454 kilograms (kg)
	1 ton (US) (t). 2000 lb	0.907 tonnes (metric tons) (T)
Volume	1 tablespoon (tbsp), 3 teaspoon (tsp)	14.79 milliliters (ml)
	1 fluid ounce (fl oz), 2 tbsp	29.57 ml
	1 cup (c). 8 fl oz	0.237 liters (l)
	1 pint (pt), 2 c	0.473 l
	1 quart (qt), 2 pt, 4 c	0.946 l
	1 gallon (US) (gal), 4 qt	3.8 l
	1 cubic foot (ft^3), 7.45 gal	28.3 l
Ratios	1 ton/acre (t.ac)	$2.241702 T.ha^{-1}$
	1 pound/acre (lb.ac)	$0.001121 T.ha^{-1}$
	1 pound/gallon (lb.gal)	$119.826427 g.l^{-1}$
	1 ounce.ft^2	$305.1517 g.m^{-2}$
	1 ounce/ft^3	$1001.1539 g.m^{-3}$
Energy	1 btu	1055.05585 joules (J)
	1 btu	1.05505585 kilojoules (kJ)
Temperature	°F (Fahrenheit)	$(1.8 \times °C) + 32°$

WEIGHTS AND MEASURES FOR BLACKBERRY FRUIT

	U.S. Units	S.I. Units
Volume of fruit	1/2 pint (6 oz)	170.5 g
	1 pint (12 oz)	341 g
	1 quart (24 oz)	681.6 g
Packaging of fruit	1 master (12 pints or 6 quarts in 2 layers)	4.0896 kg
	1 flat (12 pints or 12 half pints in one layer)	4.0896 kg or 2.0448 kg
	S.I. Units	U.S. Units Volume of fruit
Weight of fruit	125 g	4.4 ounces (oz)
	250 g	8.8 oz
	500 g	1 lb 1.6 oz
	1 kilogram (kg), 1000 g	2 lb 3.2 oz

Abbreviations (S.I. units): mm = millimeter; cm = centimeter; m = meter; km = kilometer; ha = hectare; mg = milligram; g = gram; kg = kilogram; ml = milliliter; l = liter; btu = British thermal unit
Abbreviations (U.S. units): in = inch; ft = foot; yd = yard; mi = mile; ac = acre; oz = ounce; lb = pound; t = ton; T = metric ton (tonne); tbsp = tablespoon; tsp = teaspoon; c = cup; pt = pint; qt = quart; gal = gallon

BLACKBERRIES: AN INTRODUCTION

Kim E. Hummer*

*U.S. Department of Agriculture, Agricultural Research Service,
National Clonal Germplasm Repository, Corvallis, Oregon, USA*

What was known as a favorite seasonal specialty fruit crop for many in the northern hemisphere can now be purchased in grocery stores throughout the year. What was historically gathered from the wild in Europe and North America has expanded into commercial cultivation and is now available globally.

Enhanced germplasm is providing a wealth of traits to create new growth and production opportunities. New cultivars have been developed with unusual qualities and great changes in crop production have occurred recently. In addition, production regions have expanded internationally in areas where innovative methods must be used to produce a crop for commercial production.

CLASSIFICATION AND DISTRIBUTION

Blackberries are members of *Rubus* subgenus *Rubus* (previously called subgenus *Eubatus*), while raspberries, their close relatives, are grouped in *Rubus* subgenus *Idaeobatus*. From a horticultural standpoint, each blackberry fruit is an aggregation of drupelets. Each drupelet is derived from one ovary that produces one hard-coated seed (pyrene). The seed is incased in a fleshy mesocarp with a surrounding exocarp (fruit skin). The drupelets are attached to a receptacle (torus). When the fruit is ripe, the drupelets remain attached to the receptacle, breaking away (dehiscing) from the stem (petiole) as a single unit ready for consumption. In contrast, the drupelets of raspberries separate from the receptacle. Raspberry drupelets are held together by small hairs and form a hollow, edible 'cap.'

This horticultural distinction of the fruit of blackberries and raspberries, while being an easily determined visual characteristic, is botanically arbitrary. This is because of the nature of species and cultivars within the genus, and the

---

* Corresponding author: Kim.Hummer@ars.usda.gov

great propensity of divergent groups to outcross, even those that differ in the number of chromosome sets. Many fruits that have a blackberry type fruit may contain raspberry genes in their evolutionary history, e.g. the western dewberry, *R. ursinus* (Alice and Campbell, 1999).

Blackberry plants grow in the temperate zone and have many forms. Most blackberries are 'armed.' That means that they have 'thorns,' or what botanists term 'prickles.' Besides having thorns or prickles on their stems and petioles, blackberries may also have bristles. Frequently, thorny blackberries also have thorns on the mid-rib on the underside of the leaves.

Blackberry plants can grow as trailing or upright plants or something in between. The trailing form, called 'sarmentous' by botanists, grows prostrate to the ground, similar to a grapevine. The prickles on the stem allow the blackberry vine to climb on other tree and shrub species and grow on top of them. In this case the vines are referred to as 'lians.'

For the other extreme, some blackberries have stout, thick stems and can grow upright to as much as twice a person's height, particularly in the first year of growth. This type of cane is called the primocane. Some blackberry plants have a growth habit in between prostrate and upright.

Since the time of the ancient Greeks, thornless blackberries were observed and described (Hummer, 2010; Hummer and Janick, 2007). Blackberries were cultivated in Europe for over 2000 years, initially for providing a protective barrier around areas of human habitation, then for medical purposes and for consumption. The first cultivated blackberry mentioned by name in European literature was the cut-leaf blackberry *R. laciniatus*, in Plukenet's Phytographia in 1691 (Jennings, 1988). This species was introduced into the Pacific Northwest of the USA and Canada before 1860 (Jennings, 1988).

The crowns of blackberries are perennial. In many blackberries the stems or canes that grow out of the crown are biennial: they grow one year, and die the second year. The first-year cane is called a primocane. In late summer, flower buds are initiated in a terminal raceme. A raceme has two qualities: (i) it is an arrangement of flowers on a stem where the youngest flower bud is at the tip; and (ii) the new flowers can keep growing out in an 'indeterminate' growth pattern. These primocanes and their buds overwinter. The buds receive chilling (between 4 and 6 weeks of temperatures near freezing) and break in the spring when the temperatures warm. The second-year cane is termed a 'floricane.' The flower buds on the floricane bloom and develop into fruit. The floricane dies at the end of the summer, after the blackberry fruits are harvested.

HOW MANY BLACKBERRY SPECIES ARE THERE?

More than 700 *Rubus* species have been described (Lu and Boufford, 2003). Focke prepared the first global monograph of *Rubus* (1910, 1911, 1914; see also 1894). He suggested 12 subgenera. Watson (1958) modified the subgenera and placed the blackberries into *Rubus* subgenus *Rubus*. This subgenus contains 12 sections (Table 1.1) of which species from most have been used in breeding programs to improve the cultivation of wild blackberry fruit.

The subgenus *Idaeorubus* describes a specific hybrid species: the presumptive hybrid *R. loganobaccus* L. H. Bailey, which was likely *R. idaeus* × *R. ursinus*. This subgenus represents one type of hybrid blackberry but by no means encompasses the diversity of raspberries *Rubus* subgen. *Idaeobatus* crossed with blackberry species *Rubus* subgen. *Rubus*.

Table 1.1. Subgenera, sections, and species of blackberries, raspberries, and hybrid berries.

Crop type	Subgenus	Section	Range	Some typical species
Blackberry	*Rubus*	*Allegheniensis*	Eastern North America	*R. allegheniensis* Porter, *R. alumnus* L.H. Bailey
		Arguti	Eastern North America	*R. argutus* Link, *R. frondosus* Bigelow, *R. pennsylvanicus* Poir.
		Caesii	Europe	*R. caesius* L.
		Canadenses	Eastern North America	*R. canadensis* L.
		Corylifolii	Scandinavia	*R. dumetorum* Weihe
		Cuneifolii	Eastern North America	*R. cuneifolius* Pursh
		Flagellares	Eastern North America	*R. flagellaris* Willd.
		Hispidi	Eastern North America	*R. hispidus* L.
		Persistentes	Eastern North America	*R. trivialis* Michaux
		Rubus	Europe and North America	*R. anglocandicans* A. Newton, *R. armeniacus* L., *R. axillaris* Lej., *R. fruticosus* agg.
		Setosi	Eastern North America	*R. setosus* Bigelow, *R. vermontanus* Blanch.
		Ursini	Western North America	*R. ursinus* Cham. et Schltdl.
Raspberry	*Idaeobatus*		Global	*R. idaeus* L. and many other species
Hybrid berry	*Idaeorubus*		Cultivated	*R. loganobaccus* L.H. Bailey

WHERE ARE BLACKBERRIES FOUND?

Rubus species occur on every continent except Antarctica, and are most diverse in the northern hemisphere. Many species occur in Europe and North and South America but only a few are found in Africa, Australia, and Oceana. Considerable phenotypic variation is found among the European blackberry species. Jennings (1988) refers to nearly 5000 Latin names having been recorded. He refers to these names as the tendency to recognize 'microspecies.'

HOW DOES THE LATIN NAME OF BLACKBERRY SPECIES AFFECT INTERNATIONAL TRADE?

Many people want to cultivate plants from different regions throughout the world. While this is great for new business, the challenge is to import clean plants, excluding novel diseases from the new production area. Quarantine regulations have been enacted by regional governments to insure that plants can move safely across country or regional borders without spreading diseases, Biosecurity regulations severely restrict or prohibit movement into some countries, especially Australia due to the 'weediness potential' of some species and some commercial cultivars. Plant pathologists report the diseases that have been observed in specific localities. Quarantine regulations are based on those reports. The report of a plant diseases includes the:

1. Latin name, usually genus species, of the plant affected by the disease, and
2. Locality where the disease occurs.

Plant health permits, such as plant import permits or phytosanitary certifications, are defined by quarantine regulations, and are written with those two particulars in mind. Thus, if a nursery wants to produce and ship plants across a border to another country or region, the plant species and origin must be known and specified. Plants grown in an area where diseases occur, or to be shipped to a country where certain diseases are absent, must be tested for or certified against the particular disease agents to be allowed entry.

The definition of species names for blackberries are problematic. Particularly, recent cultivars for this crop have complex pedigrees that include multiple blackberry species. Different cultivars have differing percentages of these species in their background. The botanists haven't kept up with the blackberry breeders and geneticists in describing new binomials for all possible blackberry species hybrid combinations. Because the backgrounds and pedigrees of new blackberry cultivars must be described in detail for intellectual property right protection, inventors have been creative with their reference species designation for the blackberry. A review (Table 1.2) of the USA Plant Patent database for blackberries shows that over the past 38 years, breeders have used differing binomials when referring to new blackberry cultivars, including:

Table 1.2. USA Plant Patent database for blackberries.

Number	Year	Patent title	Designated Latin name	Background or pedigree	Inventor(s)
PP25,864	2015	Blackberry plant named 'A-2312'	*Rubus* L. subgenus *Rubus*	Ark. Selection APF-1 x A-2002	Clark; John Reuben (Fayetteville, AR)
PP25,532	2015	Blackberry plant named 'Columbia Star'	*Rubus* L. subgenus *Rubus*	NZ 9629-1 x ORUS 1350-2	Finn; Chad E. (Corvallis, OR)
PP25,502	2015	Blackberry plant named 'DrisBlackSix'	*Rubus* L. subgenus *Rubus*	BF785-1 x Driscoll Cowles	Sills; Gavin R. (Gilroy, CA), Pabon; Andrea M. (Gilroy, CA), Moyles; Stephen B. (Soquel, CA)
PP25,433	2015	Blackberry plant named 'Black Jack'	*Rubus* spp.	Loch Ness x Sweet Peter (*Rubus* L. subgenus *Rubus*, including *R. allegheniensis* and *R. argutus*, the eastern upright blackberries)	Swartz; Harry J. (Oakland, MD), Vinson; Peter Edward (Hernhill, Faversham, Kent, GB)
PP24,878	2014	Blackberry plant named 'DrisBlackSeven'	*Rubus* L. subgenus *Rubus*	BJ111-2 x BH917-6	Sills; Gavin R. (Gilroy, CA), Mesa; Jose Maurilio Rodri (Michoacan, Mexico), Alcazar; Jorge Rodriguez (Texcoco, MX), Pabon; Andrea M. (Gilroy, CA)
PP24,701	2014	Blackberry plant named 'DrisBlackFive'	*Rubus* L. subgenus *Rubus*	BG837-2 x BH917-6	Sills; Gavin R. (Watsonville, CA), Rodriguez Alcazar; Jorge (Texcoco, MX), Rodriguez Mesa; Jose Maurilio (Michoacan, MX), Pabon; Andrea M. (Watsonville, CA)
PP24,609	2014	Blackberry plant named 'DrisBlackFour'	*Rubus* L. subgenus *Rubus*	Sleeping Beauty x BH917-6	Sills; Gavin R. (Watsonville, CA), Rodriguez Alcazar; Jorge (Texcoco, MX), Rodriguez Mesa; Jose Maurilio (Michoacan, MX), Pabon; Andrea M. (Watsonville, CA)

continued

Table 1.2. *continued.*

Number	Year	Patent title	Designated Latin name	Background or pedigree	Inventor(s)
PP24,298	2014	Blackberry plant named 'HJ-7'	*Rubus ursinus* Cham. et Schltdl.	Obsidian x Eaton	Johnson, Jr.; Harold A. (Aromas, CA), Johnson; Judith E. (Aromas, CA)
PP24,249	2014	Blackberry plant named 'APF-77'	*Rubus* L. subgenus *Rubus*	APF-12 x Arapaho	Clark; John Reuben (Fayetteville, AR)
PP23,725	2013	Blackberry plant named 'DrisBlackThree'	*Rubus* L. subgenus *Rubus*	Driscoll Carmel x Zorro	Sills; Gavin R. (Gilroy, CA), Pabon; Andrea M. (Gilroy, CA), Fear; Carlos D. (West Malling, England, UK)
PP23,497	2013	Blackberry plant named 'Reuben'	*Rubus* subgenus *Eubatus* sect. *Moriferi & Ursini* hybrid	A-2292T x APF-44	Clark; John Reuben (Fayetteville, AR), Fairlie; Jane (Spalding, GB)
PP23,270	2012	Blackberry plant named 'HJ-6'	*Rubus ursinus* Cham. et Schltdl.	Obsidian x Eaton	Johnson, Jr.; Harold A. (Aromas, CA), Johnson; Judith E. (Aromas, CA)
PP22,449	2012	Blackberry plant named 'APF-45'	*Rubus* L. subgenus *Rubus*	Ark. Selection APF-1 x APF-12	Clark; John Reuben (Fayetteville, AR)
PP22,358	2011	Blackberry plant named 'ONYX'	*Rubus* L. subgenus *Rubus*	Although blackberries (*Rubus* subgenus *Rubus*) are highly heterogeneous and outcrossing, and most clones contain genes from more than one species, the new cultivar and its progenitor lines phenotypically exhibit characters predominantly of the trailing western United States species, *Rubus ursinus* Cham. et Schltdl. (western trailing blackberry). *Rubus idaeus* L. (red raspberry), *R. armeniacus* Focke (Himalaya blackberry). *R. baileyanus* Britton, and *R. argutus* Link Porter (highbush blackberry) can all be found in the pedigree of this new cultivar.	Finn; Chad Elliott (Corvallis, OR)

PP22,002	2011	Blackberry plant named 'DrisBlackTwo'	*Rubus* L. subgenus *Rubus*	BH936-6 x Driscoll Cowles	Sills; Gavin R. (Watsonville, CA), Moyles; Stephen B. (Soquel, CA), Pabon; Andrea M. (Watsonville, CA)
PP20,891	2010	Blackberry plant named 'Natchez'	*Rubus* spp.	Ark. 2005 x Ark. 1857	Clark; John Reuben (Fayetteville, AR)
PP17,983	2007	Blackberry plant named 'Driscoll Thornless Sleeping Beauty'	*Rubus* L. subgenus *Rubus*	Developed from a spineless mutant of the patented cultivar 'Sleeping Beauty.'	Cabrera Avalos; Reynaldo (Jacona, MX)
PP17,162	2006	Blackberry plant named 'Ouachita'	*Rubus* sp.	Navaho x Arkansas selection 1506	Clark; John Reuben (Fayetteville, AR), Moore; James Norman (Fayetteville, AR)
PP16,989	2005	Blackberry–APF-12 cultivar	*Rubus* sp.	The new variety and its progenitor lines phenotypically exhibit characters predominantly of the erect eastern United States species, *Rubus allegheniensis* Porter (highbush blackberry) possibly introgressed with *R. argutus* Link. (tall blackberry).	Clark; John Reuben (Fayetteville, AR), Moore; James Norman (Fayetteville, AR)
PP15,788	2005	Blackberry – APF-8 cultivar	*Rubus* sp.	Its progenitor lines phenotypically exhibit characters predominantly of the erect eastern United States species, *Rubus allegheniensis* Porter (highbush blackberry) possibly introgressed with *R. argutus* Link. (tall blackberry).	Clark; John Reuben (Fayetteville, AR), Moore; James Norman (Fayetteville, AR)
PP15,058	2004	Blackberry plant named 'Driscoll Carmel'	*Rubus* L. subgenus *Rubus*	The variety is a complex *Rubus* hybrid, which can be characterized as an erect tetraploid with considerable *R. allegheniensis* background with other species such as *R. trivialis*, *R. argutus*, *R. procerus*, and *R. ulmifolius* also appearing in its background	Fear; Carlos D. (Aptos, CA), Sills; Gavin (Watsonville, CA), Cook; Fred M. (Aptos, CA), Harrison; Richard E. (Aptos, CA)

continued

Table 1.2. *continued.*

Number	Year	Patent title	Designated Latin name	Background or pedigree	Inventor(s)
PP14,935	2004	Blackberry plant named 'Clark Gold'	*Rubus trivialis*	Spontaneous mutation of the southern dewberry, *Rubus trivialis*, a wild blackberry species	Clark; John William (Palacios, TX)
PP14,780	2004	Blackberry plant named 'Driscoll Cowles'	*Rubus* L. subgenus *Rubus*	Sonoma x Loch Ness	Fear; Carlos D. (Aptos, CA), Sills; Gavin (Watsonville, CA), Cook; Fred M. (Aptos, CA), Harrison; Richard E. (Aptos, CA)
PP14,765	2004	Blackberry plant named 'Driscoll Eureka'	*Rubus* L. subgenus *Rubus*	Zorro x BY45.1	Fear; Carlos D. (Aptos, CA), Sills; Gavin (Watsonville, CA), Cook; Fred M. (Aptos, CA), Harrison; Richard E. (Aptos, CA)
PP14,682	2004	Blackberry plant named 'Driscoll Sonoma'	*Rubus* L. subgenus *Rubus*	Navaho x Hull Thornless	Carlos Fear
PP13,878	2003	Blackberry plant named 'Chesapeake'	*Rubus* hybrid	*Rubus argutus* x *R. cuneifolius* L.	Swartz; Harry J. (Laurel, MD), Fiola; Joseph A. (Keedysville, MD), Stiles; Herbert D. (Blackstone, VA), Smith; Brian R. (River Falls, WI)
PP13,759	2003	Blackberry plant named 'Zorro'	*Rubus* hybrid	The variety is a complex *Rubus* hybrid, which can be characterized as an erect tetraploid with considerable *R. allegheniensis* background with other species such as *R. trivialis, R. argutus* and *R. ulmifolius* also appearing in its background.	Carlos Fear

PP13,758	2003	Blackberry plant named 'Sleeping Beauty'	*Rubus* hybrid	The variety is described as a complex *Rubus* hybrid. It can be characterized as an erect tetraploid with considerable *R. allegheniensis* background with other species such as *R. trivialis*, *R. argutus*, *R. ulmifolius* and *R. procerus* also appearing in its background.	Carlos Fear
PP13,525	2003	Blackberry plant named 'Pecos'	*Rubus* hybrid	The variety is a complex *Rubus* hybrid, which can be characterized as an erect tetraploid with considerable *R. allegheniensis* background with other species such as *R. trivialis*, *R. argutus* and *R. ulmifolius* also appearing in its background	Carlos Fear
PP11,865	2001	Blackberry plant named 'Apache'	*Rubus* L. subgenus *Rubus*	Although blackberries (*Rubus* subgenus *Rubus*) are highly heterogeneous and outcrossing, and most clones contain genes from more than one species, the new variety and its progenitor lines phenotypically exhibit characters predominantly of the erect eastern United States species, *Rubus allegheniensis* Porter (highbush blackberry) possibly introgressed with *R. argutus* Link. (tall blackberry). Its genes for thornlessness were derived fim the British cultivar 'Merton Thornless' (non-patented), a derivative of *Rubus ulmifolius* Schott.	Clark; John Reuben (Fayetteville, AR), Moore; James Norman (Fayetteville, AR)
PP11,861	2001	Blackberry plant named 'Chickasaw'	no species given	Exhibits characters predominantly of the erect eastern United States species, *Rubus allegheniensis* Porter (highbush blackberry) possibly introgressed with *R. argutus* Link. (tall blackberry).	Clark; John Reuben (Fayetteville, AR), Moore; James Norman (Fayetteville, AR)

continued

Table 1.2. *continued.*

Number	Year	Patent title	Designated Latin name	Background or pedigree	Inventor(s)
PP9,861	1997	Blackberry – Kiowa cultivar	*Rubus* subgenus *Eubatus*	Although blackberries (*Rubus* subg. *Eubatus*) are highly heterogeneous and outcrossing, and most clones contain genes from more than one species, the new variety and its progenitor lines phenotypically exhibit characters predominantly of the erect eastern United States species, *Rubus allegheniensis* Porter (highbush blackberry) possibly introgressed with *R. argutus* Link. (tall blackberry).	Moore; James N. (Fayetteville, AR), Clark; John R. (Fayetteville, AR)
PP9,407	1995	'Everthornless' blackberry	*Rubus laciniatus* Willd.	*Ex vitro* somaclonal variant of the 'Thornless Evergreen' cultivar of *Rubus laciniatus* Willd.	McPheeters; Kenneth D. (Champaign, IL), Skirvin; Robert M. (Champaign, IL)
PP8,510	1993	Blackberry – Arapaho cultivar	no species given	Arkansas Selection 631 x Arkansas Selection 883	Moore; James N. (Fayetteville, AR)
PP8,423	1993	Blackberry-Douglass cultivar	no species given	Sander x Lawrence	Douglass; Bernard S. (Hillsboro, OR)
PP8,333	1993	Blackberry plant named Illini Hardy	no species given	Chester Thornless x NY 95	Skirvin; Robert M. (Champaign, IL), Otterbacher; Alan G. (Urbana, IL)
PP7,251	1990	Thornless blackberry named 'Per Can'	*Rubus canadensis*	A selection derived from *Rubus canadensis*; the cultivar is thus not a hybrid. The wild-growing Canadian parent was collected on the Appalachian plateau, in southern Quebec. This parent grows in thornless or nearly thornless populations.	Huber; Tony (Laval, Canada)

PP6,782	1989	Blackberry plant – Loch Ness cultivar	no species given	Two unnamed blackberry selections of complex parentage	Jennings; Derek L. (Dundee, Scotland, UK)
PP6,679	1989	Blackberry – Navaho Cultivar	no species given	Arkansas Selection 583 x Arkansas Selection 631	Moore; James N. (Fayetteville, AR)
PP6,678	1989	Choctaw Blackberry	no species given	Arkansas Selection 526 x Rosborough	Moore; James N. (Fayetteville, AR)
PP6,101	1988	Exel's everbearing blackberry plant	no species given	The plant was originated by my taking pollen from a flower of a Treeform blackberry and placing it on the stigmas of a flower of a Thornfree blackberry.	Smith; Exel R. (Green Forest, AR)
PP5,686	1986	Blackberry-Shawnee cultivar	no species given	Cherokee (non-patented) and Arkansas Selection 586 (non-patented)	Moore; James N. (Fayetteville, AR)
PP4,094	1977	Doyle's Blackberry	'wild blackberry,' no species name given	Growing in the same berry patch where the new variety was discovered were many Wild Blackberries, Tame Blackberries, Youngberries, Everbearing Raspberries, Boysenberries and Grapes. The new variety likely resulted from open pollination and is of unknown parentage. The new variety is believed to most nearly resemble the Wild Blackberry.	Doyle; Thomas E. (Washington, IN)

Rubus allegheniensis
Rubus canadensis
Rubus hybrid
Rubus spp.
Rubus fruticosis aggr.
Rubus trivialis
Rubus ursinus

Breeders in other countries have also used these binomials. The Latin binomials are appropriate for sport mutations or selections directly from the wild species but not for multiple species pedigrees. The name *R.* hybrid, is indistinct, and could be applied to any species cross within the genus, not necessarily to only blackberries species. The *R.* spp. or *R.* sp. choice implies that unknown species within the genus were the parents, and would not necessarily imply that the subject cultivar is a blackberry. The fourth, *R. fruticosis*, is a name invented by Linnaeus, and now refers to a group or aggregate of European blackberry species. Several countries, such as Australia and the United States, have specific noxious weed restrictions against '*R. fruticosus*,' so that reference may bring additional undesired constraints on a cultivar named in that category.

Recently another approach has been used for designating the complex taxonomy of blackberry cultivars. As previously mentioned (Table 1.1), blackberry species from Europe, Scandinavia, and eastern and western North America are included in *Rubus* subgenus *Rubus*. Breeders who have used a complex combination of those blackberry species have begun using this naming convention. Plant pathologists who are reporting diseases for blackberries are also using that convention.

The hybrid berries are an increasingly greater portion of the economically important blackberry type of fruit production. Breeders will continue to use available genetic flexibility to draw on the genes from both blackberries and raspberries to produce new cultivars adapted to broadening climate niches.

In this volume, reports on developments in the blackberry industry have been described by international experts in 17 chapters. This book provides a comprehensive yet concise reference for horticulture students, blackberry growers, producers, and fruit industry personnel looking for the latest production information. A summary of each section of the book follows.

CHAPTERS 2, 3, AND 4

These chapters explain growth and development, climatic requirements, and blackberry fruit quality components/health benefits. In Chapter 2, the physiology of the plant and its roots both for primocane and floricane production are described. The chapter details the mystery of primocane growth and development, cold hardiness and chilling requirements, the critical elements of flower bud development, and the complicated floricane development and fruiting.

As discussed in Chapter 3, blackberries as a species are widespread in western Europe and in the east and west of North America and with the spread of European colonization in the southern hemisphere. New cultivars have been developed for these conditions, and more recent advances in both cultural methods and varieties have allowed the expansion of blackberries into the subtropics and tropics, especially at higher elevations.

Blackberry fruit contains dietary fiber, vitamins, and minerals, along with phenolic metabolites that are likely to promote good health in humans (Chapter 4). The components, such as sugars and organic acids, anthocyanins, phenolic monomers and polymers, as well as additional quality components that are found in this fruit, are discussed. Despite the lack of a direct relationship between antioxidants and human health, the dietary contribution of blackberry fruit to the human diet is positive.

CHAPTERS 5, 6, AND 7

Chapter 5 augments previous detailed reviews of blackberry breeding with the latest releases and developments. European and North American programs initiated the crop development over the past century, however, the expansion of this crop in Asia, and Oceania, and the recent meteoric rise of Mexican and Central American production are also discussed. Breeding programs across the United States, Brazil, New Zealand, and elsewhere are noted. Selection of plants for specific architecture, and qualities such as thornlessness, primocane fruiting, chilling, disease resistance, yield, and genomic analysis are reviewed.

In Chapter 6, the clean plant program is discussed. Many pathogens infect *Rubus*. These diseases may kill plants or at the least reduce yields. New plantations should begin with certified pathogen-negative stocks. This section describes the procedure for establishing a clean foundation plant material and guidelines for blackberry nursery stock.

Traditional propagation methods are acknowledged but micropropagation protocols of blackberry species and cultivars, which are primarily used today, are defined in detail in Chapter 7. Nursery companies using these techniques produce millions of blackberry plants annually. Micropropagation protocols include several stages of plant growth: initiation, propagation, and rooting. This section describes the step-by-step procedures for micropropagation of *Rubus* plants. Recent adjustments to standard growth media and plant growth regulators are mentioned.

CHAPTERS 8, 9, 10, AND 11

While the majority of blackberry production occurs in a band of temperate latitudes and climates ranging from subtropical to Mediterranean, the expansion of future blackberry production is expected in locations previously considered impossible or inappropriate. Chapter 8, on site selection, describes new concepts of the soil, water, topography, exposure, layout, adaptation, climate, and socio-economic factors for potential new blackberry regions.

Chapter 9, site preparation, includes the soil details, physical, biological, and chemical properties, to consider in preparation for planting blackberries, as well as aspects to field layout, and planting. This information provides the key to successful planting in traditional blackberry growing regions.

For cultivation of blackberries, effective soil and water management (Chapter 10) is a key for plant health, growth, and productivity. Planning infrastructure development is important for long-term commercial success, and the correct functioning of a system designed to deliver water for blackberries both for growth and production. Careful choice of the soil for growing blackberries will assist quality production, plant health, and productivity.

Chapter 11 addresses the assessment of the nutrition status of soils for blackberry production. The levels of nutrients in the plants is also examined in detail, giving guidance on the desirable levels of each macro- and micronutrient, sampling methods, and interpretation of analyses. This will provide a framework for understanding fertilizer management for blackberry growers and for key personnel managing blackberry production on a commercial property.

CHAPTER 12

This chapter examines in detail the approach to pruning, training, and vegetative management in blackberries, differentiating between trailing types, semi-erect, and erect blackberries, including management for every-year and alternate year production and management of primocane fruiting types.

CHAPTER 13

This chapter examines the main pests and diseases of blackberries, as well as looking at weeds of blackberries and blackberries as weeds. The major fungal and bacterial pathogens are examined as are more than 30 viruses or virus-like diseases that infect *Rubus*. Frequently a complex of virus diseases infect plants in regions where blackberry are produced. This section describes significant viruses and suggests operational procedures protocols for preventing epidemics or reducing spread to new plantings.

CHAPTERS 14, 15, AND 16

Production systems for blackberries around the world are examined in Chapter 14. Machine harvest, hand harvest for fresh market, and systems for You-Pick operations are discussed. Protected culture and organic production methods are factors for the success of these operations.

The science of postharvest production of blackberries is complex (Chapter 15). Blackberries are a soft fruit, so maintaining maximum fruit quality is critical during transportation to major markets over long distances. Considerations, such as harvesting for fresh market, processing, training for hand harvesting, sanitation, packaging, cooling, and storage, are described in detail. The protocols for calculations for field heat, amount of cooling time, and refrigeration are provided. The section finishes with a discussion of the extension of shelf life.

Production protocols for production and supply of blackberries for marketing commercially are discussed in Chapter 16. Quality control and quality standards for presenting the fruit to market are important marketing factors.

CHAPTER 17

This chapter gives an overview of the decision making required in setting up and running a blackberry production operation. A carefully planned and executed enterprise will do much for the commercial success of a blackberry farm or production enterprise. An examination of risk mitigation is also discussed and draws attention to the many requirements for commercial success.

CHAPTER 18

In this chapter, the changing face of blackberry production is examined and the effects of innovation and novel varieties are evident in shaping the worldwide industry that is seen in the 21st century.

The appendices and glossaries provide additional information regarding blackberry culture.

REFERENCES

Alice, L.A. and Campbell, C.S. (1999) Phylogeny of *Rubus* (Rosaceae) based on nuclear ribosomal DNA internal transcribed spacer region sequences. *American Journal of Botany* 86(1), 81–97.

Focke, W.O. (1894) Rosaceae. In: Engler, A. and Prantl, K. (eds.) *Die natürlichen pflanzenfamilien* 3(3), 1–60. Abhandlungen Naturwissenschaftlicher Verein zu Bremen, Leipzig.

Focke, W.O. (1910) *Species Ruborum monographiae generis Rubi prodromus*. *Bibliotheca Botanica* 17, 1–120.

Focke, W.O. (1911) *Species Ruborum monographiae generis Rubi prodromus*. *Bibliotheca Botanica* 17, 121–223.

Focke, W.O. (1914) *Species Ruborum monographiae generic Rubi prodromus*. *Bibliotheca Botanica* 17, 1–274.

Hummer, K. (2010) *Rubus* Pharmacology: Antiquity to the Present. *HortScience* 45(11), 1587–1591.

Hummer, K. and Janick, J. (2007) *Rubus* iconography: antiquity to the Renaissance. *Acta Horticulturae* 759, 89–106.

Jennings, D.L. (1988) *Raspberries and Blackberries: Their Breeding, Diseases and Growth*. Academic Press, London.

Lu, L. and Boufford, D.E. (2003) *Rubus*. In: Wu, Z.Y., Raven, P.H. and Hong, D.Y. (eds.) *Flora of China* (Pittosporaceae through Connaraceae). Science Press, Beijing, and Missouri Botanical Garden Press, St. Louis, Missouri.

Watson, C.W.R. (1958) *Handbook of the Rubi of Great Britain and Ireland*. Cambridge University Press, Cambridge, UK.

2

GROWTH AND DEVELOPMENT

Bernadine C. Strik*

Oregon State University, Corvallis, Oregon, USA

INTRODUCTION

Blackberry (*Rubus* sp.) plants belong to the family Rosaceae and are closely related to the raspberry. This group or genus (*Rubus*) is collectively called 'brambles' in eastern North America, 'caneberries' in western North America and in other parts of the world simply 'blackberries' or 'hybrid blackberries.' Plants have an expected life of 15–50 years depending on the production region.

Blackberry plants typically have perennial root systems and crowns (plant base) and biennial canes. In 'thorny' cultivars, canes have spines, although spine density can vary considerably among cultivars. Thornlessness (spinelessness) may be a homogeneous genetic trait or it may be derived from a periclinal chimera (e.g. The Loganberry 'L654' and 'Thornless Evergreen'), where only the outer cell layer (L1) contains thornless genes. Genetically, thornless cultivars will not develop thorny (spiny) canes, whereas the others may develop thorny canes over time, especially 'Thornless Evergreen.' While blackberry plants are deciduous, many trailing cultivars in temperate climates do not lose all their leaves in winter. The leaves that do senesce (die) in trailing blackberries typically senesce at the leaf blade–petiole juncture rather than the petiole–cane juncture; this often leads to petioles left hanging on the cane (Fig. 2.1). In thorny cultivars, like 'Marion,' over-the-row machine harvesters equipped with brushing heads have been used to remove these petioles in winter reducing the potential for thorn contamination in machine-harvested fruit destined for the processed market (thorns are a major liability issue) (Strik and Buller, 2002).

In biennial-fruiting or 'summer-bearing' blackberries, normal flowering requires cessation of growth, bud dormancy, and sufficient chilling. The canes, in this case, are vegetative during the first year of growth (primocanes), and

---

* Corresponding author: Bernadine.Strik@oregonstate.edu

Fig. 2.1. Primocane about to break bud in late winter showing non-senescent petioles in 'Marion' blackberry (B. Strik).

flower, fruit, and then senesce in the second year (floricanes). Primocane- or annual-fruiting blackberry cultivars are different from floricane-fruiting cultivars, as they have no low temperature requirement for flower bud initiation or development (Strik, 2012). Primocane-fruiting cultivars produce fruit on the tip portion of the primocane or primocane branches in late summer through autumn. The portion of the cane that fruited senesces and is then typically removed by pruning (see Chapter 12). The base of this cane may be left in the field after pruning in winter and allowed to fruit in early summer of the second year (floricane). Producing a crop on the primocanes in autumn, followed by a crop on the floricanes the following year is called 'double-cropping' (see Chapter 12). Primocanes and floricanes exist together on mature biennial-fruiting plants and on double-cropped, primocane-fruiting plants. The extent of competition or support between the two types of canes varies with the type of blackberry grown and provides the grower with challenges unique to *Rubus* production.

There are generally three classifications or types of blackberries, based on cane architecture, including erect (e.g. 'Navaho', 'Ouachita'), semi-erect (e.g. 'Chester Thornless', 'Loch Ness'), and trailing (e.g. 'Marion', 'Obsidian', 'Columbia Star') cultivars (see Table 5.1 in Chapter 5). The three types ripen at different times of the season and require different pruning methods. Erect blackberry cultivars produce stiff, generally self-supporting primocanes, whereas the primocanes in trailing types typically grow along the ground. Erect blackberries produce primocanes from buds on the crown (at the base of floricanes) or from buds on roots (Fig. 2.2), whereas trailing and semi-erect types usually only produce new primocanes from buds on the crown (Fig. 2.3). Trailing and semi-erect blackberry cultivars are biennial fruiting, producing a crop only on the second-year floricane. However, erect blackberries include biennial- and annual- or primocane-fruiting cultivars. All types of blackberries include genetically thornless cultivars. Hybrids (hybridberry) between the red raspberry and blackberry are grown commercially (e.g. 'Boysen', 'Logan', and 'Tayberry'). These are included in the trailing blackberry types due to their growth habit.

The various blackberry types differ markedly in many characteristics, including growth habit, the timing of fruiting, cold hardiness, yield potential, and production systems. Vegetative and reproductive growth is influenced by

Fig. 2.2. Erect blackberry row in spring showing new primocanes emerging from buds on roots and the crown (foreground) (B. Strik).

Fig. 2.3. Trailing blackberry in spring showing new primocane growth at crown (B. Strik).

environment, particularly photoperiod and temperature, soil type, and cultural practices, such as fertilization, irrigation, and pruning.

ROOTS

The root system of blackberry plants is perennial. Blackberry roots are much more tolerant of heavy, more poorly drained soils than those of red or black raspberries. While over 70% of the raspberry root system is located in the top 30 cm (1 ft) of soil and the roots grow throughout much of the year (Atkinson, 1973; Colby, 1936), there has been little published on the morphology and timing of blackberry root growth to date. When 2-m-deep (6-ft) trenches were excavated alongside mature trailing blackberry plants in Oregon, blackberry roots as much as 2 m deep and extending as far as the row middle (center of aisle) were found; however roots were more concentrated in the upper 1 m (3 ft) of soil and in the row area (between plants) (Valenzuela *et al.*, unpublished). The deep root system of trailing blackberries does mean plants are able to source water at deep depths – this is why some growers in western Oregon are able to grow 'Marion' blackberry successfully without irrigation. Dixon *et al.* (2015) found no impact of withholding irrigation after fruit harvest on yield the following year.

The roots of erect blackberry cultivars have vegetative buds. These buds can produce new shoots (primocanes) in a new or established planting thus filling in the row or the space between established plants (Fig. 2.2). The roots of these cultivars can be used for vegetative propagation. As plants get older, many of the root buds remain dormant and the number of primocanes that emerge from the roots declines. In contrast, the roots of semi-erect or trailing cultivars contain few or no vegetative buds, and thus, do not produce primocanes from roots, but only from the crown.

Blackberry roots are strong sinks for nutrients and carbohydrates and these 'reserves' are important for bud break in the spring and for early-season floricane growth (Malik *et al.*, 1991; Mohadjer *et al.*, 2001; Naraguma *et al.*, 1999).

PRIMOCANE GROWTH AND DEVELOPMENT

The rate of primocane emergence and growth is dependent upon soil and air temperature, respectively. In primocane-fruiting blackberries, use of row covers from late winter through early primocane growth advanced the fruiting season by up to 2 weeks and increased yield (Strik *et al.*, 2012), but had no effect when placed over plants later in spring (Fernandez and Ballington, 2010; Strik *et al.*, 2012). The primocanes of many blackberry cultivars emerge in 'flushes' where a new group of primocanes will emerge in spring, significantly later than the first flush (Bell *et al.*, 1995a; Strik *et al.*, 2012; Takeda *et al.*, 2003). These primocanes are important for next year's yield in floricane-fruiting cultivars, especially where primocane suppression is used to limit interference of the first flush of primocanes with machine harvest (see Chapter 12). However, later flushes of primocane growth can interfere with hand harvest in primocane-fruiting blackberries and often do not produce fruit in temperate climates as they grow too late (Strik *et al.*, 2012). Delaying primocane growth by cutting all canes back to crown height has delayed fruiting or extended the season in primocane-fruiting blackberries (Thompson *et al.*, 2009) or increased cold hardiness in alternate year production of trailing blackberries (Bell *et al.*, 1995b).

The number of primocanes per plant (or per meter of row in erect types) is dependent upon plant age and health, but also on plant resources. For example, the presence of floricanes reduces primocane number and growth in trailing blackberries (Cortell and Strik, 1997a,b) and growing primocanes without floricanes present increases the subsequent cold hardiness of these canes (Bell *et al.*, 1995b; Cortell and Strik, 1997a); this response is used to advantage in alternate year production systems (see Chapter 12). In some erect cultivars (e.g. 'Natchez'), growers have observed reduced primocane growth when floricane growth and yield are particularly high. Primocane growth and the number of branches are also reduced in dense canopies (Strik *et al.*, 2012; Swartz *et al.*, 1984).

Erect and semi-erect blackberries produce primocanes that grow upright (strong apical dominance), with vigor dependent on cultivar and growing conditions (Fig. 2.4). Primocanes in these types of blackberries may reach heights of 3 m (10 ft) or more; however, they are typically summer pruned to induce branching (see Chapter 12). The primocanes of trailing blackberries are not self-supporting and after reaching about 1 m (3 ft) tall will fall over and grow along the ground (Fig. 2.5) unless they are caught in the canopy of the floricanes. Primocane length in trailing blackberries varies with cultivar, ranging from 2 to 6 m (6–20 ft). Canes do not form a terminal bud. In trailing blackberries, many primocanes often branch naturally near their base early in the growing season, thus producing strong basal branches. Branching will also occur when canes are trained onto the trellis in August ('August-training'; see Chapter 12).

Internode length depends on the blackberry type and cultivar and may be affected by growing environment and nutrition. At each node there is a compound leaf with 3 or 5 leaflets and a cluster of 1–4 mixed buds: a primary bud, a secondary bud, and one or more tertiary buds. The most developed of these, the 'primary,' typically breaks the following year, producing a fruiting lateral. However, the secondary and tertiary buds may also break (see 'Flower Bud Development,' below).

Some blackberry cultivars are adapted to higher temperatures than raspberry cultivars. For example, 'Arapaho' had a higher primocane leaf

Fig. 2.4. Upright primocane growth in 'Eureka', Watsonville, CA (B. Strik).

Fig. 2.5. 'Marion' trailing blackberry during harvest showing primocanes trailing along ground (B. Strik).

photosynthetic rate than most raspberry cultivars tested, particularly at high temperatures (35°C; 95°F); however, evapotranspiration or water loss did increase significantly from 20°C (68°F) to 35°C (95°F) (Stafne *et al.*, 2001). Many cultivars, particularly trailing ones, are not well adapted to regions that have hot, dry summers.

Primocanes will continue to grow in length until cold weather in the fall limits their development. In primocane-fruiting blackberries, primocanes or primocane branches will terminate with floral/fruit production (see 'Flower Bud Development,' below). In trailing and semi-erect blackberries, primocanes that remain on the ground or arch to the ground from tipped primocanes in the autumn will often root at their tip forming a new plant (tip layering).

In primocane-fruiting blackberries, sections of the primocane or branches that have developed fruit or flower buds will senesce naturally or after the first frost. These sections are commonly removed by pruning in winter (see Chapter 12).

COLD HARDINESS AND CHILLING

In the autumn, shortening day length and cooling temperatures slow the rate of primocane growth and lead to the onset of dormancy; this is called the acclimation phase. Not all blackberry cultivars lose all their leaves in autumn. Cultivars that have a long or late acclimation phase are more sensitive to cold temperatures in autumn than those that have a quick or early acclimation phase prior to going fully dormant. Fertilization with high rates of nitrogen, late fertilization with nitrogen (see Chapter 11), or late irrigation in some regions, may delay dormancy, increasing the risk of cold injury.

Once dormant, primocanes require a period of cold ('chilling requirement') before they start to deacclimate (leave dormancy). Prior to the chilling requirement being satisfied, a dormant plant cannot be made to grow, even when it's placed in a suitable environment. There is considerable range in chilling requirement among the blackberry types and cultivars grown, with a range of about 200–900 hours of chilling. Chilling is often estimated in the cumulative number of hours between 0 and 7°C (32–45°F) after plants are dormant; however, chilling models that are more complicated are often better predictors (Warmund and Krumme, 2005). In primocane-fruiting blackberries, once the primocanes enter dormancy, they have a chilling requirement for bud break on the floricane (Carter *et al.*, 2006). Plants that have received adequate chilling will grow (break bud) once conditions are favorable in the spring. Plants with insufficient chilling will have poor bud break on the floricane. In mild production regions, dormancy may be broken by treating primocanes with hydrogen cyanamide (Dormex®) or another growth stimulant. However, cultivars with a high chilling requirement are not well adapted to very warm climates. In contrast, cultivars with a low chilling requirement will likely be injured by cold temperatures in late winter in a cold climate. For example, 'Marion' blackberry, that has an estimated chilling requirement of 300 hours, often experiences cold damage from late-winter low temperatures when grown in Oregon (Bell *et al.*, 1992; Strik *et al.*, 1996b).

Blackberry cultivars differ in their absolute cold hardiness – the temperature threshold for damage to canes or buds when the plant is fully dormant. In general, erect, semi-erect, and trailing have the most to least cold hardy cultivars, respectively. However, a plant's susceptibility to cold injury not only depends on its genetic traits (e.g. 'Black Diamond' is more cold hardy than 'Marion') (Warmund *et al.*, 1986, 1992), but also on environmental (e.g. how quickly it got cold; whether warm temperatures preceded the cold spell; whether there was a lot of wind) and cultural factors that affect rate of growth or bud break (Warmund and George, 1990). Training time in trailing blackberries affects cold hardiness, likely a response to cane exposure to cold temperature and the impact of manipulating canes just prior to a cold temperature event (Bell *et al.*, 1992; Dixon *et al.*, 2015).

After the primocanes receive sufficient chilling and bud break occurs, the cane is then called a floricane. Scheduling floricane-fruiting blackberries for year-round production using artificial chilling and forcing in greenhouses is possible (Brennan *et al.*, 1999). Use of pruning techniques and application of chemicals to stimulate bud break and flowering of floricane-fruiting blackberries to extend the fruiting season is common in Mexico (Strik *et al.*, 2007) (see Chapter 12).

FLOWER BUD DEVELOPMENT

Annual- or primocane-fruiting blackberries will produce fruit on the tip portion of the primocane or primocane branches in the current season. The primocanes of this type of erect blackberry initiate flower buds after a short period of growth (Lopez-Medina *et al.*, 1999) and floral development is likely unaffected by day length (day-neutral). Chilling of root cuttings enhanced primocane growth and flowering (Lopez-Medina and Moore, 1999). However, the factors that influence flower bud initiation and development in primocane-fruiting blackberries are relatively unknown at this time. Still, factors which hasten primocane growth in spring, such as rate of nitrogen fertilization and air temperature, may advance the fruiting season. Similarly, in primocane-fruiting red raspberries, flowering will occur on primocanes in the absence of chilling (Dale *et al.*, 2005), but chilling improves and advances flowering (Carew *et al.*, 2001; Takeda 1993; Vasilakakis *et al.*, 1980). In addition, the proportion of the primocane that fruits in the autumn is affected by day length and air temperature (Carew *et al.*, 2003; Sønsteby and Heide, 2009). While summer tipping of the primocane delays harvest in primocane-fruiting raspberries (DeGomez *et al.*, 1986), this pruning practice has no impact on fruiting season in primocane-fruiting blackberries (Thompson *et al.*, 2009). When summer pruning primocane-fruiting blackberries, the buds below the point of tipping must still be vegetative – if they have already initiated into floral buds, yield will be greatly reduced by tipping (see Chapter 12).

In biennial- or floricane-fruiting blackberry cultivars, flower bud initiation occurs under short days and low temperatures. However, the time of flower bud initiation and pattern of development on the cane can vary with growing location and blackberry type or cultivar (Stanley *et al.*, 1999; Takeda *et al.*, 2002a,b). In 'Marion' and 'Boysen' trailing blackberries, time of flower bud initiation was first observed microscopically in November–December when sepal primordia were first evident; flower bud development continued through winter in Oregon. In Arkansas, flower bud initiation and development in the 'Cherokee' erect blackberry started later than in Oregon, due to colder average temperatures. 'Chester Thornless' semi-erect blackberry did not show evidence of flower bud development until spring (Takeda *et al.*, 2002a). While flower bud initiation and development proceeds from the tip of the cane to the base in red raspberries (Robertson, 1957), this is not as apparent in blackberries (Takeda *et al.*, 2002b).

Bud break and perhaps flower bud initiation may be resource limited in biennial-fruiting blackberries as it is in raspberries. Yield for August-trained 'Marion' was 45% greater than for February-trained plants, due to a greater percent bud break and more fruit per lateral with canes more exposed to light in autumn and winter (Bell *et al.*, 1995a). However, there was no effect of training time on bud break or yield of 'Marion' and 'Black Diamond' when grown in a certified organic production system (Dixon *et al.*, 2015). In trailing blackberries, shorter canes have a higher percent bud break than longer canes in 'Marion' (Bell *et al.*, 1995a).

Loss of primary buds on floricanes, either through winter cold injury (Strik, personal observation; Warmund and George, 1990) or pruning (Strik *et al.*, 1996b), leads to compensation from later emerging laterals from secondary buds and a delay in flowering and production. This effect has also been observed in erect blackberries in North Carolina (Shires, 2015). Flower bud development occurs later in secondary buds than in primary buds, and may extend from late winter into spring during growth of the floricane. In the semi-erect cultivar 'Loch Ness,' fruiting laterals from secondary buds could be forced by pruning the floricane, thus leading to a 'second crop' (Pitsioudis *et al.*, 2009).

Primocane removal during the fruiting season did not increase yield of the floricanes in 'Marion' (Strik, unpublished) likely because most primocane growth in trailing blackberries occurs after the fruiting season (Cortell and Strik, 1997a) and flower bud initiation occurs late in the season; also primo-canes are trained under the canopy of the floricanes (unlike in red raspberries where primocanes can shade the floricanes) (Fig. 2.5). However, re-cutting primocanes to crown height, in the non-production year of alternate year pro-duction to encourage a later flush of canes, reduced fruit per lateral if done too late (Bell *et al.*, 1995a). Primocanes are an important source of carbohydrates (Bell *et al.*, 1995b; Cortell and Strik, 1997b), nitrogen (Dixon *et al.*, 2016a,b; Harkins *et al.*, 2014; Malik *et al.*, 1991; Mohadjer *et al.*, 2001; Naraguma *et al.*, 1999) and other nutrients needed for fruiting lateral and fruit growth the following spring (Dixon *et al.*, 2016a,b; Harkins *et al.*, 2014).

FLORICANE DEVELOPMENT AND FRUITING

The percentage of bud break on the floricane varies by blackberry type, culti-var, plant vigor, and year. Blackberry canes rarely have 100% bud break. In trailing blackberries, percent bud break typically ranges from 20 to 75%, depending on cultivar, and is reduced in particularly long canes and with a high number of canes per plant; there is evidence that bud break is limited by resources in the cane (Bell *et al.*, 1995a; Cortell and Strik, 1997b; Dixon *et al.*, 2015).

The growth on the floricane at each node is called a fruiting lateral (Fig. 2.6). Blackberry plants normally produce only one fruiting lateral per node if a bud at that node breaks in spring. However, it is possible to have multiple buds break at a node (e.g. the primary and secondary buds break). Fruiting laterals

Fig. 2.6. Fruiting lateral growth in 'Marion' (B. Strik).

produce leaves and from 5 to more than 40 flowers, depending on the cultivar and production system (Takeda, 1987; Thompson *et al.*, 2007). The order of flower opening varies within an inflorescence or on the fruiting lateral leading to a range in ripening time. Laterals will differ in fruitfulness with location on the floricane. In trailing blackberries, laterals at the very tip of the cane are often shorter, with fewer flowers, than those in the mid- or basal-section of the cane. In erect or semi-erect blackberry cultivars, fruiting laterals that develop toward the base of the cane often have fewer flowers, perhaps a response of less light exposure during flower bud development. Lateral length is dependent on position on the cane, cultivar, and nitrogen nutrition.

Blackberry flowers and fruit have a similar morphology to raspberries. Flowers commonly have 5 sepals, 5 white or pink petals, and many stamens (male part) and pistils (female part) (Fig. 2.7). The pistils are arranged on a raised core in the center of the flower, called the receptacle. For pollination to occur, pollen must be released from the anthers (sacs at the end of each stamen) and be deposited on the tip of styles (attached to each pistil). Pollen transfer may occur within a flower (self-pollination) or between plants (within the same cultivar) or between cultivars (cross-pollination). Blackberry plants are self-fertile – good yield may be achieved when growing only one cultivar. Native or domesticated bees are important for pollination (transfer of pollen). Blackberry flowers are attractive to bees as they contain a high amount of nectar.

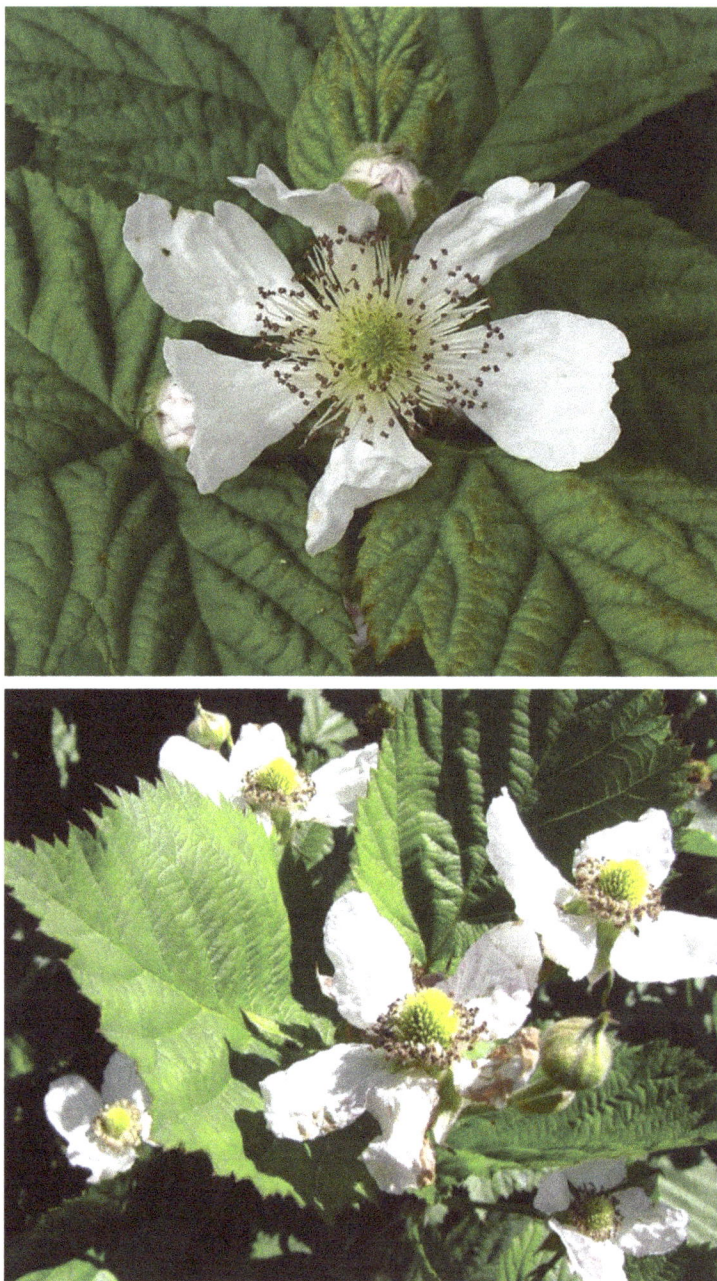

Fig. 2.7. *Top:* 'Marion' blackberry flowers (B. Strik). *Bottom:* 'Triple Crown' blackberry flowers (B. Strik).

Blackberry flowers contain as many as 250 pistils per flower, depending on the order of the flowering within the inflorescence or lateral and the cultivar (Strik *et al.*, 1996a). Pollen grains germinate after landing on the style and the pollen tube grows down the style. If the pollen grain reaches the ovary while the ovule is still receptive (during the effective pollination period), fertilization will occur. Each fertilized pistil/ovule will develop into a fleshy drupelet containing one seed (a pyrene). Erect and semi-erect blackberry cultivars produce fruit with relatively large pyrenes compared to those of trailing blackberries.

The percentage of ovules that set to become drupelets is often called drupelet set. Berries with a higher percentage of drupelet set are larger or weigh more (Strik *et al.*, 1996a) and have a more uniform berry shape. In addition to lack of pollination, a reduction in drupelet set may be caused by infection with *Raspberry bushy dwarf virus* (Strik and Martin, 2003), boron or copper deficiency which may reduce pollen tube germination or growth, and high temperature during bloom in some cultivars of primocane-fruiting blackberries (Stanton *et al.*, 2007). However, even under mild temperatures and adequate plant nutrition, percent drupelet set ranged from 50 to 75% depending on cultivar and flower order (Strik *et al.*, 1996a).

The 'berry' is an aggregate fruit consisting of many drupelets arranged on a central core or receptacle. In all blackberry fruit the fleshy receptacle or 'torus' separates from the plant when picked and is part of the fruit that is consumed. Hybridberry cultivars like 'Tayberry,' with 30% or more raspberry content, have a less fleshy receptacle and this may separate from the drupelets and float on the surface of jam or other processed products when cooked. The ease of fruit removal when ripe is cultivar-dependent, with those suited for machine harvest requiring an easy fruit removal. Drupelet size varies among cultivars (e.g. Strik *et al.*, 1996a) and may impact fruit firmness. While cultivars with larger drupelets may be softer than those with smaller drupelets, skin toughness is also an important trait related to shelf life or fresh market potential.

Blackberry fruits usually ripen from 35 to 60 days after pollination (Shires, 2015; Strik *et al.*, 2012; Thompson *et al.*, 2007). The rate of increase in size is not uniform. The fruit increase in size rapidly right after fruit set (cell division stage), then size expansion slows while the embryos are developing and the seed coat of each pyrene hardens (lag phase), followed by a rapid increase in size as each drupelet expands (cell expansion stage); growth thus follows an S-shaped curve.

Individual fruit typically weigh from 3 to 15 g. Primocane-fruiting blackberry plants growing under a tunnel tended to produce heavier fruit (32%, on average) than those grown in the open field (Thompson *et al.*, 2009). Maintaining good plant water status is important for maximizing fruit size, and growing blackberries in an environment with low humidity results in reduced berry size. Berry weight was reduced in 'Marion' and 'Black Diamond' trailing blackberry plants grown without weed control in the row (Dixon *et al.*, 2015; Harkins *et al.*, 2013). Plants that had competition from weeds also produced fruit with a lower water content, but high percent soluble solids. In well-watered 'Marion,' floricanes did not compete with primocanes for water (Bryla and Strik, 2008). However, water status of floricanes was lower during fruiting, perhaps due to greater resistance to water transport or accumulation of solutes during fruiting. In addition, xylem development may be poor in fruiting laterals limiting efficiency of water movement.

Many blackberry fruit are not at their optimum flavor until they change from glossy black to dull black in color. Acid content (mainly citric acid in blackberries) decreases during the later stages of ripening, especially in an environment with hot day and warm night temperatures, while sugar content increases. The sugar-to-acid ratio varies considerably among blackberry types and cultivars and may vary within a cultivar across different environments. For maximum productivity, flavor and sweetness, fruit must reach full maturity and full size before harvest. However, fruit firmness often decreases in the later stages of fruit maturation. Blackberry fruit are not considered climacteric because fruit respiration does not increase during ripening, even though ethylene is produced (Perkins-Veazie and Nonnecke, 1992; Walsh *et al.*, 1983). During fruit ripening, some cultivars are quite sensitive to either heat damage (softening or drying of drupelets) or sunburn/white drupelet, thought to be caused by excessive exposure to ultraviolet light (Fig. 2.8).

Fig. 2.8.   Drupelet damage in trailing blackberry caused by heat and/or ultraviolet light (B. Strik)

REFERENCES

Atkinson, D. (1973) Seasonal changes in the length of white unsuberized root on raspberry plants grown under irrigated conditions. *Journal of Horticultural Science & Biotechnology* 48(4), 413–419.

Bell, N., Nelson, E., Strik, B.C. and Martin, L. (1992) *Assessment of Winter Injury to Berry Crops in Oregon, 1991.* Agricultural Experiment Station Special Report, 902. Oregon State University, Corvallis, Oregon.

Bell, N., Strik, B.C. and Martin, L. (1995a) Effect of date of primocane suppression on 'Marion' trailing blackberry. I. Yield components. *Journal of the American Society for Horticultural Science* 120(1), 21–24.

Bell, N., Strik, B.C. and Martin, L. (1995b) Effect of date of primocane suppression on 'Marion' trailing blackberry. II. Cold hardiness. *Journal of the American Society for Horticultural Science* 120(1), 25–27.

Brennan, R.M., McNicol, R., Gillespie, T. and Raffle, S. (1999) Factors affecting out-of-season *Rubus* production. *Acta Horticulturae* 505, 115–120.

Bryla, D. and Strik, B.C. (2008) Do primocanes and floricanes compete for soil water in blackberry? *Acta Horticulturae* 777, 477–482.

Carew, J.G., Mahmood, K., Darby, J., Hadley, P. and Battey, N.H. (2001) The effects of low temperature on vegetative growth and flowering of the primocane fruiting raspberry 'Autumn Bliss' raspberry. *Journal of Horticultural Science & Biotechnology* 76(3), 264–270.

Carew, J.G., Mahmood, K., Darby, J., Hadley, P. and Battey, N.H. (2003) The effect of temperature, photosynthesis photon flux density, and photoperiod on the vegetative growth and flowering of 'Autumn Bliss' raspberry. *Journal of the American Society for Horticultural Science* 128(3), 291–296.

Carter, P.M., Clark, J.R., Drake Particka, C. and Yazzetti Crowne, D. (2006) Chilling response of Arkansas blackberry cultivars. *Journal of the American Pomological Society* 60(4), 187–197.

Colby, A.S. (1936) Preliminary report on raspberry root systems. *Proceedings of the American Society of Horticultural Science* 34, 372–376.

Cortell, J. and Strik, B.C. (1997a) Effect of floricane number in 'Marion' trailing blackberry. I. Primocane growth and cold hardiness. *Journal of the American Society for Horticultural Science* 122(5), 604–610.

Cortell, J. and Strik, B.C. (1997b) Effect of floricane number in 'Marion' trailing blackberry. II. Yield components and dry mass partitioning. *Journal of the American Society for Horticultural Science* 122(5), 611–615.

Dale, A. Pirgozliev, S., King, E.M. and Sample, A. (2005) Scheduling primocane-fruiting raspberries (*Rubus idaeus* L.) for year-round production in greenhouses by chilling and summer-pruning of canes. *Journal of Horticultural Science & Biotechnology* 80(3), 346–350.

DeGomez, T.E., Martin, L.W. and Breen, P.J. (1986) Effect of nitrogen and pruning on primocane fruiting red raspberry 'Amity'. *HortScience* 21(3), 441–442.

Dixon, E.K., Strik, B.C., Valenzuela-Estrada, L.R. and Bryla, D.R. (2015) Weed management, training, and irrigation practices for organic production of trailing blackberry: I. Mature plant growth and fruit production. *HortScience* 50(8), 1165–1177.

Dixon, E.K., Strik, B.C. and Bryla, D.R. (2016a) Weed management, training, and irrigation practices for organic production of trailing blackberry: III. Accumulation and removal of aboveground biomass, carbon, and nutrients. *HortScience* 51(1), 51–66.

Dixon, E.K., Strik, B.C. and Bryla, D.R. (2016b) Weed management, training, and irrigation practices for organic production of trailing blackberry: II. Soil and aboveground plant nutrient concentrations. *HortScience*, 51(1), 36–50.

Fernandez, G.E. and Ballington, J.R. (2010) Performance of primocane-fruiting experimental blackberry cultivars in the Southern Appalachian Mountains. *HortTechnology* 20(6), 996–1000.

Harkins, R.H., Strik, B.C. and Bryla, D.R. (2013) Weed management practices for organic production of trailing blackberry: I. Plant growth and early fruit production. *HortScience* 48(9), 1139–1144.

Harkins, R.H., Strik, B.C. and Bryla, D.R. (2014) Weed management practices for organic production of trailing blackberry: II. Accumulation and loss of biomass and nutrients. *HortScience* 49(1), 35–43.

Lopez-Medina, J. and Moore, J.N. (1999) Chilling enhances cane elongation and flowering in primocane-fruiting blackberries. *HortScience* 34(4), 638–640.

Lopez-Medina, J., Moore, J.N. and Kim, K.-S. (1999) Flower bud initiation in primocane-fruiting blackberry germplasm. *HortScience* 34(1), 132–136.

Malik, H., Archbold, D. and MacKown, C.T. (1991) Nitrogen partitioning by 'Chester Thornless' blackberry in pot culture. *HortScience* 26(12), 1492–1494.

Mohadjer, P., Strik, B.C., Zebarth, B.J. and Righetti, T.L. (2001) Nitrogen uptake, partitioning and remobilization in 'Kotata' blackberries in alternate year production. *Journal of Horticultural Science & Biotechnology* 76(6), 700–708.

Naraguma, J., Clark, J.R., Norman, R.J. and McNew, R.W. (1999) Nitrogen uptake and allocation by field-grown 'Arapaho' thornless blackberry. *Journal of Plant Nutrition* 22(4–5), 753–768.

Perkins-Veazie, P. and Nonnecke, G. (1992) Physiological changes during ripening of raspberry fruit. *HortScience* 27(4), 331–333.

Pitsioudis, F., Odeurs, W. and Meesters, P. (2009) Early and late production of raspberries, blackberries, and red currants. *Acta Horticulturae* 838, 33–37.

Robertson, M. (1957) Further investigations of flower-bud development in the genus *Rubus*. *Journal of Horticultural Science* 32(4), 265–273.

Shires, D. (2015) Impact of bud/lateral thinning on subsequent growth and yield of erect blackberry. Presentation, tour *XI International Rubus and Ribes Symposium*, Asheville, North Carolina, June 21.

Sønsteby, A. and Heide, O.M. (2009) Effects of photoperiod and temperature on growth and flowering in the annual (primocane) fruiting raspberry (*Rubus idaeus* L.) cultivar 'Polka'. *Journal of Horticultural Science & Biotechnology* 84(4), 439–446.

Stafne, E.T., Clark, J.R. and Rom, C.R. (2001) Leaf gas exchange response of 'Arapaho' blackberry and six red raspberry cultivars to moderate and high temperatures. *HortScience* 36(5), 880–883.

Stanley, C.J., Harris-Virgin, P.M., Morgan, C.G.T. and Snowball, A.M. (1999) Boysenberry primocane management for improved productivity. *Acta Horticulturae*, 505, 79–86.

Stanton, M.A., Scheerens, J.C., Funt, R.C. and Clark, J.R. (2007) Floral competence of primocane-fruiting blackberries Prime-Jan and Prime-Jim grown at three temperature regimes. *HortScience* 42(3), 508–513.

Strik, B.C. (2012) Flowering and fruiting on command in berry crops. *Acta Horticulturae* 926, 197–214.

Strik, B. and Buller, G. (2002) Reducing thorn contamination in machine-harvested 'Marion' blackberry. *Acta Horticulturae* 585, 677–681.

Strik, B. and Martin, R. (2003) Impact of Raspberry Bushy Dwarf Virus on 'Marion' blackberry. *Plant Disease* 87(3), 294–296.

Strik, B.C., Mann, J. and Finn, C.E. (1996a) Drupelet set varies among blackberry genotypes. *Journal of the American Society for Horticultural Science* 121(3), 371–373.

Strik, B., Cahn, H., Bell, N., Cortell, J. and Mann, J. (1996b) What we've learned about 'Marion' blackberry – potential alternative production systems. *Proceedings of the Oregon Horticultural Society* 87, 131–136.

Strik, B.C., Clark, J.R., Finn, C.E. and Bañados, P. (2007) Worldwide production of blackberries, 1995 to 2005 and predictions for growth. *HortTechnology* 17(2), 205–213.

Strik, B.C., Clark, J.R., Finn, C.E. and Buller, G. (2012) Management of primocane-fruiting blackberry – impacts on yield, fruiting season, and cane architecture. *HortScience* 47(5), 593–598.

Swartz, H.J., Gray, S.E. Douglass, L.W., Durner, E., Walsh, C.S. and Galletta, G.J. (1984) The effect of a divided canopy trellis design on thornless blackberry. *HortScience* 19(4), 533–535.

Takeda, F. (1987) Some factors associated with fruit maturity range in cultivars of the semi-erect, tetraploid thornless blackberry. *HortScience* 22(3), 405–408.

Takeda, F. (1993) Chilling affects flowering of primocane-fruiting 'Heritage' red raspberry. *Acta Horticulturae* 352, 247–252.

Takeda, F., Strik, B.C., Peacock, D. and Clark, J.R. (2002a) Cultivar differences and the effect of winter temperature on flower bud development in blackberry. *Journal of the American Society for Horticultural Science* 127(4), 495–501.

Takeda, F., Strik, B.C. Peacock, D. and Clark, J.R. (2002b) Patterns of floral bud development in canes of erect and trailing blackberry. *Journal of the American Society for Horticultural Science* 128(1), 3–7.

Takeda, F., Hummell, A.K. and Peterson, D.L. (2003) Primocane growth in 'Chester Thornless' blackberry trained to the rotatable cross-arm trellis. *HortScience* 38(3), 373–376.

Thompson, E., Strik, B.C., Clark, J.R. and Finn, C.E. (2007) Flowering and fruiting patterns of primocane-fruiting blackberries. *HortScience* 42(5), 1174–1176.

Thompson, E., Strik, B.C., Finn, C.E., Zhao, Y. and Clark, J.R. (2009) High tunnel vs. open field: Management of primocane-fruiting blackberry using pruning and tipping to increase yield and extend the fruiting season. *HortScience* 44(6), 1581–1587.

Vasilakakis, M.D., McCown, B.H. and Dana, M.H. (1980) Low temperature and flowering of primocane-fruiting red raspberry. *HortScience* 15(6), 750–751.

Walsh, C.S., Popenoe, J. and Solomos, T. (1983) Thornless blackberry is a climacteric fruit. *HortScience* 18(3), 482–483.

Warmund, M.R. and George, M.F. (1990) Freezing survival and supercooling in primary and secondary buds of *Rubus* spp. *Canadian Journal of Plant Science* 70(2), 893–904.

Warmund, M.R. and Krumme, J. (2005) A chilling model to estimate rest completion in erect blackberries. *HortScience* 40(5), 1259–1262.

Warmund, M.R., George, M.F. and Clark, J.R. (1986) Bud mortality and phloem injury of six blackberry cultivars subjected to low temperature. *Fruit Varieties Journal* 40(4), 144–146.

Warmund, M.R., Takeda, F. and Davis, G.A. (1992) Supercooling and extracellular ice formation in differentiating buds of eastern thornless blackberry. *Journal of the American Society for Horticultural Science* 117(6), 941–945.

3

CLIMATIC REQUIREMENTS

Fumiomi Takeda*

Appalachian Fruit Research Station, USDA-ARS, Kearneysville, West Virginia, USA

INTRODUCTION

The effect of several environmental factors on the above-ground parts of the blackberry plant, including seasonal changes, the climatic requirements, and the effects of light and temperature on blackberry plant survival and fruit quality, along with the author's own research (Takeda et al., 2008; Takeda et al., 2013; Takeda and Glenn, 2016), will be discussed in this chapter. Abiotic factors, such as temperature and water extremes, which can cause plant stress, are difficult to analyze individually because stress seldom occurs in the absence of some other stressor. However, separate effects of temperature extremes have been examined for blackberries under specific environmental circumstances that occur in the eastern USA (Takeda et al., 2013; Takeda and Glenn, 2016). These stressors occasionally cause permanent injury from which blackberry plants cannot recover, or they place a strain on the biological processes that produce chemical and/or physical changes that lead to loss of the plant's overall photosynthetic capacity and reduce economic yield (Warmund et al., 2008).

Weather, especially temperature, determines where blackberries can be grown (Crandall, 1995; Magness and Traub, 1941; Strik et al., 2007). High temperatures in the summer, usually associated with high solar radiation, contribute to stress-related plant responses such as poor pollination and white drupelet disorder (Takeda et al., 2013). In winter and spring, unseasonably low temperature can kill canes and damage flower buds and open flowers (Warmund et al., 2008). The low-temperature limit for blackberries is about −21°C (−6°F) in mid-winter for eastern thornless (USA) blackberries (Bushway et al., 2008; Kraut et al., 1986) and only −11°C (12°F) for trailing blackberries (Crandall, 1995). In the spring, temperatures ≤−3°C (27°F) during bloom can injure flowers and buds (Takeda and Glenn, 2016; Warmund et al., 2008). It is

---

* Corresponding author: Fumi.Takeda@ars.usda.gov

clear that damage related to freezing temperatures can occur during the fall, winter, and spring in many of the regions where blackberries are grown commercially.

With primocane fruiting (PF) blackberries, the signal for the development of flower buds, flowers, and fruit is physiological and occurs after the plant has reached the critical stage of growth, per a specific variety, allows differentiation to occur. If this occurs early enough in the growing season, the plant is able to carry fruit through to ripening and harvest within the current growing season. With flowering and fruiting occurring on current year's growth, in the case of PF blackberries, minimizing winger damage is less of a concern to growers.

In the case of floricane-fruiting (FF) blackberries, shortening days (hours of light) and the cool temperatures of autumn are needed for the differentiation of flower buds that will grow and produce fruit in the following season (Crandall, 1995; Takeda et al., 2002). At the same time, the canes begin to slow in growth and acquire cold hardiness. Excessive fertilization or irrigation in the latter half of the growing season promotes vegetative growth and interferes with the normal hardening-off process, making the plants more susceptible to cold injury even at $-7°$ to $-10°C$ ($19°–14°F$) (Crandall, 1995).

In northern climates, flower bud differentiation in FF blackberries occurs as early as late August and continues until mid-December (Takeda and Wisniewski, 1989; Takeda et al., 2002). At the same time, cane and bud hardiness is reached and the blackberry plant enters into a rest period or endo-dormancy. An extended period of time at temperatures below $-7°C$ ($19°F$) is needed to overcome the rest period. Once the chill requirement is met and favorable climatic conditions (e.g. $\geq -10°C$ ($14°F$)) are present, the blackberry plant can resume normal growth (Crandall, 1995).

Chilling requirement is cultivar dependent (Carter et al., 2006; Drake and Clark, 2000; Warmund and Krumme, 2005, 2008). A study in Arkansas elucidated the response to chilling in whole plants and stem cuttings. In using the stem-cutting technique, 'Kiowa,' 'Ouachita,' and 'Prime-Jim' needed 100–300 chill-hours (below $7°C$ ($45°F$)) to break rest with an intense bud break, while 'Arapaho,' 'Choctow,' and 'Shawnee' needed 300–600 hours (h) and 'Navaho,' 'Chicksaw,' and 'Apache' needed 700 h or more (Drake and Clark, 2000). The time of rest completion among blackberry cultivars has been estimated by Warmund and Krumme (2005, 2008). 'Kiowa' and 'Arapaho' blackberry had the shortest rest periods, while those for 'Shawnee,' 'Navaho,' and 'Chickasaw' buds were intermediate and those for 'Apache' and 'Darrow' buds needed the longest rest periods. This study elucidated that a chilling model which accounted for chilling inception temperature of $-2.2°C$ ($28°F$) was better at estimating rest completion in erect blackberries. They also suggested using a model that weighted temperatures between 0 and $9.1°C$ ($32–48°F$) differently from temperatures below and above this range for estimating rest completion.

COLD HARDINESS AND WINTER INJURY

Cold hardiness in blackberries is less than the hardy raspberry varieties developed in the northeast USA, Canada, and in continental Eastern Europe and Russia (Magness and Traub, 1941). Breeding for hardiness in New York State, produced the variety Darrow, which has deeper dormancy and requires more cold accumulation for chilling requirement to be satisfied. At the same time it is able to handle colder temperatures in the depths of winter in this region (Warmund and George, 1990). Breeding programs have resulted in the variety 'Illinois Hardy' from Illinois, 'Gazda' and other new cultivars from Poland, and the development of hardier genetics from Nova Scotia, Canada (Wojcik-Seliga and Wokcik-Gront, 2013). Among the five eastern thornless (USA) blackberries evaluated for cold hardiness, 'Smoothstem' was most susceptible, 'Hull Thornless' and 'Chester Thornless' were intermediate, and 'Dirksen' was the least susceptible to cold injury (Kraut *et al.*, 1986; Warmund and George, 1990; Warmund *et al.*, 1992). The time of floral bud differentiation varies in these cultivars; some develop floral meristem in the autumn, whereas in others, the floral differentiation is not observed until spring (Takeda and Wisniewski, 1989); however, these authors reported that differential thermal analysis experiments showed that both differentiated and undifferentiated buds exhibited exotherms when exposed to freezing temperatures and the amount of damage was not related to the stage of floral development (Warmund and George, 1990; Warmund *et al.*, 1992).

In a study conducted in Poland (Wojcik-Seliga and Wokcik-Gront, 2013), 13 cultivars of blackberries and their hybrids were evaluated for resistance to adverse winter conditions. Following a winter in which the minimum temperature dropped to −23°C (−9.4°F), 'Chester Thornless' and 'Black Satin,' two eastern thornless blackberries screened in Illinois and Ohio were the only cultivars that produced a satisfactory yield the following summer. All blackberry hybrids developed in western USA ('Boysenberry,' 'Kotata,' 'Loganberry,' 'Oregon Thornless,' and 'Silvan') and the United Kingdom (UK) ('Loch Ness,' 'Lock Tay,' and 'Tayberry') produced a small crop. These findings strongly suggested that cultivars developed in western USA and the UK lack the sufficient winter hardiness to escape the damaging effects of extremely low winter temperatures <−20°C (−9.4°F). Blackberry cultivars from the University of Arkansas breeding program were thought to possess a less or equivalent winter hardiness level compared to the eastern thornless cultivars from the United State Department of Agriculture (USDA) blackberry breeding program (Takeda *et al.*, 2013). For these reasons, commercial blackberry production has not been recommended for the northern part of the eastern USA (latitude 40–45°N), from Massachusetts in the east to the Midwest and Kansas and Oklahoma in the central part of the USA, where low winter temperatures of −25°C (−13°F) are common. In these areas, commercial blackberry production has been limited to small acreage and a pick-your-own market (Takeda,

2012). The main factor contributing to lack of commercial plantings in this region was either the lack of sufficient winter hardiness or poor post-harvest fruit quality in existing cultivars. However, in the early 21st century, there has been an expansion of blackberry production in this region. The increase in land area (150 ha (375 acre)), starting in 2011 is attributed to the combination of (i) the introduction of cultivars (e.g. 'Natchez' and 'Ouachita' from the University of Arkansas) with superior fresh-market quality suitable for long-distance transport, and (ii) a means for improved protection of blackberry plants from severe winter conditions with the use of the Rotating Cross-Arm (RCA) trellis and cane training system (Takeda and Peterson, 1999; Takeda *et al.*, 2008; Takeda, 2012; Takeda *et al.*, 2013) and the application of a rowcover on canes lowered close to the ground to improve winter survival of blackberry plants (Takeda *et al.*, 2008; Takeda and Phillips, 2011).

The use of the RCA trellis system has helped to mitigate the effect of severe cold, winter temperatures where partial to full crop losses have been reported to occur in three out of four years. The details of the RCA trellis and the specific cane training protocol for this trellis are discussed in Chapter 12 (Takeda, 2012; Takeda and Peterson, 1999). When the primocanes are trained properly (Takeda *et al.*, 2003a,b), the rotation of the articulating cross-arm can be performed with little effort and cane breakage and allows the canopy to be positioned close to the ground. With the canopy close to the ground, a winter row cover can be applied on top of the canopy (Takeda, 2012; Takeda *et al.*, 2013). There is an economic impetus to growing blackberries in a northern Midwest region (e.g. Illinois, Kansas, and Iowa, USA), where winter temperatures as low as $-26°C$ ($-15°F$) are recorded annually, and the potential for winter injury is high (Takeda *et al.*, 2008). If winter injury is avoided, then the fruit on surviving floricanes ripens later than anywhere else in the USA. The growers see a financial gain from harvesting in late summer and early fall, when the price for fresh blackberries at the Chicago Terminal Market is approximately 200% greater than in June (USDA AMS, 2015).

SPRING FROST

In April 2007, the Midwest, central and southern plains and southeast portions of the USA experienced a record-breaking freezing event that caused unprecedented damage to many economically important crops, including blackberries (Warmund *et al.*, 2008). The cold event experienced across much of these regions was an advective type of freeze. Advective freezes are characterized by the movement of a large-scale cold air mass into a region with freezing temperatures and a relatively low dew point (Biel, 1961). Wind speeds were usually greater than $2.2 \, m·s^{-1}$ (5 mph) and temperature inversions did not develop, thus most efforts to provide frost protection effective in a radiation frost event were of limited effectiveness. As much as 70% of the USA blackberry crop was eliminated by the April freeze (Warmund et *al.*, 2008). In Bailey, North Carolina, where the minimum temperature recorded was −5.6°C (22°F), when flower buds had recently emerged on the new lateral growth, all visible buds of 'Apache,' 'Arapaho,' and 'Ouachita' exhibited oxidative browning. In contrast, only 19% of the visible 'Navaho' buds appeared injured (Fernandez, 2007; Warmund et *al.*, 2008). The loss of reproductive potential from the fruit developing on primary lateral shoots was compensated by the latent growth and development from secondary buds (Takeda and Glenn, 2016; Warmund et *al.*, 2008).

Spring temperatures at which crop injury occurs have been studied in other small fruits, such as strawberry (Woo and Warmund, 1992), grape (Poling, 2008), and blueberry (Rowland et *al.*, 2013), but little is known about the survival of blackberry flower buds from exposure to sub-freezing temperatures at growth stages following bud break. Vincent and Garcia (2011) and Warmund et *al.* (2008) reported that in several commercial blackberry plantings, all open flower buds were killed when exposed to −5.0°C (23°F) and all new growth from buds on the floricanes were necrotic when the temperature dropped to −7.3°C (19°F). In these studies, ambient temperatures were recorded, but no information was provided on flower tissue temperatures. In most frost resistance studies, frost simulation chambers have been used in which the plant is cooled by conduction or by convection (Spiers, 1978).

According to Fuller and Le Grice (1998), conduction or convection chambers do not reproduce radiation frost events and ice crystal growth. Consequently, the test would not be a true reflection of radiation frost as it occurs in the field and may not give a correct indication of plant survival. A freezing chamber design based on radiative cooling which mimics overnight radiative freezing can be constructed (Fuller and Le Grice, 1998). It has unique design features, such as the radiative cooling plate at the top of the chamber which acts as a cold black body. The sides of the chamber could be cooled to variable temperatures in order to prevent the chamber wall radiating to the plant material during testing. Thermocouple measurements showed that air and plant temperatures simulated the radiative cooling conditions found during spring (Takeda and Glenn, 2016).

There are several methods to resolve how plants freeze, where freezing is initiated in a plant, and how the freezing process is propagated. Freeze injury can be determined with a simple visual inspection (Takeda and Glenn, 2016). To determine ice nucleation has occurred, thermocouples, which rely on the detection of the heat of fusion as water changes from a liquid to a solid state, are used (Warmund and George, 1990). High-resolution infrared thermography is available to study the freezing process, rates of ice propagation, and the effects of plant structure on the freezing process (e.g. SC8000 HD Series infrared camera, FLIR Systems, Wilsonville, Oregon, USA, www.flir.com). A thermographic camera detects radiation in the infrared range and allows one to see variations in surface temperature and produces images of that radiation and how the pattern of freezing in plants relates to visual tissue damage (Wisniewski *et al.*, 2008).

Takeda and Glenn (2016) determined injuries in blackberry flower buds and flowers of whole plants. They used a freezing chamber based on radiative cooling designed to mimic overnight radiative freezing. Along with tissue-based viability tests on flowers, they recorded exothermic events using thermocouples and monitored the spread of ice formation with a high-resolution infrared thermographic video camera. They showed that internal temperatures of open flowers and tight buds followed closely to the chamber air temperature. Exothermic vents in reproductive organs at distinct developmental stages occurred at about −2.5 to −3.0°C (27.5–27°F), a narrow range of temperatures. These findings suggested that it was unlikely that less developed reproductive organs were more resistant to freezing temperatures (Takeda and Glenn, 2016).

Furthermore, infrared thermography (Fig. 3.1) revealed additional information about freezing in blackberry plants that supported the data that had been accumulated with thermocouples (Warmund and George, 1990; Warmund *et al.*, 1992; Takeda and Glenn, 2016). The images from the thermographic video recording showed that (i) freezing in long-cane plants was initiated in the canes, propagated into individual flower shoots, and then into flower buds and open flowers, and (ii) the entire process of water freezing in the cane and ice propagating into individual flower buds and flowers occurred

Fig. 3.1. Infrared thermography for studying where ice formation is initiated and avenue of ice propagation into flower in a blackberry plant. *Left:* A close-up photograph of blackberry flower shoots on a floricane. Note thermocouple wires are inserted into a number of flowers on the shoot for detecting exothermic events. *Right:* An infrared (IR) thermographic image of the same two flower shoots captured with a FLIR IR camera. Plant tissues in which an exothermic event has occurred are illustrated in yellow color. Here the ice formation occurred initially in the floricane and now has propagated half-way up the flower shoots (arrows). The flowers and subtending leaves are yet to freeze as indicated by their black color.

quickly. For example, ice propagated from a section of the floricane, approximately 20 cm (8 in) from the juncture of the floricane, spread into several flowers on the 20-cm-long flower shoot in less than one minute. Infrared (IR) imaging clearly revealed that once ice formation had reached the flower shoot the freezing of several buds and flowers occurred essentially at the same time (Takeda and Glenn, 2016). Assessment of individual floral parts for oxidative browning following the freezing tests showed that little or no tissue browning developed when flower buds, open flowers and green/red fruit were exposed to temperatures above −2°C (28°F). However, after exposure to −3°C (27°F) and −4°C (25°F), injury as indicated by tissue browning was observed in all floral organs. All samples from tight bud stage to green and red fruit were completely brown after exposure to −4°C (25°F) for 1 hour. In the case of the corolla, damage was apparent as either a water-soaked or necrotic appearance (oxidative browning) (Takeda and Glenn, 2016). The findings showed that injury of floral parts developed at temperatures between −2.5°C (27°F) and −3.0°C (26°F). The damage occurred first in the gynoecium (the ovules, followed by injuries in the ovaries and styles) and then spread to the vascular tissues in the receptacle, the entire receptacle, and finally to the corolla and the androecium (anther sacs and anther filaments).

Previously, studies on spring frost injury were conducted in small fruit crops (Carter *et al.*, 2001; Spiers, 1978; Vincent and Garcia, 2011) to determine the relationship between bud development stage and sensitivity to low temperatures. Findings from these studies indicated that flowers at more advanced stages of development were more sensitive to frost injury than earlier stages of development. It was clearly shown that the pistil became increasingly vulnerable to low temperatures in blueberries (Rowland *et al.*, 2013; Spiers, 1978), blackcurrant (Carter *et al.*, 2001), and blackberries (Vincent and Garcia, 2011). Vincent and Garcia (2011) reported that in flower clusters of PF blackberries, bud mortality was higher in flowers at the 'popcorn' stage than those still at tight-bud stage after a spring frost in which ambient temperatures reached −5°C (23°F).

The results of the Takeda and Glenn (2016) study using infrared thermography, thermocouples, and visual inspection were not in agreement with previous reports on blackberries (Vincent and Garcia, 2011). A controlled temperature radiation frost chamber study suggested that freezing of reproductive organs in blackberries from tight-bud to green fruit development stage occurs over a narrow range of temperatures. In fact, freezing in open flowers and tight buds occurred within a 0.3°C range. This implied that it is unlikely that reproductive organs at tight-bud stage are more resistant to freeze damage than those of more developed flower buds. Also, the findings of Takeda and Glenn (2016) showed that buds and flowers in a single inflorescence froze over a narrow temperature range and ice formation spread into the cane and flowers at the same rate of 1 cm·s^{-1} (3/8 inch·s^{-1}). It appears that in the case of blackberry plants at the time of bloom, the barrier to movement of ice was absent or had become inactive at the junction of the cane (e.g. woody tissue) and the lateral reproductive appendages, such as flower shoots (e.g. herbaceous tissue) (Carter et al., 2011; Takeda and Glenn, 2016; Wisniewski et al., 2008).

Factors related to the genotype, the stage of development, the formation of ice, and biological status of the reproductive organ play a part in the vulnerability or resistance of flowers to spring frosts (Woo and Warmund, 1992). The capacity of flowers to supercool at blooming time, flower bud density, and presence of leaves subtending the reproductive appendages and uniformity of floral development appears to affect frost tolerance or the actual temperatures of these plants (Carter et al., 2011; Wisniewski et al., 2008).

Strawberry flowers and buds surrounded by nearby leaves may escape frost damage because the leaves provide some degree of insulation from the ambient temperature and even shorten the exposure to near lethal temperatures (Olney, 1958). Thus, research studies can determine whether the location of blackberry flowers relative to the plant canopy and the ground and their close proximity to leaves can affect the tissue temperatures and possibly aid in avoiding exposures to lethal temperatures. The blackberry plants trained to the RCA trellis (Fig. 3.2) are preferred to test the hypothesis. With this trellis system, the flower shoots can be positioned either above (Fig. 3.2a) (Takeda and Peterson, 1999; Takeda et al., 2013) or below (Fig. 3.2b) the canes that are oriented horizontally during the bloom period or on either side of vertically oriented canes (Fig. 3.2c). Additionally, it is possible to use a rowcover to insulate the flowers from the damaging temperatures (Figs. 3.2d,e) and take advantage of heat radiating from the soil, which can be trapped by the rowcover, during calm, cold nights.

SUMMER HEAT

Blackberry plants are grown under diverse environmental conditions; however, they perform best in areas with mild winter and cool summer temperatures (Magness and Traub, 1941; Crandall, 1995). Trailing blackberries are grown effectively in Oregon and California where both winter and summer temperatures are milder than in northeast USA and Canada. Varieties well adapted to these conditions have been produced in the USDA breeding program in Oregon and private programs in California. Winter hardiness is also an issue for them as production is attempted further north or east of the Rocky Mountains (Takeda et al., 2008, 2013), but they can also be grown effectively in the temperate regions of South America, New Zealand, Australia, and South Africa, as well as in southern Europe (Strik et al., 2007). Other varieties have been developed that can be grown effectively in Mexico and Brazil at higher elevations, where temperatures are not too high for fruit set to occur or for fruit damage to occur during ripening. In western Europe and the UK, winter

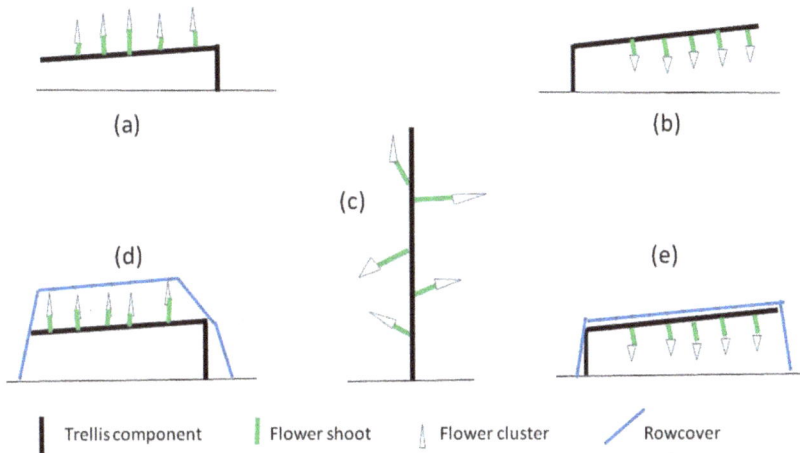

Fig. 3.2. Blackberry trained to the Rotating Cross-Arm (RCA) trellis system (a and b) and on a standard, upright, 'I') trellis (c). For additional information regarding the RCA trellis and cane training system see Chapter 12. The RCA trellis has a long rotatable cross-arm and the plant canopy at anthesis can be rotated about the pivot point at the top of the short post. Normally, the cross-arm is rotated to the left (a) so that the flower shoots will point up. However, the cross-arm can be rotated up to vertical and downward to the right (b) to orient the developed flower shoots downward. In the upright trellis (c) system, the flower shoots develop to the right and left depending on which way the buds were pointed at bud break. A rowcover can be applied during frost events to protect the buds and flowers on plants with a low canopy configuration (d and e).

temperatures are mild but summer temperatures are significantly lower than in the USA. In the UK, varieties have been developed that will set fruit effectively at low temperatures and ripen within their relatively short summer. Varieties from Oregon, Australia, and New Zealand can also be grown there, especially when grown under tunnels for protection from rainfall and cooler temperatures.

Some production areas experience dramatic seasonal variations in temperatures and heat that affect fruit ripening time and quality. Even under temperate-zone growing conditions, the fruit of some cultivars become susceptible to solar injury when fruit is exposed to intense sunlight and high temperatures (Takeda, 2012; Takeda et al., 2013). In particular, the white drupelet disorder in 'Apache' blackberry (Fig. 3.3a) can cause as much as a 30% reduction in fresh-market quality pack-out. The white drupelet disorder has been attributed to high solar radiation and high ultraviolet (UV) radiation. Reducing both photosynthetic active radiation and UV are effective means of reducing sunburn damage. Although, there is a genetic factor attributable to this disorder as it develops in some cultivars and not in others, horticultural practices

Fig. 3.3. High heat and ultraviolet radiation cause fruit damage. *Left:* In 'Apache' and some western trailing blackberries, high heat and ultraviolet radiation can cause some drupelets of ripe fruit to change from black to white (arrow). With the RCA trellis system, fruit can be positioned away from direct sunlight depending on the row orientation and which way the long cross-arm is positioned after bloom. On the *right*, a blackberry field is shown with the rows established in a northeast–southeast direction approximately at a 70°/250° axis and the trellis system installed to have the wall of fruit towards the northwest. With these row and trellis configurations the developing fruit were maintained in partial shade. Note that in the morning, with the sun rising from the left, the plant canopy casts a shadow to its right side. In late mornings, when the sun has moved to the right, the left side of the rows become shaded and the fruit will be in the shade.

to reduce the solar radiation transmission to the fruit were shown to reduce white drupelet disorder in 'Apache' blackberry (Takeda *et al.*, 2013). The use of RCA trellis technology allowed the fruit to be positioned only on the north side of a row oriented east–west and not be exposed to direct sunlight in the morning and afternoon (Fig. 3.3b). High light intensity either in the morning or in the afternoon was sufficient to cause white drupelet disorder in 'Apache' blackberry (Table 3.1). The fruit skin temperature of a fruit exposed to direct sunlight was higher by as much as 8°C (46°F) compared to a fruit shaded by leaves (Takeda *et al.*, 2013).

Similar reduction in white drupelet disorder can be achieved by decreasing solar light transmission with a placement of a shade fabric over the plants (Fig. 3.4). White drupelet disorder was observed the entire harvest period in July and August. Both the morning and afternoon irradiation was sufficient to cause one to five drupes on the exposed side of the fruit to become white (Takeda *et al.*, 2013). Reducing the light intensity with an overhead shading frame reduced white drupelet disorder by as much as 21%. All three fabric colors (black, red, or blue) and shade levels (40% and 60% light transmission) were equally effective in reducing white drupelet disorder in 'Apache' blackberry.

Table 3.1. The influence of fruit location (east side or west side) and shade fabrics on blackberry plants growing in rows oriented north and south on the percentage of 'Apache' blackberry fruit with white drupelet disorder over the entire season and on four selected harvest dates in July and August, 2012.

	Percentage (%) of fruit with white drupelet disorder				
	Over season	18 July	25 July	2 August	9 August
Fruit location					
East side	11a	11a	14a	9a	11a
West side	14a	14a	8a	17a	16a
Shade					
Present	6a[z]	9a	4a	3a	5a
Absent	19b	16a	18b	24b	21b
P > F					
Location	0.4776	0.5752	0.0983	0.1185	0.2286
Shade	0.0009	0.0755	0.0002	0.0005	0.0004
Location × Shade	0.8546	0.9964	0.3842	0.6633	0.5626

[z]Mean within a column and fruit location or shade treatment followed by the same letter do not differ according to T-test ($P < 0.05$).

Fig. 3.4. A portable shade structure (*left*) and permanent shade structure for covering an expansive area (*right*). Portable shade structures can either be flat-top or have a gable roof. Large-scale shade structures either consist of tall poles supporting cables on which a shade cloth rests or sometimes multi-bay high tunnels. Shade cloths (black, red, or white) that reduce solar radiation by 30–50% are sufficient to prevent white drupelet formation.

SUMMARY

A number of new cultural techniques are available to mitigate low winter temperature, spring frost, and high light intensity in the summer that can cause high cane and flower mortality and loss of fruit quality (Takeda *et al.*, 2008; Takeda, 2012; Takeda *et al.*, 2013). These new production techniques have enabled blackberry production to expand into areas with more severe winter conditions than those that occurred in warmer production areas, and allow blackberries to be grown in areas with high solar radiation, especially during the time the fruit is reaching maturity (e.g. turning black). New production technology has fostered niche market blackberry production and has helped create opportunities for growers to produce blackberries in areas where fruit do not ripen until late summer.

The blackberry plant has a dynamic structure. Research has shown that trellis and cane training systems can be used to change the vigor and viability of canes to survive harsh winter conditions and minimize the exposure of fruit to direct sunlight in the summer (Takeda *et al.*, 2013). The interaction of the plant and its environment determines the economic success or failure of a planting. Selecting cultivars, growing method, and canopy manipulation will help avoid plant failure and insure maximum yield under prevailing environmental conditions.

REFERENCES

Biel, E.R. (1961) Microclimate, bioclimatology, and notes on comparative dynamic climatology. *American Scientist* 49(3), 327–357.

Bushway, L., Pritts, M. and Handley, D. (eds.) (2008) Plant selection. In: *Raspberry & Blackberry Production Guide for the Northeast, Midwest, and Eastern Canada*. Northeast Regional Agricultural Engineering Service, Ithaca, New York, pp. 17–27.

Carter, J., Brennan, R. and Wisniewski, M. (2001) Patterns of ice formation and movement in blackcurrant. *HortScience* 36(6), 1027–1032.

Carter, P.M., Clark, J.R., Drake, C.A., Particka, C. and Crowne, D.Y. (2006) Chilling response of Arkansas blackberry cultivars. *Journal of the American Pomological Society* 60(4), 187–197.

Crandall, P.C. (1995) *Bramble Production: The Management and Marketing of Raspberries and Blackberries*. Haworth Press, Binghamton, New York.

Drake, C.A. and Clark, J.R. (2000) Determination of the chilling requirements of Arkansas thornless blackberry cultivars. *Journal of Dale Bumpers College of Agricultural Food and Life Sciences* 1, 15–19.

Fernandez, G. (2007) Freeze hits North Carolina blackberry growers in 2007. *Southern Region Small Fruit Consortium* 7, 2–4.

Fuller, M.P. and Le Grice, P. (1998) A chamber for the simulation of radiation freezing of plants. *Annals of Applied Biology* 133(1), 111–121.

Kraut, J., Walsh, C.S. and Ashworth, E.N. (1986) Acclimation and winter hardiness patterns in eastern thornless blackberry. *Journal of the American Society for Horticultural Science* 111(3), 347–352.

Magness, J.R. and Traub, H.P. (1941) Climatic adaption of fruit and nut crops in *Climate and Man – 1941 Yearbook of Agriculture*. Washington, DC: USDA. pp. 400–420.

Olney, A.S. (1958) Frost resistance in strawberries. *Fruit Varieties and Horticultural Digest* 13, 25.

Poling, E.B. (2008) Spring cold injury to winegrapes and protection strategies and methods. *HortScience* 43(6), 1652–1662.

Rowland, L.J., Ogden, E.L., Takeda, F., Glenn, D.M. and Ehlenfeldt, M.K. (2013) Variation among highbush blueberry cultivars for frost tolerance of open flowers. *HortScience* 48(6), 692–695.

Spiers, J.M. (1978) Effect of stage of bud development on cold injury in rabbiteye blueberry. *Journal of the American Society for Horticultural Science* 103, 452–455.

Strik, B.C., Clark, J.R., Finn, C.E. and Banados, P. (2007) Worldwide production of blackberries, 1995 to 2005 and predictions for growth. *HortTechnology* 17(2), 205–213.

Takeda, F. (2012) Innovating blackberry production system. *Small Fruit Newsletter* 12(3), 2–5.

Takeda, F. and Glenn, D.M. (2016) Susceptibility of blackberry flowers to freezing temperatures. *European Journal of Horticultural Science* 81(2), 115–121.

Takeda, F. and Peterson, D.L. (1999) Considerations for machine harvesting fresh-market eastern thornless blackberries: trellis design, cane training systems, and mechanical harvester development. *HortTechnology* 9(1), 16–21.

Takeda, F. and Philips, J. (2011) Horizontal cane orientation and rowcover application improve winter survival and yield of trailing 'Siskiyou' blackberry. *HortTechnology* 21(2), 170–175.

Takeda, F. and Wisniewski, M. (1989) Organogenesis and patterns of floral bud development in two eastern thornless blackberry cultivars. *Journal of the American Society for Horticultural Science* 114(4), 528–531.

Takeda, F., Strik, B.C., Peacock, D. and Clark, J.R. (2002) Cultivar differences and the effect of winter temperature on flower bud development in blackberry. *Journal of the American Society for Horticultural Science* 127(4), 495–501.

Takeda, F., Hummell, A.K. and Peterson, D.L. (2003a) Primocane growth in 'Chester Thornless' blackberry trained to the rotatable cross-arm trellis. *HortScience* 38(3), 373–376.

Takeda, F., Hummell, A.K. and Peterson, D.L. (2003b) Effects of cane number on yield components in 'Chester Thornless' blackberry on the rotatable cross-arm trellis. *HortScience* 38(3), 377–380.

Takeda, F., Demchak, K., Warmund, M.R., Handley, D.T., Grube, R. and Feldhake, C. (2008) Rowcovers improve winter survival and production of western thornless 'Siskiyou' blackberry in the eastern United States. *HortTechnology* 18(4), 75–82.

Takeda, F., Glenn, D.M. and Tworkoski, T. (2013) Rotating cross-arm trellis technology for blackberry production. *Journal of Berry Research* 3(1), 25–40.

USDA AMS (2015) USDA Agricultural Marketing Service fruit and vegetable market news. Available at: www.ams.usda.gov/mnreports/wa_fv101.txt (accessed October 6, 2015).

Vincent, C.I. and Garcia, M.E. (2011) A system of defined phenological stages for cold tolerance and development of floricane inflorescences of primocane-fruiting blackberries. *Journal of the American Pomological Society* 65(1), 54–60.

Warmund, M.R. and George, M.F. (1990) Freezing survival and supercooling in primary and secondary buds of *Rubus* spp. *Canadian Journal of Plant Science* 70(3), 893–904.

Warmund, M.R. and Krumme, J. (2005) A chilling model to estimate rest completion of erect blackberries. *HortScience* 40(5), 1259–1262.

Warmund, M.R. and Krumme, J. (2008) A chilling model to estimate rest completion of erect blackberries. *Acta Horticulturae* 777, 275–280.

Warmund, M.R., Takeda, F. and Davis, G.A. (1992) Supercooling and extracellular ice formation in differentiating buds of eastern thornless blackberry. *Journal of the American Society for Horticultural Science* 117(6), 941–945.

Warmund, M.R., Guinan, P. and Fernandez, G. (2008) Temperature and cold damage to small fruit crops across the eastern United States associated with the April 2007 freeze. *HortScience* 43(6), 1643–1647.

Wisniewski, M., Glenn, D.M., Gusta, L. and Fuller, M.P. (2008) Using infrared thermography to study freezing in plants. *HortScience* 43(6), 1648–1651.

Wojcik-Seliga, J. and Wokcik-Gront, E. (2013) Evaluation of blackberry and hybrid berry cultivars new to Polish climate – short communication. *Horticultural Science* (Prague) 40(2), 88–91.

Woo, K.K. and Warmund, M.R. (1992) Low-temperature injury to strawberry floral organs at several stages of development. *HortScience* 27(12), 1302–1304.

4

BLACKBERRY FRUIT QUALITY COMPONENTS, COMPOSITION, AND POTENTIAL HEALTH BENEFITS

Jungmin Lee*

USDA-ARS-HCRU Worksite, Parma, Idaho, USA

INTRODUCTION

Small-fruit quality is closely related to its production of primary and secondary metabolites (i.e. type, concentration). Blackberry metabolites continue to undergo anabolism and/or catabolism within the fruit until harvest, and these components that remain, and are not degraded by the time the fruit (or fruit product) reaches consumers, determine blackberry fruit quality. The primary and secondary metabolite composition of a blackberry defines its characteristic appearance, taste, and texture. Blackberries are typically purchased for consumption as fresh, individually quick frozen (IQF), or as a further processed product incorporating them into jams, syrups, wines, teas, juices, concentrates, and purees. Many potential health benefits from consuming blackberries or blackberry products are attributed to their metabolites. Metabolites also directly and indirectly influence processing regimes, shelf life, and consumer likeability.

Blackberries contain dietary fiber, vitamin C (ascorbic acid), vitamin A, vitamin E, potassium, and calcium (for additional nutrition facts, see USDA, 2015), along with the phenolic metabolites that are a source of possible health benefits. Sensory attributes, typically used to describe the taste and flavor of blackberries, include fresh fruit, cooked fruit, cooked berry, strawberry, raspberry, vegetal, stemmy, and earthy (Du *et al.*, 2010). In this chapter, a summary of distinct primary and secondary metabolites crucial to blackberry quality will be presented focusing on fruit, although all parts of the plant (leaves, canes, and roots) have historically been used as foods or herbal remedies (Arnason *et al.*, 1981; Hummer, 2010). The concentration ranges for compounds related to blackberry quality are summarized in Table 4.1.

---

* Corresponding author: Jungmin.Lee@ars.usda.gov

Table 4.1. Blackberry and blackberry hybrid fruit quality components and reported concentration ranges (in fresh weight).

Fruit quality component	Reported ranges (n = sample size)	References
Fruit mass	3.8–28.3 g (n = 19)	Finn et al., 2014; Vrhovsek et al., 2008; H.K. Hall, personal observation
Calories	43–64 kcal/100 g (n = 3)	USDA, 2015
% soluble solids	6.9–16.8 (n = 90)	Fan-Chiang and Wrolstad, 2010; Finn et al., 2014; Mertz et al., 2007; Thomas et al., 2005; Vrhovsek et al., 2008; Wang et al., 2008
Titratable acidity	0.08–2.7 g/100 g (n = 82)	Fan-Chiang and Wrolstad, 2010; Finn et al., 2014; Mertz et al., 2007; Thomas et al., 2005; Veberic et al., 2014; Wang et al., 2008
pH	2.6–3.9 (n = 73)	Fan-Chiang and Wrolstad, 2010; Finn et al., 2014; Mertz et al., 2007; Thomas et al., 2005; Veberic et al., 2014
Simple sugars	2.6–13.9 g/100 g (n = 63)	Fan-Chiang and Wrolstad, 2010; Mikulic-Petkovsek et al., 2012; Veberic et al., 2014
Organic acids*	0.5–2.9 g/100 g or (n = 73)	Fan-Chiang and Wrolstad, 2010; Mikulic-Petkovsek et al., 2012; Vrhovsek et al., 2008
Vitamin C (ascorbic acid)	1.2–11.9 mg/100 g (n = 12)	Thomas et al., 2005; Veberic et al., 2014
Anthocyanins	28–366 mg/100 g (n = 1,306)	Conner et al., 2005; Fan-Chiang and Wrolstad, 2005; Finn et al., 2014; Scalzo et al., 2008; Sellappan et al., 2002; Vasco et al., 2009; Veberic et al., 2014; Wang et al., 2008
Phenolic monomers	0.7–555 mg/100 g (n = 22)	Acosta-Montoya et al., 2010; Bilyk and Sapers, 1986; Gancel et al., 2011; Sellappan et al., 2002; Vasco et al., 2009; Veberic et al., 2014
Ellagic acid conjugates	17–27 mg/100 g (n = 5)	Gasperotti et al., 2010
Ellagitannins (phenolic polymers)	85–390 mg/100 g (n = 23)	Gancel et al., 2011; Gasperotti et al., 2010; Vasco et al., 2009; Vrhovsek et al., 2008
Carotenoids	0.44–0.59 mg/100 g (n = 2)	Curl, 1964; Marinova and Ribarova, 2007

*Excluding ascorbic acid, listed separately.

SUGARS AND ORGANIC ACIDS

Two simple quality measurements of blackberries and their hybrids assess the most influential categories of their taste perception: sugar and organic acid content. Typically sweetness is described as percent (%) soluble solids, while acidity is reported as titratable acidity (for concentration ranges, see Table 4.1). Sugars reported in blackberry fruits are fructose, glucose, sucrose, and occasionally exceedingly low levels of sorbitol (Fan-Chiang and Wrolstad, 2010; Lee, 2015; Mikulic-Petkovsek *et al.*, 2012; Wrolstad *et al.*, 1980, 1981). Minute concentrations of sorbitol (sugar alcohol) in processed blackberry products (e.g. juice) likely originated from processing enzymes or immature (under ripe) fruit (Fan-Chiang and Wrolstad, 2010; Lee, 2015). Since sorbitol is seldom found in ripe blackberries (Lee, 2015), the detection of sugar alcohol may be an indicator of accidental or fraudulent adulteration of products with cheaper fruit (i.e. apples, pears) juices or concentrates (Lee, 2015; Lee *et al.*, 2012; Wrolstad *et al.*, 1981). Lee (2015) clarified that recent United States media claims of blackberries containing high levels of sugar alcohol were inaccurate and actually the opposite of scientific findings.

Blackberry tartness is due to nonvolatile organic acids, including ascorbic acid (vitamin C), citric acid, isocitric acid, lactoisocitric acid, malic acid, shikimic acid, fumaric acid, and succinic acid (Fan-Chiang and Wrolstad, 2010; Mikulic-Petkovsek *et al.*, 2012; Veberic *et al.*, 2014; Vrhovsek *et al.*, 2008). Lactoisocitric acid can be a useful organic acid marker for blackberries, with the caveat that the amounts in blackberry hybrids (e.g. 'Loganberry,' 'Boysenberry') may be too low to act as an effective indicator. Two distinctive patterns were observed in the acid makeup of blackberry samples examined by Fan-Chiang and Wrolstad (2010). The samples' organic acid profiles resembled either that of 'Marion' (higher citric acid levels) or of 'Evergreen' (higher isocitric acid levels), suggesting this distinguishing factor might allow the identification of cultivars used in commercial blackberry products.

Famiani and Walker (2009) investigated changes in blackberry primary metabolites during fruit ripening, where they found soluble solids increased while titratable acidity decreased during the final growth stage prior to harvest. Ratios of sugars to acids will not be discussed here, since ratios are misleading in terms of apparent flavor; equivalent sugar–acid ratios do not equal similar taste assessments.

ANTHOCYANINS

Blackberry fruit phenolics have been thoroughly reviewed by Lee *et al.* (2012) and Kaume *et al.* (2012). Unlike other dark-colored *Rubus* fruit (i.e. black raspberry and red raspberry), blackberry pigments are chiefly cyanidin-based anthocyanins (Lee *et al.*, 2012). Though the characteristic black color of intact

fresh blackberry fruit is actually from its concentration and types of anthocyanins (natural red pigments). Since anthocyanin color is pH dependent, color
linked to its structural form that undergoes transformation with changes in
pH, a slight change in pH makes the red anthocyanin within blackberries turn
a deep purple to black color. However, there are some rare blackberries lacking
anthocyanins, including 'Snowbank' (*R. allegheniensis*; Hummer et *al.*, 2015)
and 'Clark Gold' (*R. trivalis*; US PP14935 P2). While these uncommon fruits
are white to yellow, consumers gravitate towards more commercially available
dark blackberries thought to have high pigment concentrations. Blackberry
anthocyanin levels (see Table 4.1) are actually in the lower ranges of what can
be found in black raspberries (anthocyanin levels ranging from 39 to
996 mg/100ml, n > 1,000; Dossett et *al.*, 2012), or blueberries (anthocyanin
levels ranging from 101 to 400 mg/100 g, n = 37; Lee et *al.*, 2004), but higher
than red raspberries (anthocyanin levels ranging from 6 to 98 mg/100 g,
n = 644; Scalzo et *al.*, 2008).

Blackberries contain acylated and non-acylated anthocyanins, but most
are the non-acylated form. The major anthocyanins that have been reported
are cyanidin-glucoside, cyanidin-rutinoside, cyanidin-xyloside, cyanidin-
malonylglucoside, and cyanidin-dioxalylglucoside (or possibly cyanidin-
hydroxymethylglutaroylglucoside) (Conner et *al.*, 2005; Fan-Chiang and
Wrolstad, 2005; Finn et *al.*, 2014; Jordheim et *al.*, 2011; Lee et *al.*, 2012;
Stintzing et *al.*, 2002a; Veberic et *al.*, 2014, 2015). While the most predomi-
nate pigment in blackberries is cyanidin-glucoside (44–95% of total), the ratios
of subsequent anthocyanins vary with cultivar and genotype. Blackberry
hybrids contain different anthocyanin profiles compared to non-hybrids. For
example, hybrids of raspberry and blackberry (i.e. 'Boysenberry' and 'Loganberry') contain cyanidin-sophoroside, as found in red raspberry, but not in any
non-hybrid blackberry (Fan-Chiang and Wrolstad, 2005; Lee et *al.*, 2012).

Some blackberries including 'Marion,' 'Waldo,' 'Evergreen,' 'Black
Douglass,' 'Hull Thornless,' 'Chester Thornless,' and 'Shawnee,' contain cyanidin-dioxalylglucoside (Fan-Chiang and Wrolstad, 2005; Kolniak-Ostek et *al.*,
2015; Stintzing et *al.*, 2002a). This identification may be, at least
partially, disputed, as an independent group has claimed that the
accepted identity of cyanidin-dioxalylglucoside is actually cyanidin-
hydroxymethylglutaroylglucoside (unconfirmed; Jordheim et *al.*, 2011). It
should be noted that while both of those anthocyanins are unique to blackberries (Fan-Chiang and Wrolstad, 2005; Jordheim et *al.*, 2011; Veberic et *al.*,
2014), they are not necessarily found in all varieties, and neither was detected
in the new 'Columbia Star' (Finn et *al.*, 2014).

The prevailing minor anthocyanin in many blackberries is pelargonidin-
glucoside, as Veberic et *al.* (2014) found in the cultivars 'Black Satin,' 'Čačanska
Bestrna,' 'Chester Thornless,' 'Thornless Evergreen,' 'Loch Ness,' and 'Thornfree.' Although one previous study did not find pelargonidin-glucoside when
identical cultivars were tested (Fan-Chiang and Wrolstad, 2005).

Blackberry anthocyanins increase in concentration with fruit maturity (Acosta-Montoya et al., 2010; Famiani and Walker, 2009). While darker fruit is an indication of ripeness, post-processing modifications in appearance are expected. Freezing, thawing, or storage induced visual color changes of dark black to hues of red, yellow, or blue are due to slight alterations in pH and degradation of ascorbic acid, anthocyanins, etc. (Stintzing et al., 2002b; Veberic et al., 2014). Anthocyanin-based color is also affected by the physical form of water (liquid versus ice) and the chemical state of the fruit itself.

Anthocyanin profiles, as well as the presence of sorbitol, mentioned earlier, can point to adulteration in blackberry and non-blackberry based products (Lee, 2015; Lee et al., 2012; Wrolstad et al., 1981). For instance, a black raspberry freeze-dried powder (~$19 per 100 g), sold as a dietary supplement, was found to actually contain blackberry powder (~$14 per 100 g); Lee (2014) confirmed this with repeated purchases over time, and by analyzing blackberry and black raspberry powdered products from the same anonymous vendor. Although the fruit powders are similar visually, they have distinct anthocyanin profiles (Lee et al., 2012).

NON-ANTHOCYANIN PHENOLIC MONOMERS

The non-anthocyanin phenolic monomers (range shown in Table 4.1) found in blackberry fruit are the phenolic acids: ellagic acid, gallic acid, p-coumaric acid esters, caffeic acid, caffeic acid esters (like neochlorogenic acid), ferulic acid, ferulic acid esters; the flavanols: catechin and epicatechin; and the flavonol-glycosides: quercetin-, kaempferol-, isorhamnetin-, and myricetin-glycosides (Bilyk and Sapers, 1986; Kolniak-Ostek et al., 2015; Lee et al., 2012; Mertz et al., 2007; Sellappan et al., 2002; Veberic et al., 2014). Acylated flavonol-glycosides have also been reported in blackberries (Veberic et al., 2014). A more detailed list of blackberry phenolic monomers can be found in Lee et al. (2012), but additional work is needed to clarify these phenolic classes in blackberries.

While ellagic acid is the main phenolic acid seen in blackberries, it is a challenging compound to analyze, since it has poor solubility in water; although improved in alcohol, its solubility is enhanced best by increasing solution pH well above what is normally found in foods (Bala et al., 2006). At least one study reported flavonol-glycoside levels decreased with fruit ripening (Acosta-Montoya et al., 2010).

PHENOLIC POLYMERS

Blackberries, and other Rubus fruit (i.e. red raspberries, cloudberries), are a rich source of ellagitannins (also known as hydrolyzable tannins, and distinct

from the more extensively studied condensed tannins). Red raspberry ellagitannin concentrations, at 94–172 mg/100 g, were found lower than for blackberries (for range, see Table 4.1), but within the same study their single black
raspberry sample tested in the blackberry range at 330 mg/100 g (Vrhovsek
et al., 2008). The main intact (non-hydrolyzed) blackberry ellagitannins have
been recognized as lambertianin C and sanguiin H-6 (Acosta-Montoya et al.,
2010; Gancel et al., 2011; Gasperotti et al., 2010; Mertz et al., 2007;
Sangiovanni et al., 2013). Kool et al. (2010) did not find lambertianin C in
their 'Boysenberry' samples, but found sanguiin H-6 as the primary ellagitannin, along with three other supplementary ellagitannins.

Some researchers have broken down (hydrolyzed) ellagitannins before
analysis and reported the ellagitannin subunits as methyl gallate, ellagic acid
derivative, ellagic acid, and methyl sanguisorboate, with mean degree of
polymerization (indication of size) ranging from 1.59 to 1.92 (Mertz et al.,
2007; Vrhovsek et al., 2006, 2008). This points to ellagitannin values determined by hydrolysis prior to high performance liquid chromatography (HPLC)
separation as offering the closest probable approximation of ellagitannin
concentrations within unprocessed fruit. These remain a difficult group of
compounds to extract, separate, and identify. Beside conventional challenges
in investigating a naturally complex class of compounds, work with them is
further hampered by the lack of available pure commercial standards, no confirmed identifications as of 2016, and problems keeping these compound in
their native states (Acosta et al., 2014; Acosta-Montoya et al., 2010; Aripitsas,
2012; Gasperotti et al., 2010; Lee et al., 2012; Sangiovanni et al., 2013;
Vrhovsek et al., 2006, 2008).

Ellagitannins are found in all fractions of *Rubus* fruit, but the highest concentrations are in seed fractions (Hager et al., 2008). Ellagitannin levels can
decrease during the fruit-ripening period from red to fully ripe (Acosta-Montoya
et al., 2010). They also are reduced during food processing by high temperature degradation (e.g. pasteurization), precipitation out of solution, and
hydrolysis to ellagic acid (Gancel et al., 2011). Some reports have likely under-
estimated ellagitannins by incomplete extraction due to inappropriate solvents
and non-optimized extraction techniques (Lee et al., 2012; Lei et al., 2001).

Ellagitannins play several roles within plants, including protecting them
from pathogen attack and inhibiting premature seed germination (Lee et al.,
2012; Lei et al., 2001). To humans, these are the compounds seen as sediments
during wine and juice processing, and they can contribute to turbidity issues
for food products requiring clarity as a quality assessment (Lee et al., 2012).
However, these same sediments (i.e. ellagic acid and ellagitannin) from post-
processing waste are also potential future ingredients for value-added products
(Acosta et al., 2014). No work has yet been published on blackberry ellagitannins' taste, but work done with wood tannins (same phenolic class as found in
blackberries, but different types of ellagitannins) shows their contributions
range from no detectable flavor to added bitterness and astringency, with

threshold concentration dependent to each specific compound examined (Glabasnia and Hofmann, 2006). Additional work is needed for identification, quantification, and sensory evaluation of this phenolic class from blackberries.

ADDITIONAL QUALITY COMPONENTS

Other quality constituents of blackberries are carotenoids, vitamins, minerals, proteins, fiber, and aroma compounds. Aroma compounds reported in black-berries are esters (ethylacetate), aliphatic alcohols (heptanol, hexenol, hex-anol, and octanol), terpenes (carveol), aldehydes (hexanal, hexenal, and benzaldehyde), and ketones (heptanone) were found in *R. ulmifolius* Schott (D'Agostino et al., 2015; Perez-Gallardo et al., 2015). Blackberry fruit carote-noids are lutein, β-carotene, zeaxanthin, and β-cryptoxanthin (Marinova and Ribarova, 2007). Additional work needs to be conducted on the quality of components listed in this section to provide a better understanding of similari-ties and differences among cultivars, field treatments, etc. Novelty products made from food processing by-products, such as juice presscake and black-berry seed oils, are currently available and are a source of vitamin E (Bushman et al., 2004; Van Hoed et al., 2011).

POTENTIAL HEALTH BENEFITS

Besides the flavor and color blackberries provide to foods, their naturally high level of phenolics could have potential health benefits that may help protect their consumers from some chronic diseases. Blackberry fruit phenolics have been implicated in providing anticancer, antiproliferative, antineurodegenera-tive, anti-inflammatory, antidiarrheal, antidiabetic, antimicrobial, and antivi-ral activities (Bakkalbasi et al., 2009; Landete, 2011, 2012; Lee et al., 2012; Pojer et al., 2013). Phenolics bioavailability, metabolism, and potential health benefits have been previously well reviewed (Bakkalbasi et al., 2009; Landete, 2011, 2012; Pojer et al., 2013), although the exact mechanisms of how black-berries may impart protection after ingestion remain unclear. It is clear that diets rich in fruits and vegetables are valuable in preventing some cancers and reducing the risk of cardiovascular disease (Basu et al., 2010; Van Duyn and Pivonka, 2000; Pojer et al., 2013), and blackberries can be an element of that diet.

Blackberry fruit anthocyanins are found in their native forms at low con-centrations after digestion; the majority are found as protocatechuic acid and its derivatives (i.e. ferulic acid, hippuric acid, vanillic acid), phenylacetic acid, phenylpropenoic acid, methylated conjugates, glucuronidated conjugates, and many other metabolites (Czank et al., 2013; de Ferrars et al., 2014; Felgines *et al.*, 2005; Pojer *et al.*, 2013). Fang (2014) reviewed the absorption route of cyanidin-glucoside, the chief blackberry anthocyanin. Czank *et al.* (2013) demonstrated isotopically labeled cyanidin-glucoside (500 mg) remained in circulation in male subjects for over 48 hours. Cyanidin-glucoside has been shown to inhibit proliferation of human lung carcinoma cells, and cancer cell migration in mice (Ding *et al.*, 2006). Blackberries have also been linked to providing neuroprotective effects in human neuroblastoma (extracranial solid cancer) cells (Tavares *et al.*, 2013).

As with the previously mentioned challenges to analyzing blackberry phenolic polymers (i.e. ellagitannins), their size and poor solubility limit their bioavailability as well (Garcia-Munoz and Vaillant, 2014). Ellagitannins' health beneficial effects were well summarized recently (Landete, 2011; Garcia-Munoz *et al.*, 2014). A post-ingestion assessment of blackberry ellagitannins in human urine found that the metabolites had been converted into urolithins by gut microbiota (Garcia-Munoz *et al.*, 2014). Urolithins are associated with preventing or controlling colon, breast, esophageal, and prostate cancers (Garcia-Munoz and Vaillant, 2014; Landete, 2011). As each of us has heterogeneous gut microbiota, some researchers have classified individuals into urolithin A, urolithin B, or non-urolithin (unidentified metabolites) excreters (Tomas-Barberan *et al.*, 2014). A review of a variety of gut microbiota metabolizing assorted classes of phenolics was well summarized by Selma *et al.* (2009). Additional data on this topic will become available as the number of identified gut bacteria grows, and they come to be further studied.

The exact mechanism of how blackberry dietary phenolics benefit human health is not fully elucidated and more work needs to be conducted to clarify this. While phenolics are attributed in disease prevention, they are also antinutritive and hinder absorption of certain minerals and proteins (Landete, 2012). Human clinical studies on blackberries alone, not mixed berries, are limited. A USA human clinical study on the influence of blackberries on cancer processes has been completed, but results are not yet available (US clinical trial identifier NCT01293617). Additional work is needed to clarify the contradicting reports among *in vitro* and *in vivo* work, animal models versus human trials, length of intervention, types of cells used, etc. (Garcia-Munoz and Vaillant, 2014; Landete, 2011, 2012).

Antioxidant claims have been deemed scientifically uncorroborated (Hollman *et al.*, 2011), and will not be discussed due to the controversial limitations surrounding *in vitro* methods (Carocho and Ferreira, 2013; Frankel and Meyer, 2000). Numerous studies and reviews are available on the lack of evidence for a relationship between antioxidant activity and human health (Hollman *et al.*, 2011; Lee *et al.*, 2012). In 2010, the European Food Safety Authority (EFSA), analogous in the European Union (EU) to the US Food and Drug Administration (FDA), rejected a petition for labeling food packages with health claims related to antioxidant activity, citing the lack of scientific data from human trials to substantiate such a claim (Gilsenan, 2011).

SUMMARY

As many become convinced that blackberries' quality compounds offer desirable potential health benefits, efforts have been conducted to further enhance the dietary phenolic content of blackberry fruit. Various techniques, including stimulation of metabolite production with field management factors via biotic elicitors (e.g. *Pseudomonas fluorescens*) and plant hormones (e.g. methyl jasmonate), have been attempted (Garcia-Seco *et al.*, 2013; Ramos-Solano *et al.*, 2014, 2015; Wang *et al.*, 2008). Nutrient regimes have also been shown to alter blackberry phenolics (Ali *et al.*, 2011). The high perishability of fruit sold in the fresh market has also created a demand to prolong blackberry shelf life; positive reports have used calcium in combination with pectin spray and starch-beeswax coating to prolong the shelf life (Perez-Gallardo *et al.*, 2015; Sousa *et al.*, 2007).

Growers, processors, and consumers should remain conscious that the quality components discussed here vary with a plant's genus, species, cultivar/genotype, and age; along with environment and management practices such as growing region and conditions, harvest decisions, fruit maturity indices, processing methods, and storage. Beside their dietary phenolic content, blackberries also have other important contributors to nutrition, including vitamin C, vitamin A, vitamin E, vitamin B6, folic acid, dietary fiber, potassium, phosphorus, magnesium, calcium, and iron.

REFERENCES

Acosta, O., Vaillant, F., Perez, A.M. and Dornier, M. (2014) Potential of ultrafiltration for separation and purification of ellagitannins in blackberry (*Rubus adenotrichus* Schltdl.) juice. *Separation and Purification Technology* 125, 120–125.

Acosta-Montoya, A., Vaillant, F., Cozzano, S., Mertz, C., Perez, A.M. and Castro, M.V. (2010) Phenolic content and antioxidant capacity of tropical highland blackberry (*Rubus adenotrichus* Schltdl.) during three edible maturity stages. *Food Chemistry* 119(4), 1497–1501.

Ali, L., Alsanius, B.W., Rosberg, A.K., Svensson, B., Nielsen, T. and Olsson, M.E. (2011) Effects of nutrition strategy on the levels of nutrients and bioactive compounds in blackberries. *European Food Research and Technology* 234(1), 33–44.

Aripitsas, P. (2012) Hydrolyzable tannin analysis in food. *Food Chemistry* 135(3), 1708–1717.

Arnason, T., Heba, R.J. and Johns, T. (1981) Use of plants for food and medicine by native peoples of eastern Canada. *Canadian Journal of Botany* 59(11), 2189–2325.

Bakkalbasi, E., Mentes, O. and Artik, N. (2009) Food ellagitannins – occurrence, effects of processing and storage. *Critical Reviews in Food Science and Nutrition* 49(3), 283–298.

Bala, I., Bhardwaj, V., Hariharan, S. and Ravi Kumar, M.N.V. (2006) Analytical methods for assay of ellagic acid and its solubility studies. *Journal of Pharmaceutical and Biomedical Analysis* 40(1), 206–210.

Basu, A., Rhone, M. and Lyons, T.J. (2010) Berries: emerging impact on cardiovascular health. *Nutrition Reviews* 68(3), 168–177.

Bilyk, A. and Sapers, G.M. (1986) Varietal differences in the quercetin, kaempferol, and myricetin contents of highbush blueberry, cranberry, and thornless blackberry fruits. *Journal of Agricultural and Food Chemistry* 34(4), 585–588.

Bushman, B.S., Phillips, B., Isbell, T., Ou, B., Crane, J.M. and Knapp, S.J. (2004) Chemical composition of caneberry (*Rubus* spp.) seeds and oils and their antioxidant potential. *Journal of Agricultural and Food Chemistry* 52(26), 7982–7987.

Carocho, M. and Ferreira, C.F.R. (2013) A review on antioxidants, prooxidants and related controversy: natural and synthetic compounds, screening and analysis methodologies and future perspectives. *Food and Chemical Toxicology* 51, 15–25.

Conner, A.M., Finn, C.E., McGhie, T.K. and Alspach, P.A. (2005) Genetic and environmental variation in anthocyanins and their relationship to antioxidant activity in blackberry and hybridberry cultivars. *Journal of the American Society for Horticultural Science* 130(5), 680–687.

Curl, A.L. (1964) The carotenoids of several low-carotenoid fruits. *Journal of Food Science*, 29(3), 241–245.

Czank, C., Cassidy, A., Zhang, Q., Morrison, D.J., Preston, T., Kroon, P.A., Botting, N.P. and Kay, C.D. (2013) Human metabolism and elimination of the anthocyanins, cyanidin-3-glucoside: a ^{13}C-tracer study. *American Journal of Clinical Nutrition* 97(5), 995–1003.

D'Agostino, M.F., Sanz, J., Sanz, M.L., Giuffre, A.M., Sicari, V. and Soria, A.C. (2015) Optimization of a solid-phase microextraction method for the gas chromatography-mass spectrometry analysis of blackberry (*Rubus ulmifolius* Schott) fruit volatiles. *Food Chemistry* 178, 10–17.

de Ferrars, R.M., Czank, C., Zhang, Q., Botting, N.P., Kroon, P.A., Cassidy, A. and Kay, C.D. (2014) The pharmacokinetics of anthocyanins and their metabolites in humans. *British Journal of Pharmacology* 171(13), 3268–3282.

Ding, M., Feng, R., Wang, S.Y., Bowman, L., Lu, Y., Qian, Y., Castranova, V., Jiang, B. and Shi, X. (2006) Cyanidin-3-glucoside, a natural product derived from blackberry, exhibits chemopreventive and chemotherapeutic activity. *Journal of Biological Chemistry* 281(25), 17359–17368.

Dossett, M., Lee, J. and Finn, C.E. (2012) Anthocyanin content of wild black raspberry germplasm. *Acta Horticulturae*, 946, 43–47.

Du, X.F., Kurnianta, A., McDaniel, M., Finn, C.E. and Qian, M.C. (2010) Flavour profiling of 'Marion' and thornless blackberries by instrumental and sensory analysis. *Food Chemistry* 121(4), 1080–1088.

Famiani, F. and Walker, R.P. (2009) Changes in abundance of enzymes involved in organic acid, amino acid and sugar metabolism, and photosynthesis during the ripening of blackberry fruit. *Journal of the American Society for Horticultural Science* 134(2), 167–175.

Fan-Chiang, H. and Wrolstad, R.E. (2005) Anthocyanin pigment composition of blackberries. *Journal of Food Science* 70(3), C198–C202.

Fan-Chiang, H. and Wrolstad, R.E. (2010) Sugar and nonvolatile acid composition of blackberries. *Journal of AOAC International* 93(3), 956–965.

Fang, J. (2014) Some anthocyanins could be efficiently absorbed across the gastrointestinal mucosa: extensive presystemic metabolism reduces apparent bioavailability. *Journal of Agricultural and Food Chemistry* 62(18), 3904–3911.

Felgines, C., Talavera, S., Texier, O., Gil-Izquierdo, A., Lamaison, J. and Remesy, C. (2005) Blackberry anthocyanins are mainly recovered from urine as methylated and glucuronidated conjugates in humans. *Journal of Agricultural and Food Chemistry* 53(20), 7721–7727.

Finn, C.E., Strik, B.C., Yorgey, B.M., Peterson, M.E., Lee, J., Martin, R.R. and Hall, H.K. (2014) 'Columbia Star' thornless trailing blackberry. *HortScience* 49(8), 1108–1112.

Frankel, E.N. and Meyer, A.S. (2000) The problems of using one-dimensional methods to evaluate multifunctional food and biological antioxidants. *Journal of the Science of Food and Agriculture* 80(13), 1925–1941.

Gancel, A., Feneuil, A., Acosta, O., Perez, A.M. and Vaillant, F. (2011) Impact of industrial processing and storage on major polyphenols and the antioxidant capacity of tropical highland blackberry (*Rubus adenotrichus*). *Food Research International* 44(7), 2243–2251.

Garcia-Munoz, C. and Vaillant, F. (2014) Metabolic fate of ellagitannins: implications for health, and research perspectives for innovative functional foods. *Critical Reviews in Food Science and Nutrition* 54(12), 1584–1598.

Garcia-Munoz, C., Hernandez, L., Perez, A. and Vaillant, F. (2014) Diversity of urinary excretion patterns of main ellagitannins' colonic metabolites after ingestion of tropical highland blackberry (*Rubus adenotrichus*) juice. *Food Research International* 55, 161–169.

Garcia-Seco, D., Bonilla, A., Algar, E., Garcia-Villaraco, A., Manero, J.G. and Ramos-Solano, B. (2013) Enhanced blackberry production using *Pseudomonas fluorescens* as elicitor. *Agronomy for Sustainable Development* 33(2), 385–392.

Gasperotti, M., Masuero, D., Vrhovsek, U., Guella, G. and Mattivi, F. (2010) Profiling and accurate quantification of *Rubus* ellagitannins and ellagic acid conjugates using direct UPLC-Q-TOF HDMS and HPLC-DAD analysis. *Journal of Agricultural and Food Chemistry* 58(8), 4602–4616.

Gilsenan, M.B. (2011) Nutrition and health claims in the European Union: a regulatory overview. *Trends in Food Science and Technology* 22(10), 536–542.

Glabasnia, A. and Hofmann, T. (2006) Sensory-directed identification of taste-active ellagitannins in American (*Quercus alba* L.) and European oak wood (*Quercus robur* L.) and quantitative analysis in bourbon whiskey and oak matured red wines. *Journal of Agricultural and Food Chemistry* 54(9), 3380–3390.

Hager, T.J., Howard, L.R., Liyanage, R., Lay, J.O. and Prior, R.L. (2008) Ellagitannin composition of blackberry as determined by HPLC-ESI-MS and MALDI-TOP-MS. *Journal of Agricultural and Food Chemistry* 56(3), 661–669.

Hollman, P.C.H., Cassidy, A., Comte, B., Heinonen, M., Richelle, M., Richling, E., Serafini, M., Scalbert, A., Sies, H. and Vidry, S. (2011) The biological relevance of direct antioxidant effects of polyphenols for cardiovascular health in humans is not established. *Journal of Nutrition*, 141(5), 989S–1009S.

Hummer, K.E. (2010) *Rubus* pharmacology: antiquity to the present. *HortScience*, 45(11), 1587–1591.

Hummer, K.E., Finn, C.E. and Dossett, M. (2015) Luther Burbank's best berries. *HortScience*, 50(2), 205–210.

Jordheim, M., Enerstvedt, K.H. and Andersen, O.M. (2011) Identification of cyanidin 3-*O*-β-(6"-hydroxy-(3-methylglutaroyl)glucoside) and other anthocyanins from wild and cultivated blackberries. *Journal of Agricultural and Food Chemistry* 59(13), 7436–7440.

Kaume, L., Howard, L.R. and Devareddy, L. (2012) The blackberry fruit: a review on its composition and chemistry, metabolism and bioavailability, and health benefits. *Journal of Agricultural and Food Chemistry* 60(23), 5716–5727.

Kolniak-Ostek, J., Kucharska, A.Z., Sokol-Letowska, A. and Fecka, I. (2015) Characterization of phenolic compounds of thorny and thornless blackberries. *Journal of Agricultural and Food Chemistry* 63(11), 3012–3021.

Kool, M.M., Comeskey D.J., Cooney J.M. and McGhie, T.K. (2010) Structural identification of the main ellagitannins of a Boysenberry (*Rubus loganbaccus* x *baileyanus* Britt.) extract by LC-ESI-MS/MS, MALDI-TOF-MS and NMR spectroscopy. *Food Chemistry* 119(4), 1535–1543.

Landete, J.M. (2011) Ellagitannins, ellagic acid and their derived metabolites: a review about source, metabolism, functions and health. *Food Research International* 44(5), 1150–1160.

Landete, J.M. (2012) Updated knowledge about polyphenols: functions, bioavailability, metabolism, and health. *Critical Reviews in Food Science and Nutrition* 52(10), 936–948.

Lee, J. (2014) Marketplace analysis demonstrates quality control standards needed for black raspberry dietary supplements. *Plant Foods for Human Nutrition* 69(2), 161–167.

Lee, J. (2015) Sorbitol, *Rubus* fruit, and misconception. *Food Chemistry* 166, 616–622.

Lee, J., Finn, C.E. and Wrolstad, R.E. (2004) Anthocyanin pigment and total phenolic content of three *Vaccinium* species native to the Pacific Northwest of North America. *HortScience*, 39(5), 959–964.

Lee, J., Dossett, M. and Finn, C.E. (2012) *Rubus* fruit phenolic research: the good, the bad, and the confusing. *Food Chemistry* 130(4), 785–796.

Lei, Z., Jervis, J. and Helm, R.F. (2001) Use of methanolysis for the determination of total ellagic and gallic acid contents of wood and food products. *Journal of Agricultural and Food Chemistry* 49(3), 1165–1168.

Marinova, D. and Ribarova, F. (2007) HPLC determination of carotenoids in Bulgarian berries. *Journal of Food Composition and Analysis* 20(5), 370–374.

Mertz, C., Cheynier, V., Gunata, Z. and Brat, P. (2007) Analysis of phenolic compounds in two blackberry species (*Rubus glaucus* and *Rubus adenotrichus*) by high-performance liquid chromatography with diode array detection and electrospray ion trap mass spectrometry analysis. *Journal of Agricultural and Food Chemistry* 55(21), 8616–8624.

Mikulic-Petkovsek, M., Schmitzer, V., Slatnar, A., Stampar, F. and Veberic, R. (2012) Composition of sugars, organic acids, and total phenolics in 25 wild or cultivated berry species. *Journal of Food Science* 77(10), C1064–C1070.

Perez-Gallardo, A., Garcia-Almendarez, B., Barbosa-Canovas, G., Pimentel-Gonzalez, D., Reyes-Gonzalez, L.R. and Regalado, C. (2015) Effect of starch-beeswax coating on quality parameters of blackberries (*Rubus* spp.). *Journal of Food Science and Technology* 52(9), 5601–5610.

Pojer, E., Mattivi, F., Johnson, D. and Stockley, C.S. (2013) The case for anthocyanin consumption to promote human health: a review. *Comprehensive Reviews in Food Science and Food Safety* 12(5), 483–508.

Ramos-Solano, B., Garcia-Villaraco, A., Gutierrex-Manero, F.J., Bonilla, L.A. and Garcia-Seco, D. (2014) Annual changes in bioactive contents and production in field-grown blackberry after inoculation with *Pseudomonas fluorescens*. *Plant Physiology and Biochemistry* 74, 1–8.

Ramos-Solano, B., Algar, E., Gutierrez-Manero, F.J., Bonilla, A., Lucas, J.A. and Garcia-Seco, D. (2015) Bacterial bioeffectors delay postharvest fungal growth and modify total phenolics, flavonoids and anthocyanins in blackberries. *LWT- Food Science and Technology* 61(2), 437–443.

Sangiovanni, E., Vrhovsek, U., Rossoni, G., Colombo, E., Brunelli, C., Brembati, L., Truvulzio, S., Gasperotti, M., Mattivi, F., Bosisio, E. and Dell'Agli, M. (2013) Ellagitannins from *Rubus* berries for the control of gastric inflammation: *in vitro* and *in vivo* studies. *PLoS One* 8(8), p. e71762.

Scalzo, J., Currie, A., Stephens, J., McGhie, T. and Alspach, P. (2008) The anthocyanin composition of difference *Vaccinium, Ribes* and *Rubus* genotypes. *Biofactors* 34(1), 13–21.

Sellappan, S., Akoh, C.C. and Krewer, G. (2002) Phenolic compounds and antioxidant capacity of Georgia-grown blueberries and blackberries. *Journal of Agricultural and Food Chemistry* 50(8), 2432–2438.

Selma, M.V., Espin, J.C. and Tomas-Barberan, F.A. (2009) Interaction between phenolics and gut microbiota: role in human health. *Journal of Agricultural and Food Chemistry* 57(15), 6485–6501.

Sousa, M.B., Canet, W., Alvarez, M.D. and Fernandez, C. (2007) Effect of processing on the texture and sensory attributes of raspberry (cv. Heritage) and blackberry (cv. Thornfree). *Journal of Food Engineering* 78(1), 9–21.

Stintzing, F.C., Stintzing, A.S., Carle, R. and Wrolstad, R.E. (2002a) A novel zitterionic anthocyanin from evergreen blackberry (*Rubus laciniatus* Willd.). *Journal of Agricultural and Food Chemistry* 50(2), 396–399.

Stintzing, F.C., Stintzing, A.S., Carle, R., Frei, B. and Wrolstad, R.E. (2002b) Color and antioxidant properties of cyanidin-based anthocyanin pigments. *Journal of Agricultural and Food Chemistry* 50(21), 6172–6181.

Tavares, L., Figueira, I., McDougall, G.J., Vieira, H.L.A., Stewart, D., Alves, P.M., Ferreira, R.B. and Santos, C.N. (2013) Neuroprotective effects of digested polyphenols from wild blackberry species. *European Journal of Nutrition* 52(1), 225–236.

Thomas, R.H., Woods, F.M., Dozier, W.A., Ebel, R.C., Nesbitt, M., Wilkins, B. and Himelrick, D.G. (2005) Cultivar variation in physicochemical and antioxidant activity of Alabama-grown blackberries. *Small Fruits Review* 4(2), 57–71.

Tomas-Barberan, F.A., Garcia-Villalba, R., Gonzalez-Sarrias, A., Selma, M.V. and Espin, J.C. (2014) Ellagic acid metabolism by human gut microbiota: consistent observation of three urolithin phenotypes in intervention trials, independent of food source, age, and health status. *Journal of Agricultural and Food Chemistry* 62(28), 6535–6538.

USDA (2015) United States Department of Agriculture, National Nutrient database. Available at: http://ndb.nal.usda.gov (accessed July 10, 2015).

Van Duyn, M.A.S. and Pivonka, E. (2000) Overview of the health benefits of fruit and vegetable consumption for the dietetics professional: selected literature. *Journal of the American Dietetic Association* 100(12), 1511–1521.

Van Hoed, V., Barbouche, I., De Clercq, N., Dewettinck, K., Slah, M., Leber, E. and Verhe, R. (2011) Influence of filtering of cold pressed seed oils on their antioxidant profile and quality characteristics. *Food Chemistry* 127(4), 1848–1855.

Vasco, C., Riihinen, K., Ruales, J. and Kemal-Eldin, A. (2009) Phenolic compounds in Rosaceae fruits from Ecuador. *Journal of Agricultural and Food Chemistry* 57(4), 1204–1212.

Veberic, R., Stampar, F., Schmitzer, V., Cunja, V., Zupan, A., Koron, D. and Mikulic-Petkovsek, M. (2014) Changes in the contents of anthocyanins and other compounds in blackberry fruits due to freezing and long-term frozen storage. *Journal of Agricultural and Food Chemistry* 62(29), 6926–6935.

Veberic, R., Slatnar, A., Bizjak, J., Stampar, F. and Mikulic-Petkovsek, M. (2015) Anthocyanin composition of different wild and cultivated berry species. *LWT – Food Science and Technology* 60(1), 509–517.

Vrhovsek, U., Palchetti, A., Reniero, F., Guillou, C., Masuero, D. and Mattivi, F. (2006) Concentration and mean degree of polymerization of *Rubus* ellagitannins evaluated by optimized acid methanolysis. *Journal of Agricultural and Food Chemistry* 54(12), 4469–4475.

Vrhovsek, U., Giongo, L., Mattivi, F. and Viola, R. (2008) A survey of ellagitannin content in raspberry and blackberry cultivars grown in Trentino (Italy). *European Food Research and Technology* 226, 817–824.

Wang, S.Y., Bowman, L. and Ding, M. (2008) Methyl jasmonate enhances antioxidant activity and flavonoid content in blackberries (*Rubus* sp.) and promotes antiproliferation of human cancer cells. *Food Chemistry* 107(3), 1261–1269.

Wrolstad, R.E., Culbertson, J.D., Nagaki, D.A. and Madero, C.F. (1980) Sugars and nonvolatile acids of blackberries. *Journal of Agricultural and Food Chemistry* 28, 553–558.

Wrolstad, R.E., Cornwell, C.J., Culbertson, J.D. and Reyes, F.G.R. (1981) Establishing criteria for determining the authenticity of fruit juice concentrates. In: Teranishi, R. and Barrera-Benitez, H. (eds.) *Quality of Selected Fruits and Vegetables of North America*, Chapter 7. ACS symposium series 170, 77–93.

5

CULTIVAR DEVELOPMENT AND SELECTION

Chad E. Finn[1,*] and John R. Clark[2]

[1]USDA-ARS, HCRL, Corvallis, Oregon, USA; [2]University of Arkansas, Fayetteville, Arkansas, USA

INTRODUCTION

Successful blackberry production and marketing depends on planting cultivars that are adapted to the region, efficiently produce high yields, and have the fruit quality the market, whether local or distant, demands. Blackberry breeding programs have developed cultivars that consumers like to eat and these have provided the basis for the expansion of the blackberry industry. No cultivar is ever perfect, and so breeding programs are on a constant quest to meet the changing demands of the industry and market place. The background of the most important non-proprietary blackberry cultivars with a focus on more modern cultivar is given in Table 5.1.

Excellent reviews of blackberry breeding have been written that go into much more depth than will be covered here and may be useful for those looking for that level of detail. These include Jennings (1988), Hall (1990), Daubeny (1996), Clark *et al.* (2007), Clark and Finn (2011), and Finn and Clark (2012). See Chapter 1 for patented cultivars.

ORIGIN AND REGIONAL ADAPTATION

Blackberry breeding is a relatively recent phenomenon dating back to the late 1800s and early 1900s. Until that time, most cultivars were selections from the wild. 'Dorchester' and 'Lawton' (synonym 'New Rochelle') are generally considered to be the first cultivars selected from wild species in North America (Hedrick, 1925), although the European wild-selected cultivars 'Himalaya,' from *Rubus armeniacus* Focke, and 'Evergreen' from *R. laciniatus* Willd. were introduced in the United States Pacific Northwest (PNW) during the mid-1800s by pioneers on the Oregon Trail. Private breeding efforts, as those of

---

* Corresponding author: Chad.Finn@ars.usda.gov

Table 5.1. Background of most important non-proprietary blackberry cultivars. For older or less common cultivars see Clark et al. (2007).

Cultivar	Origin	Type	Thorny or source of thornlessness	Year patented or released if not patented	Patent number	Primary use	Major strengths	Major weaknesses
Apache	Univ. of Arkansas	Erect	MT Thornless	2001	PP11,865	Wholesale and local fresh	Large firm fruit	Prone to white drupe
Arapaho	Univ. of Arkansas	Erect	MT Thornless	1993	PP8,510	Wholesale and local fresh	Earliest ripening erect type	Low yield
Black Diamond	USDA-ARS, OR	Trailing	AT Thornless	2004	No	Processing; local fresh	High yields; very good-quality fruit; machine harvestable; good cold hardiness	Open canopy increases incidence of sun damage
Black Magic™ (APF-77)	Univ. of Arkansas	Primocane fruiting, Erect	Thorny	2014	PP24,249	Home garden/ local fresh	Some heat tolerance on primocane fruiting	Soft fruit
Black Pearl	USDA-ARS, OR	Trailing	AT Thornless	2004	No	Processing; local fresh	Moderate yields; very good quality fruit for fresh or processed; machine harvestable	Inconsistent yield
Black Satin	USDA-ARS, MD/IL	Semi-erect	MT Thornless	1974	No	Local fresh	High yields	Soft, poor fruit quality
Boysen	Private, R. Boysen, CA	Trailing	Thorny	1935	No	Processing; local fresh	Unique and excellent fruit quality	Fair yield; difficult marketing

Cultivar	Origin	Habit	Type	Year	Patent	Use	Strengths	Weaknesses
Čačanska Bestrna	Cacak Res. Stat., Serbia	Semi-erect	MT Thornless	1987	No	Processing; local fresh	High yields	Poor fruit quality; very tart
Chester Thornless	USDA-ARS, MD/IL	Semi-erect	MT Thornless	1985	No	Wholesale and local fresh	Very high yields; ships well	Poor flavor
Columbia Giant	USDA-ARS, OR	Trailing	LL Thornless	2015	PPAF	Local fresh and processing	High yields; excellent-quality fruit; machine harvestable	Potential commercial yield not known
Columbia Star	USDA-ARS, OR	Trailing	LL Thornless	2015	PP25,532	Processing; local fresh	High yields; very large, very good-quality fruit; machine harvestable	Potential commercial yield not known
Columbia Sunrise	USDA-ARS, OR	Trailing	LL Thornless	2016	PPAF	Local fresh; processing	Very early production; excellent fruit quality, particularly flavor; machine harvestable	Potential commercial yield not known, but expected to be moderate
Eclipse	USDA-ARS, OR	Semi-erect	MT Thornless	2017	PAAF	Wholesale and local fresh	High yields; excellent flavor and firm in 'Loch Ness' season	Potential commercial yield not known
Everthornless	Univ. of Illinois	Trailing	Somaclonal selection of L1 chimera in 'Thornless Evergreen'	1995	PP9,407	Processing	High yield; very late season	Fair flavor; soft; prone to botrytis fruit rot
Galaxy	USDA-ARS, OR	Semi-erect	MT Thornless	2017	PPAF	Wholesale and local fresh	High yields; very good flavor, large and firm in 'Loch Ness' season	Potential commercial yield not known

continued

Table 5.1. *continued.*

Cultivar	Origin	Type	Thorny or source of thornlessness	Year patented or released if not patented	Patent number	Primary use	Major strengths	Major weaknesses
Hall's Beauty	USDA-ARS, OR	Trailing	LL Thornless	2017	PPAF	Local fresh and processing	High yields; very sweet, firm fruit in early season; extremely large, showy flowers	Potential commercial yield not known
Heaven Can Wait™ (A-1960)	Univ. of Arkansas	Erect	MT Thornless	2013	PP26,405	Home garden/ local fresh	Excellent flavor	Variable yields
Hull Thornless	USDA-ARS, MD/IL	Semi-erect	MT Thornless	1981	No	Local fresh	High yields	Soft, poor fruit quality
Illini Hardy	Univ. of Illinois	Erect	Thorny	1993	PP8,331	Local fresh	Cold hardy	Poor-quality fruit; poor yield
Karaka Black	NZ Plant and Food Res.	Trailing	Thorny	2003	PPAF	Wholesale and local fresh	Firm; early ripening	Fair flavor
Kiowa	Univ. of Arkansas	Erect	Thorny	1997	PP9,861	Local fresh	Large fruit	Soft fruit
Kotata	USDA-ARS, OR	Trailing	Thorny	1984	No	Wholesale fresh	Moderate yield; cold hardy; excellent fruit quality	Small–moderate size; can bleed
Loch Maree	James Hutton Inst., Scotland	Semi-erect	MT Thornless	2003	No	Wholesale and local fresh	Moderate yields; firm, excellent flavor	High quality; size similar to Loch Tay

Cultivar	Source	Habit	Thorns	Year	Patent	Market	Strengths	Weaknesses
Loch Ness	James Hutton Inst., Scotland	Semi-erect	MT Thornless	1989	PP6,782	Wholesale and local fresh	Moderate yields; firm	Poor fruit flavor
Loch Tay	James Hutton Inst., Scotland	Semi-erect	MT Thornless	2003	No	Wholesale and local fresh	Moderate yields; earliest ripening semi-erect	Fair quality; size drops quickly after primary fruit
Logan	Private, J. Logan, Santa Cruz, CA	Trailing	Thorny	1881	No	Processing; local fresh	Unique and excellent fruit quality	Fair yield; cannot be machine harvested, but difficult to pick for fresh
Marion	USDA-ARS, OR	Trailing	Thorny	1956	No	Processing; local fresh	Outstanding fruit quality, especially flavor; machine harvestable	Spines, marginal winter hardiness, soft turns purple after harvest.
Metolius	USDA-ARS, OR	Trailing	Thorny	2004	No	Wholesale fresh	Moderate yield; very early ripening; excellent fruit quality	Moderate yield; medium size
Natchez	Univ. of Arkansas	Erect	MT Thornless	2010	PP20,891	Wholesale and local fresh	Productive, large fruit, early	Can overcrop, berries can be tart early season
Navaho	Univ. of Arkansas	Erect	MT Thornless	1989	PP6,679	Wholesale and local fresh	Firm	Orange rust susceptible

continued

Table 5.1. *continued.*

Cultivar	Origin	Type	Thorny or source of thornlessness	Year patented or released if not patented	Patent number	Primary use	Major strengths	Major weaknesses
Newberry	USDA-ARS, OR	Trailing	Thorny	2010	No	Wholesale and local fresh	Very high yield; unique and excellent fruit quality; sold as 'non-bleeding' Boysen	Where does it fit in market?
Obsidian	USDA-ARS, OR	Trailing	Thorny	2004	No	Wholesale fresh	Very high yield; very early ripening; excellent fruit quality	Borderline firmness for wholesale fresh
Olallie	USDA-ARS, OR	Trailing	Thorny	1950	No	Local fresh	Excellent flavor	Too soft
Onyx	USDA-ARS, OR	Trailing	Thorny	2011	PP22,358	Wholesale and local fresh	Moderate yield; excellent fruit quality	Small–moderate size
Ouachita	Univ. of Arkansas	Erect	MT Thornless	2006	PP17,162	Wholesale and local fresh	High yields; firm fruit for shipping	Not fully adapted to low chill environments
Prime-Ark® 45	Univ. of Arkansas	Primocane fruiting, Erect	Thorny	2012	PP22,449	Wholesale and local fresh	Primocane fruiting; firm fruit; productive	Reversion can develop
Prime-Ark® Freedom	Univ. of Arkansas	Primocane fruiting, Erect	MT Thornless	2013	PPAF	Home garden, local fresh	Primocane fruiting/large berry	Soft; tendency for double fruit

Cultivar	Origin	Habit	Thorns	Year	PPAF	Use	Characteristics	Comments
Prime-Ark® Traveler	Univ. of Arkansas	Primocane fruiting, Erect	MT Thornless	2014	PPAF	Wholesale and local fresh	Primocane fruiting; firm; sweet	Not widely evaluated
Ranui	NZ Plant and Food Res.	Trailing	Thorny	1984	No	Local fresh	Excellent flavor	Very early, red-purple fruit
Silvan	Dept. of Agriculture, Victoria, Australia	Trailing	Thorny	1984	No	Processing; local fresh	Excellent flavor; more cold hardy by trailing standards	Very soft; prone to purple fruit
Tayberry	James Hutton Inst., Scotland	Trailing	Thorny	1979	No	Processing; local fresh	Unique and excellent fruit quality	Very difficult to pick by hand
Triple Crown	USDA-ARS, MD/OR	Semi-erect	MT Thornless	1996	No	Wholesale and local fresh	High yield; excellent fruit quality	Too soft for most shipping
Tupy	EMBRAPA, Brazil	Erect	Thorny	2003	No	Wholesale and local fresh	Low chilling; large; adapted to Central Mexico	Reversion tendency
Von	NC State Univ.	Erect	MT Thornless	2012	PPAF	Wholesale and local fresh	Later season, excellent postharvest handling	Not widely evaluated
Wild Treasure	USDA-ARS, OR	Trailing	AT Thornless	2010	No	Processing; local fresh	High quality, very small fruit	Moderate yield; very small fruit

Judge James Logan in the late 1800s and Byrnes M. Young in the early 1900s, along with the public programs with the United States Department of Agriculture-Agricultural Research Service (USDA-ARS) in Oregon and Maryland and the New York Fruit Experiment Station in Geneva, NY, USA, began to release cultivars selected from crosses. Breeding work in blackberries was also initiated at the John Innes Institute in the 1920s, resulting in the transfer of recessive spinelessness from the diploid species *R. ulmifolius* to the tetraploid John Innes and then to 'Merton Thornless,' which was released in 1934. A significant shift in the worldwide blackberry industry began in the mid-1900s, as cultivars from these breeding programs began to dominate the commercial acreage. While strong from the early 1900s, the advent of machine harvesting in the 1960s worked well with cultivars like 'Marion,' developed by the USDA-ARS in Oregon, allowed the industry in the USA Pacific Northwest to establish its predominance for the processing industry. This continued and further improved as machine-harvestable, high-yielding, thornless cultivars with high fruit quality, such as 'Black Diamond,' were developed (Finn et al., 2005b; Yorgey and Finn, 2005). The development of thornless cultivars, such as 'Thornfree' and 'Smoothstem' in the late 1960s by the USDA-ARS in Maryland, was a major improvement for local fresh-market sales. The early releases, such as 'Cherokee' and 'Comanche' in 1974, from the 1964-established program at the University of Arkansas provided the groundwork for an expanded fresh-market industry. With the exception of 'Cherokee,' these cultivars were only suited for local sales as they were too perishable for long-distance shipping. The cultivars from these programs could take the rigors of shipping. These cultivars, along with the amazing 'Tupy' developed by Empresa Brasileira de Pesquisa Agropecuária (EMBRAPA) in Brazil, but planted extensively in Mexico, allowed the fresh-market industry to blossom from a very short-season crop with questionable quality to a crop that was available year-round in the grocery store (Finn and Clark, 2011). The recent widespread commercialization of 'Prime-Ark® 45,' the first commercially viable primocane-fruiting cultivar for the wholesale fresh market from the University of Arkansas (Clark and Perkins-Veazie, 2011), allows growers flexibility in managing the crop for different harvest seasons and should only help further drive the expansion of the blackberry market.

As has been addressed in Chapters 1 and 2, there are three predominant types of blackberries, including erect (includes the primocane-fruiting types), semi-erect, and trailing. The erect and semi-erect types were largely derived from a similar germplasm pool that included eastern North American blackberry species, such as *R. allegheniensis* Porter and R. argutus Link, and share many common characteristics. The trailing types were largely developed from the western USA trailing blackberry *R. ursinus* Cham. et Schltdl. but, beginning with Judge Logan's breeding of 'Loganberry,' they also have a substantial red raspberry (*R. idaeus* L.) genetic component.

Pacific Northwest USA

The PNW USA industry in the early 1900s was based largely on 'Logan,' a blackberry×raspberry hybrid, and 'Evergreen' or its sport 'Thornless Evergreen' (Logan, 1955; Waldo, 1977). The other important raspberry/ blackberry hybrid, 'Boysen,' was also important from the early 1900s to today (Hall et al., 2002). Waldo's release of 'Pacific' and 'Cascade' blackberries, derived largely from *R. ursinus*, in 1940, marked the beginning of the use of cultivars developed in the PNW USA (Finn, 2001). 'Santiam' (synonym 'Ideal Wild'), found in the wild, was either a hermaphroditic selection of *R. ursinus* or the offspring of an *R. ursinus*×'Logan' cross, and was also commonly grown in the PNW USA. The release of 'Marion' in 1956 followed by its adoption in the 1960s put in motion a steady but strong shift from an industry of 'Logan,' 'Evergreen,' and 'Boysen' to one that by 1980 was predominantly 'Marion' (Finn et al., 1997). A mix of about 60–70% 'Marion,' 20–30% 'Thornless Evergreen,' with the remainder a variety of cultivars was typical in the region until around 2000. In 2004, the release of the thornless 'Black Diamond' led to a major transition in the industry; it has been the most popular cultivar planted each year since its release (Finn et al., 2005a). As 'Thornless Evergreen' and 'Marion' plantings are taken out due to poor performance, they are being replaced by 'Black Diamond' or the recent release 'Columbia Star' (Finn et al., 2014). 'Logan' has almost disappeared from the northwest industry and the 'Boysen' industry is becoming very small. A thornless sport of 'Marion,' 'Willamette Valley Thornless Marion,' has been identified and is being trialed, but it is unclear how widely it will be planted (Heidt, 2014).

Eastern United States

Eastern USA production was primarily only for local markets until the early 2000s. Production was limited partially due to limited cultivar choices, but also by winter hardiness limitations in the Midwest and northward, and a major disease in the South. Until the early 2000s, the markets were largely limited to fresh, on-farm sales, pick-your-own, and farmers' markets. For these markets, postharvest handling was not a major requirement of a cultivar. The advent of improved cultivars, including 'Chester Thornless' and 'Triple Crown,' provided for expanded fresh-market production options in the East and Midwest USA. The major limitation in the South, the disease double blossom/ rosette (caused by *Cercosporella rubi* (G. Wint.) Plakidas), was overcome by thornless introductions from the University of Arkansas (Clark, 2005). The first of these was 'Navaho,' followed by several others (e.g. 'Arapaho,' 'Apache,' 'Ouachita') with greatly reduced susceptibility to the disease. Further, the Arkansas thornless releases had improved fruit firmness, and served as the basis for a shipping industry that began about 2002 (Clark, 2005). Commercial plantings for shipping located in the South were established primarily in Georgia, North Carolina, South Carolina, Arkansas, and Texas, and totaled approximately 1500–2000 acres by 2015 (Clark, 2016). With the commercial adaptation of the rotatable cross-arm trellis along with row cover protection from damaging winter temperatures, these same shipping-quality cultivars began to be utilized in the Midwest (Takeda et al., 2013). This has led to a small commercial shipping industry where only local sales were possible prior (Clark, 2016). A major contributor to the domestic shipping industry has been the importance of blackberries in the retail markets in the USA brought about by Mexican imports that are available generally from October to June. Domestic shippers desired a continuous supply of berries to complete the annual marketing cycle; this has contributed to the Southern and now Midwest USA supplying fruit for the national retail fresh market.

California, USA

'Boysen,' a raspberry × blackberry hybrid that was originally found on a farm in California in the early 1900s, was the major cultivar in California for much of the 1900s (Hall et al., 2002). 'Olallie,' a 1950 release from the USDA-ARS in Oregon, USA, and a parent of 'Marion,' always performed better in California than Oregon and was a very important cultivar there until the early 1990s. While 'Olallie' has outstanding fruit quality, it is very soft and as the fresh, wholesale blackberry market rapidly expanded in the 1990s, it was very quickly replaced by semi-erect and erect, proprietary cultivars that could be shipped across the country. Since the early 2000s, blackberry production for the shipping market has greatly expanded in California, which is the leading fresh-market producer in the USA. This has been brought about by several key factors; paramount was the development of improved cultivars. Proprietary cultivars developed by Driscoll Strawberry Associates and cultivars developed by the University of Arkansas have provided for major expansion of production in California. Due to moderate summer temperatures in coastal regions of California, the Arkansas primocane-fruiting cultivars have been most successful there. This has led to major expansion in production for late summer to autumn production as improved primocane-fruiting cultivars have been released.

Mexico

The rise of the Mexican blackberry has been meteoric. In the 1990s, there was a nascent industry based on 'Brazos.' Beginning in the early 2000s, the combination of the Brazilian cultivar, 'Tupy,' with new and creative plant manipulation practices, allowed the industry to expand and become the leading blackberry production area in the world. 'Tupy' is very thorny and, while the fruit ship well, many cultivars have much better eating quality. However, in the central Mexican highlands, it can be intensively manipulated to produce in almost any month of the year. As of 2016, a limited number of other cultivars are in early stages of trial or in small commercial plantings, including primocane-fruiting developments. It is anticipated that expanded floricane- and primocane-fruiting cultivar options will further enhance this major production region.

Europe

Semi-erect cultivars have been the predominant cultivars grown in Europe. 'Loch Ness,' from the James Hutton Institute (formerly the Scottish Crop Research Institute), has been the main cultivar for the wholesale fresh market throughout the more temperate portions of Europe. In Spain, it is grown in pots for storage and sequential harvest over much of the year. Serbia has a long history of producing red raspberries and blackberries for the processing market and relied predominantly on 'Čačanska Bestrna' and 'Thornfree.' Turkey has major production areas and is showing further promise for the production of semi-erect and primocane-fruiting cultivars. While fresh market blackberries produced in mild climate regions of Europe, such as Spain and Portugal, are beginning to be explored, no specific cultivars dominate production. However, with expanded floricane- and primocane-fruiting options, production for the expanded European market is assured.

New Zealand

The New Zealand industry has been best known for its processing market. 'Boysenberry,' in the 1980s, produced over 5000 tonnes (5500T) per year, but 2015 production was around 30% of this total (Hall, personal communication). Nevertheless, 'Boysenberry' remains a key element in hybrid berry production in New Zealand. A 'Boysenberry' breeding program was started in the late 1970s, but as of 2016, no cultivar has been released that is satisfactory to replace the clones. 'Riwaka Choice,' 'Tasman,' 'Mapua,' 'McNicol's Choice' and 'LDE2' were selected in the 1990s from New Zealand Boysenberry plantings. These comprised almost all of the 2015 production. 'Youngberry' has also been grown in New Zealand as 'early Boysen' and a reselected clone of this cultivar, 'RS4,' has replaced the standard cultivar.

Other hybrid berry cultivars, such as 'Karaka Black' and 'Ranui,' developed in New Zealand, have been important for local and regional fresh sales (Hall, personal communication). 'Karaka Black' is also grown in Australia, the United Kingdom and Europe. 'Ranui' is grown in New Zealand as a fresh market very early 'Boysen,' and in Australia, where it has been renamed 'Crimson Giant' blackberry.

China

China is the center for diversity for *Rubus*, and therefore, is host to hundreds of *Rubus* species, most of which harvest like a raspberry, rather than a blackberry; many of these are harvested in the wild. Cultivated blackberry plantings, which are mainly cultivars that were developed in the eastern USA, have only been planted in the past 25 years. Originally, 'Hull Thornless' and 'Chester Thornless' were the predominant cultivars, but more recently, 'Triple Crown' and some of the Arkansas cultivars, such as 'Navaho,' have been widely trialed.

South America

Until the development of the Mexican industry, Chile was the predominant supplier of fresh fruit in the off-season in the USA and Canada. 'Cherokee,' 'Chester Thornless,' and 'Navaho' were the common cultivars planted for the fresh market; those plantings have largely been removed since 2000 or their fruit has been diverted to the processed or local fresh market. 'Boysen,' 'Marion,' and fruit harvested from wild blackberries make up most of the fruit supply for the processing market. Since 2000, production for the domestic market for fresh berries has gained interest in Chile, as well as other South American countries, including Brazil and Argentina.

Other production areas

Australian production continues to be limited as adapted cultivar options are few. Due to rigid biosecurity laws in Australia, the opportunities to import new cultivars are very limited. European blackberries are widely naturalized and many older cultivars, from which modern cultivars have been developed, were introduced and some are still grown. Production of blackberries in other areas of the South Pacific and southern Asia is limited also. South Africa has limited production that is utilized in the domestic market. Production in other countries could increase, but currently is limited due to lack of adapted cultivars or unfamiliarity with this fruit.

BREEDING PROGRAMS

United States Department of Agriculture – Agricultural Research Service (USDA-ARS)

The USDA-ARS began breeding blackberries 1919 with crosses in Georgia that were evaluated in Maryland, USA, and from which 'Brainerd' was released (Darrow, 1937). The USDA-ARS programs in Beltsville, MD, and Carbondale, Illinois, were of critical importance in bringing the 'Merton Thornless' source of thornless to commercial cultivars. The first significant releases were 'Smoothstem' and 'Thornfree' in 1966 from Beltsville, followed by 'Black Satin' and 'Dirksen Thornless' from Carbondale in 1974. While this group made commercial, semi-erect, thornless blackberries a reality, the later releases of 'Chester Thornless' in 1985 from Carbondale and 'Triple Crown' in 1996 from Beltsville would prove to be the most commercially viable.

The USDA-ARS program in Oregon began in 1927 and continues today. This program brought the native western trailing blackberry (*R. ursinus*) germplasm into commercially viable cultivars (Finn, 2001). The first releases were 'Cascade' and 'Pacific' in 1940. Since the Pacific Northwest in the early 1900s was very distant from the major population centers in much of the USA, it was a major production area for fruit that could be processed (canned or frozen) and sent back East. As a result, the breeding program was focused on trailing blackberries suited for processing. 'Olallie,' released in 1950, was often marketed as 'Olallieberries' and was commonly grown in California for the wholesale fresh market until the early 1990s. Its offspring 'Marion,' marketed as 'Marionberry,' released in 1956, became a standard by the late 1960s and the predominant cultivar grown in the Northwest USA beginning in the 1970s until 2016. While 'Marion' in 2015 is producing the most tonnage in the Northwest, the thornless 'Black Diamond' has been the number 1 blackberry being planted since 2004; the release of the thornless 'Columbia Star' surpassed 'Marion' in plant sales in 2014 (Finn *et al.*, 2014). Other historically important cultivars from this program include 'Kotata,' which has been important for the fresh and processed market, and 'Obsidian' and 'Metolius,' which are commonly grown as the earliest ripening, fresh-market blackberries (Finn *et al.*, 2005b, 2005c).

University of Arkansas

The University of Arkansas began a blackberry-breeding program in 1964. The program focused initially on large fruit size, high fruit quality, erect canes, extended harvest season (primarily earliness of ripening), thornlessness, and high productivity (Clark, 2016). Two parents, the New York-developed

'Darrow,' and 'Brazos' from Texas, plus other parents, such as the USDA-ARS thornless germplasm, were major components of the foundation germplasm. After initial releases of several thorny cultivars in the 1970s to mid-1980s, the thornless 'Navaho' was released in 1989. It has had a major impact because it was not only thornless, but also, one of the first blackberries with substantial postharvest storage potential. Subsequent releases expanded thornless options, with 'Ouachita' contributing to the expansion of the wholesale shipping-market. Primocane-fruiting focus was developed in the 1990s, with the first cultivars 'Prime-Jan®' and 'Prime-Jim®' released in 2004 (Clark et al., 2005). These cultivars were primarily recommended for home garden production, although 'Prime-Jan®' had some commercial potential. 'Prime-Ark® 45' was released in 2009, providing the first primocane-fruiting blackberry with post-harvest handling capability for shipping to commercial markets (Clark and Perkins-Veazie, 2011). The thornless 'Prime-Ark® Freedom' and 'Prime-Ark® Traveler' expanded options of this plant type, the latter cultivar recommended for shipping (Clark, 2014; Clark, 2016).

Empresa Brasileira de Pesquisa Agropecuária (EMBRAPA)

The breeding program based at EMBRAPA in Pelotas, Rio Grande de Sul, Brazil has been active since the late 1970s. This low-chill site has focused on fresh-market blackberries, using University of Arkansas cultivars and other adapted germplasm. Several cultivars have been released, the most important being 'Tupy,' which has become the most important fresh-market cultivar in the world, due primarily to its production in Central Mexico. It is also planted in many areas of Brazil.

Proprietary programs

The most important proprietary blackberry breeding program has been that of Driscoll Strawberry Associates (Watsonville, CA). The Driscoll program was very small until the early 1990s. As the blackberry industry began to expand rapidly in California, there was a need for cultivars that were firm enough to be shipped across country with truckloads of strawberries and red raspberries. 'Pecos,' 'Sleeping Beauty,' 'Zorro,' and 'Driscoll Cowles' were released in the early 2000s and were quickly planted substantially in California and Mexico (primarily 'Sleeping Beauty'). Newer releases, including the floricane-fruiting cultivars 'Dasha' and 'Victoria,' as well as the primocane-fruiting 'Elvira,' have expanded cultivar options. Other USA companies, such as Plant Sciences Inc. (Watsonville, CA), Five Aces Breeding (Oakland, MD), and Pacific Berry Breed-ing (Watsonville, CA) have active breeding programs and have released cultivars.

Other programs

A number of programs that either are no longer active or were fairly small have played a significant role in producing important cultivars. 'Bedford Giant,' a hexaploid with a raspberry × blackberry parent ('Veitchberry') was released by a private breeder, Laxton, in the 1930s and is still an important cultivar for local sales in the United Kingdom (Jennings, 1988). In the 1960s, the Scottish Horticultural Research Institute (currently the James Hutton Institute), brought together the germplasm pools of the eastern and western North American programs in the UK to produce 'Loch Ness' in 1988 and later 'Loch Tay' and 'Loch Maree.' 'Loch Ness' is probably the most widely planted cultivar in Europe outside of Serbia. In Serbia, the thornless 'Čačak Bestrna' from the Čačak Research Station has been a very important cultivar for processing. North Carolina State University has conducted blackberry breeding for many years, and maintains an effort in cultivar improvement, with 'Von,' a thornless, fresh-market cultivar, the most recent release.

The New Zealand program was in its prime during the 1970s and 1980s, a comparable large effort, with the Arkansas and Oregon programs, with efforts invested in developing new Boysenberry cultivars, improved 'Loganberry' types, and a range of hybrid blackberries and tetraploid blackberries. Advances made included the development of a new dominant gene for thornlessness (spinelessness) from the thornless Loganberry 'L654,' which provided thornlessness for 'Columbia Star.' This thornlessness was transferred to tetraploid types like the Arkansas blackberries, to black fruited hexaploid types, and also to other breeding material at each ploidy level from 3× up to 12×. A little work continues, but the majority of the germplasm generated remains and could be yet used in future.

BREEDING GOALS

The complexity of blackberry genetics, including the mixing of diverse species in the recent history of most cultivars, combined with very small number of breeding programs, has made some of the traditional heritability and other genetic studies scarce, and often impractical or impossible (Thompson, 1997; Meng and Finn, 1997). The eastern blackberries are tetraploid (2n = 4× = 48), largely derived from the *R. allegheniensis* and *R. argutus* with thornlessness from *R. ulmifolius*, through 'Merton Thornless.' The western blackberries consist of cultivars that are 6×, 7×, 8×, 9×, 10×-2, and 12× and while largely derived from *R. ursinus*, also have ancestry that traces to *R. allegheniensis*, *R. armeniacus*, *R. argutus*, *R. baileyanus* Britton, and *R. idaeus*. Breeding for most traits is quite straightforward for the traits that can be measured or scored easily. Superior parents for a trait or complementary parents for a series of traits are chosen and offspring that are improvements over their parents are selected.

Molecular markers are being developed to be able to better choose parents, to screen seedling populations for traits such as fruit quality, and to improve the efficiency of the breeding process.

FRUIT QUALITY

Flavor

Breeding for subjective traits, such as flavor, is extremely difficult. Whose taste buds does one trust? What do most consumers want? How does one compare fruit evaluated when cool in the morning vs. warm at midday? After or before rain/irrigation? Primary vs. secondary fruit?

At the seedling stage, breeders are primarily selecting for a well-balanced sweet/acid flavor with no off flavors and minimal to no bitterness. What is well-balanced varies with primary customers. If one is breeding for fruit that will be eaten fresh out of a clamshell, sweet fruit with no bitter off flavors is critical (Wang et al., 2005). For the processing market, flavors with more intensity and acidity are required, because the fruit is often a small part of the total product (i.e. ice cream), and often sugar is added. Many of the cultivars for processing have very aromatic fruit that helps define their overall flavor (Du et al., 2010). Once breeders have identified selections that meet the minimum criteria, greater levels of scrutiny can be brought to bear as the selection begins to differentiate itself as superior from other selections. This may involve evaluating the flavor before and after storage, if fresh or as a thawed frozen fruit, puree, or juice is used in the processed product. Also, as the selection is moved along in the evaluation process, more evaluators and growing locations will be involved in the flavor evaluation. Flavor evaluations usually begin with a single breeder but often expand to others in the program and then to growers, packers, and other industry members and finally the target consumers.

Shipping

Production of blackberries for shipping to fresh markets was largely non-existent in the USA until the mid- to late 1990s. In general, the eastern USA germplasm has been most widely used thus far for breeding enhanced post-harvest storage potential. 'Navaho' was one of the first cultivars identified for postharvest storage potential, as it had much-improved fruit firmness compared to most prior releases. Breeding for shipping and storage garnered substantial attention in the early 1990s and onward in the Arkansas program. A key cooperation was begun with Perkins-Veazie, formerly with the USDA-ARS based in Lane, Oklahoma and the University of Arkansas. This cooperation included storage potential evaluations of new cultivars and advanced selections developed in the program, and a standard protocol was developed to subjectively evaluate overall storage potential using ratings of fruit firmness, leak, mold, and red drupelet development in storage (often referred to as reversion) (Clark and Perkins-Veazie, 2011). This protocol was used in identifying the shipping potential of 'Ouachita,' 'Natchez,' 'Prime-Ark® 45,' 'Osage,' and 'Prime-Ark® Traveler' (Clark and Moore, 2005, 2008; Clark and Perkins-Veazie, 2011; Clark, 2013, 2016). While less effort has been devoted to the adaptability of trailing cultivars to the fresh market, some of these cultivars have excellent fresh market potential (Fernandez-Salvador *et al.*, 2015). 'Karaka Black,' derived from 'Comanche' out of the Arkansas breeding program and from 'Aurora' from Oregon, has good fruit firmness and extended shelf life.

Breeding for shipping potential improvement includes several traits of focus. Traits including firmness and reversion are discussed in other sections of this chapter.

PROCESSING

Fruit for processing are commonly processed into a variety of bulk products that are then used to manufacture products for consumers, including individually quick frozen (IQF), block frozen, juice, puree, and dried/freeze-dried. While the berries might be made into something that is mostly fruit, such as a jam or fruit spread, they may also go into a product where the berries are a small part of the overall product, such as yogurt, cereal, or smoothie. For each of these products, it is essential that the berries retain their typical flavor and color. Some of the components that can be important to evaluate include soluble solids (sweetness), titratable acidity, pH, and drip loss. The importance of sweetness to flavor has been discussed, but sweetness also impacts how a product holds up in storage and how that berry's flavor interacts with the other components of the product. In general, high soluble solids levels are valued as packers can more easily meet the soluble solids requirements for their contracts, and, because the sugar is a preservative, the product maintains a higher quality level throughout the handling of the fruit. Different measures of acidity, titratable acidity, and pH are inextricably related. The pH is a measure of the acidity of the fruit and a pH below 3.5 is good in processed fruit products as the anthocyanins stay bright and red vs. graying. Titratable acidity is a measure of the buffering capacity of the fruit. Processors need to know the titratable acidity of a new cultivar as they begin to develop product formulations from that cultivar. Drip loss is simply a measure of the loss of the liquid lost when a berry is frozen and then thawed. The more important it is in a product to have entire berries or 'chunks' of fruit, the more important low drip loss becomes.

The retention of intense color through processing in addition to being a function of pH is also a function of the inherent anthocyanin levels.

Blackberries are naturally high in anthocyanins, but there is tremendous variability present for breeding for higher anthocyanin, and corresponding higher antioxidant, levels (Connor et al., 2005a, 2005b; Siriwoharn et al., 2004).

Color/color reversion

Blackberries marketed fresh should be black when the consumer views them. However, with the expansion of the shipping industry, the occurrence of red drupelets, termed reversion, has become of high importance. This phenomenon is not well understood, and appears to differ among cultivars and fruit temperature exposure after harvest. In fresh-market blackberry breeding, evaluations of reversion are among the highest priority activities. In the Arkansas program, reversion is rated after storage for seven or more days, and multiple harvests and years are normally utilized to fully evaluate the potential for reduced reversion occurrence (Clark and Perkins-Veazie, 2011).

In evaluating parents for firmness in the Arkansas program, a 'crispy' texture was identified. This texture was also found to provide for much-reduced occurrence of reversion (Salgado, 2015). This germplasm is being used extensively in crossing to provide offspring that are not only firm, but also lower in reversion. No cultivars have been released with this improved combination of traits, however.

Color retention after freezing and thawing can be important and is evaluated during a selection's evaluation. Important older cultivars, such as 'Marion,' often turn purple when processed while newer ones, such as 'Obsidian,' 'Black Diamond,' and 'Columbia Star,' tend to retain their black color.

HEAT AND ULTRAVIOLET LIGHT (UV) DAMAGE

As blackberry production has expanded in more varied climates in recent years, environmental impacts on fruit have become more common. Damage from heat is usually manifested when temperatures are very high (approximately 32°C (90°F) and above) and can include soft berries, increased leakage, and reduction in black color retention. Damage from sunlight (UV light) is often involved with heat damage, and the most substantial sun damage results in entire sides of berries damaged, drupelets become pink or white in color. White drupelets are often seen, particularly with specific cultivars, and in these instances single drupelets, or several non-adjacent drupelets, lose color and have a white appearance. The full reason this occurs is not known. Some have hypothesized that it is due to a combination of sunlight and moisture, as it is often worse when associated with rain, or berries located nearer the soil, subject to heavier dew. In contrast, the same cultivars grown in the lower Midwest, USA, where the UV intensity is less intense due to higher humidity, suffer less damage than when grown in the Pacific Northwest, USA, where the UV light intensity is high, the humidity is very low, and the temperatures are milder. Reduction in white drupelets can be achieved by reducing light by use of shade-cloth. Breeding for reduced damage due to heat, sunlight exposure/moisture on berries, or other causes is considered difficult to achieve and routine selection away from these traits is commonly practiced. Heat effects on primocane fruiting are also substantial. These effects will be covered in the primocane-fruiting section of this chapter.

SEASON

In a climate like the Willamette Valley in Oregon where all types of blackberries can be grown, a grower could choose cultivars that ripen with mid- to late season strawberries ('Obsidian,' 'Metolius') to ones grown in tunnels and could be ripening fruit in November ('Prime-Ark® 45') and any time in between. In general, the trailing blackberries have a much earlier ripening season than the eastern USA blackberries due to a genetic background that includes red raspberry and the early ripening *R. ursinus*. USDA-ARS selections from wild collections of *R. ursinus* germplasm are being incorporated into breeding germplasm and are moving the ripening season earlier, into the early and mid-season ripening of strawberries (Finn, 2001). There is potential to move this season even earlier with further breeding with very early *R. ursinus* parents. The development of primocane-fruiting cultivars has initially greatly increased options for very late-ripening blackberries that ripen fruit from late summer to frost or beyond with protection from precipitation and cold temperatures. The potential to manipulate primocane-fruiting types will allow growers to bring black-berries into ripening in nearly any season they desire, climate permitting, and to also potentially take advantage of early spring crops on floricanes. Therefore, with the advent of these new developments, blackberry growers can have the opportunity to harvest and market fruit 5–7 months per year, compared to the prior 4–6-week harvest period utilizing floricane-fruiting cultivars only. In certain climates, such as in central Mexico, 'Tupy' can be manipulated to come into production in any month of the year.

SIZE

Fruit size has been dramatically increased through selection in breeding programs from species with small 2–3-gram (g) berries to cultivars, such as 'Natchez,' 'Kiowa,' and 'Columbia Giant' that commonly have 15–20-g berries. While there seems to be no limit to the genetic potential for increasing fruit size, from a practical standpoint larger at some point is not better. Fresh-market shippers typically pack in clamshell containers with a legally defined weight.

As berry size increases, it becomes more difficult to achieve the defined weight without having to put in one more berry that then might be crushed when the clamshell is closed. For this reason, many agree an 8–9-g berry is ideal for clamshell packaging, as it looks large but allows weights to be made without crushing fruit. For processing, a wider range of fruit sizes is acceptable and is driven by the cost of evaluating the fruit on a packing line. Small fruit on a packing line are less efficient to evaluate than larger fruit, and depending on the buyer's final use of the product, a giant berry may not work in a muffin application, but would be fine if pureed. 'Wild Treasure,' a hybrid between wild *R. ursinus* and 'Waldo,' is a thornless cultivar with small, 2–3-g fruit that can be machine harvested and is well suited to small berry applications (Finn et *al.*, 2010).

FIRMNESS

Fruit firmness is a combination of several factors including skin toughness, uniformity of the drupelet size, drupelet flesh firmness, and the ratio of pyrene to flesh in each drupelet. A berry that does not 'bleed' is one of the most critical requirements of a berry for fresh-market sales. Fruit are most firm when they first turn black, but may not be edible at this point as the acids may be high and sugars low. Growers try to pick fruit for the fresh market as soon after the fruit turn black and have developed acceptable fruit quality. The challenge for breeders is to evaluate fruit as the growers will experience it. They cannot wait to evaluate it until the fruit is at its ideal ripeness and best flavor, as typically that fruit is soft and cannot be shipped. On the other hand, if the fruit is for machine harvesting and processing, the fruit will not be picked until it is fully mature. A breeder needs to select for enough firmness to handle the harvest and packing line and have appealing flavor.

Firmness is considered a quantitative trait, and the breeding approach is similar as for other traits of this type. However, environmental influences can be substantial, particularly rainfall near harvest, as well as high temperatures. Use of objective measures of firmness can be of value in evaluating germplasm. Salgado (2015) found compression firmness and drupelet skin penetration were the measurements that better represented blackberry firmness, while receptacle penetration was less effective in firmness differentiation. Multiharvest and year measurements are required to fully assess firmness and overall shipping potential.

In the early 1990s, an unusually firm selection was identified in the Arkansas program, one which was often referred to as 'crispy' when evaluated. It was not clear why this unusual firmness occurred, however. Further research by Salgado (2015) revealed that crisp genotypes in the Arkansas program maintained greater cellular integrity in drupelet mesocarp compared to softer-fruited cultivars. Although no cultivars have resulted from this very firm germplasm, this firmness will be incorporated in future developments thus enhancing postharvest potential.

SHAPE/UNIFORMITY

Ideally a fruit has very uniformly shaped and sized drupelets. This gives the overall shape a uniform appearance and leads to fewer drupelets extending out from the others where they may be more likely to be torn. Berries from western USA trailing blackberry germplasm have historically tended to be a long conic shape and those from eastern USA germplasm more round. While trailing blackberries tend to now be very uniformly conically shaped (e.g. 'Black Diamond,' 'Columbia Star'), eastern-developed cultivars are much more uniformly shaped than in the past, but can range in overall shape from fairly round (e.g. 'Osage') to conic (e.g. 'Prime-Ark® 45'), to long-conic ('Natchez'). While breeders have selected for increased uniformity, a wide variety of overall shapes seem to be acceptable.

PLANT TRAITS

Architecture

Blackberry types largely are defined by plant architecture – erect, semi-erect, and trailing; however, what is acceptable is slowly changing. A trailing architecture offers some advantages when machine harvesting the crop, as the primocanes for next year's crop are on the ground below the catch plates of the machine harvester where they do not interfere with fruit dropping through the canopy and are not injured by the rubbing of the catch plates. However, the management costs associated with training trailing canes to a trellis are much more than those for the canes for erect or semi-erect blackberries. As a result, there has been an increasing emphasis on selecting for the oxymoron of erect, trailing blackberries. The newer thornless cultivars, such as 'Black Diamond' and 'Columbia Star,' are more erect than 'Marion' and some cultivars, such as the 'Onyx' have a very similar habit to the semi-erect blackberries.

Erect and semi-erect blackberries were developed from eastern USA germplasm. For semi-erects (e.g. 'Chester Thornless,' 'Loch Ness,' and 'Triple Crown'), a trellis is required. The Arkansas program has emphasized erect canes in its developments. Although erect cultivars (e.g. 'Ouachita,' 'Osage,' and 'Navaho') were developed to be free-standing, most commercial plantings include support wires that prevent canes from falling over in high winds or heavy crop load situations.

More recently, dwarf to semi-dwarf blackberries have begun to be a focus in breeding programs. Dwarf types that might have been discarded in the past began to be kept in the 1990s with an eye towards new architectures that might be components of novel, high-density production systems using substrate culture. They also can make attractive ornamentals for the homeowner market; 'Sharon's Delight' was recently released for the Japanese garden market as well as 'Baby Cakes'™ in the U.S. and 'Black Cascade' in the UK.

Thornlessness

While blackberry 'thorns' are botanically spines, in North America, the terms 'thorns' and 'thornless' are most commonly used and will be used here. Multiple sources of thornlessness have been utilized in blackberry breeding (Clark et al., 2007); however, the 'Merton Thornless' and 'Austin Thornless' sources have been the most important. The 'Merton Thornless' source is a recessive 4× source (s) that has been useful in the breeding of the semi-erect and erect blackberries in eastern North America and Europe. The recessive nature of this source offers the advantages that once the breeding material is predominantly thornless and crossing is among thornless parents, all of the offspring in the field are thornless. However, this characteristic is also a disadvantage when working new germplasm into this pool as it will take a couple of generations to generate thornless progeny in most instances. The 'Austin Thornless' source was discovered in an 8× species and so not surprising has been useful in breeding at the 6× and higher ploidy level in trailing blackberries. This dominant source (designated S_f) took several generations of breeding to be useful as it was often associated with brittle canes and sterility. One of the drawbacks with this source is that the plants with 'Austin Thornless' source of thornlessness are often thorny until they reach ~60 cm (24 in) tall making them functionally thornless as the catcher plates on the machine harvesters and fruiting zone are separate from the spines. Since the young seedlings may have thorny canes until they are over ~60 cm (24 in) tall, one must plant all of them into the field to tell whether they will be thornless or thorny when mature. The most important cultivar with this source of thornlessness is 'Black Diamond.'

Recently, cultivars have been released using a new source of thornlessness, the 'Lincoln Logan' source (S_{fL}). The trailing 'Columbia Star' is the first blackberry to be released with this source of thornlessness other than the original 'Lincoln Logan' and 'Waimate' that have 'Logan'-type fruit and 'Marahau' that has 'Boysen'-type fruit (Hall et al., 1986a, 1986b; Hall and Stephens, 1999; Finn et al., 2014). This source was derived from somaclonal cells derived from the chimeral mutation of the L1 layer of a thornless mutation of 'Logan' called 'Lincoln Logan.' This source is useful in trailing germplasm because the seedlings can be screened for thornlessness when they are very young, thereby the breeder only plants thornless seedlings into the field. When first used in breeding, many of the offspring had weak, erect brittle canes, greater foliar disease problems, and a tendency towards red raspberry/'Logan'-colored and

-flavored fruit. In just a few generations of breeding, these problems have largely been eliminated.

Going forward it will be very difficult to justify releasing a thorny blackberry unless it has a trait that is so unique (i.e. very early season, very firm, primocane fruiting) that the liability associated with thorns in the product is accepted.

Machine harvest

The advent of machine harvesting beginning in the 1960s has had a major impact on the production of many berry crops, especially blackberries. The Iron Wino Co. (Canby, Oregon; Weygandt, 1964), Littau (Stayton, Oregon), and BEI (South Haven, Michigan) in the 1960s, followed later by Korvan (Lynden, Washington; now part of Oxbo Intl. Corp.) developed machines that success-fully harvested blackberries and black and red raspberries. Over the past four decades, these machines have progressed from those that were rough on the plants and fruit, meaning that the fruit could only be used for processing. Cur-rently, machines harvest fruit that look undamaged. While it is very unusual to be able to do, some growers are able to harvest by machine for the fresh market and air freight to markets within a few hours of their farm. Moving to the future, it will continue to be essential to harvest fruit destined for processing by machine and, at some point, for the fresh market, as can regularly be done with blueberries. Firmness and skin toughness are essential for a berry to be machine harvested successfully. The fruit must easily abscise between the pedicel and the fruit when the fruit is fully mature. Very few blackberries have problems with this abscission zone. The primary fruit characteristic that has eliminated prom-ising selections from consideration is when the fruit break at the point the pedi-cel attaches to the plant before it breaks at the typical pedicel fruit abscission zone. When these weak pedicels occur, the harvested fruit has fruit with pedi-cels and calyxes in the product, which is unacceptable.

Primocane fruiting

One has only to look at the impact of primocane fruiting in red raspberries since the 1980s to see the value of primocane fruiting of this crop. Almost all red raspberries in retail fresh markets are produced by plants of this type. This is due to several factors; two major ones include length of harvest/market sea-son and diversification of management techniques.

Breeding for primocane fruiting in blackberry did not get substantial atten-tion until the 1990s (Clark, 2008). The first primocane-fruiting cultivar releases were the tetraploid 'Prime-Jan®' and 'Prime-Jim®' in 2004 (Clark et al., 2005) although NC-194 had been released earlier as germplasm exhibiting this trait (Ballington and Moore, 1995). These cultivars were recommended for home-garden production primarily. 'Prime-Ark® 45' was released in 2009, providing the first primocane-fruiting blackberry with postharvest handling capability for shipping to commercial markets (Clark and Perkins-Veazie, 2011). The thornless 'Prime-Ark® Freedom' and 'Prime-Ark® Traveler' expanded options of this plant type, the latter cultivar recommended for shipping (Clark, 2014, 2016). Additional cultivars were released from cooperative programs between the University of Arkansas and private partners, including 'Reuben,' released in Europe by Hargreaves Plants (Finn and Clark, 2014).

Primocane fruiting is considered a recessive trait (Lopez-Medina *et al.*, 2000). The greatest effort has been carried out in the Arkansas program, but breeding for primocane fruiting proved difficult, as selection of improved seedlings was often impossible in the high summer and fall temperatures in Arkansas, which reduced expression of the trait or damaged developing flowers and fruit. A shift in approach was made in the early 2000s in Arkansas, from evaluation of primocane fruits to that of floricane fruits of primocane-fruiting plants. It was hoped that one could extrapolate floricane-fruit performance to that of primocanes (Clark, 2008). This was proven successful when additional testing of selections in coastal California revealed the full potential of primocane fruiting of these floricane-fruiting-selected plants. Primocane fruiting is now impacting commercial production, particularly in extending the fresh-market blackberry production season in the USA, and will likely expand in use as cultivars broaden the areas they are used.

Chilling requirement

Chilling requirement of blackberries has received limited attention in breeding and related research. Historically, blackberries have been grown in high-chill areas where lack of chill was not a limitation. In the last 30 years, commercial production has expanded to areas considered not high chill (generally approx. 800 hours or more below 7°C (45°F)) to mid-chill (300–500 hours), to low-chill (100–200 hours), or to no chill as in the world's largest fresh-market production area of Central Mexico. This expansion of production has increased interest in chilling requirement as a breeding objective. One of the greatest challenges is differentiating chilling requirement in genotypes in breeding programs, particularly when the breeding program is based in a high-chill location, such as Oregon or Arkansas. The most successful method is to test selections in a reduced-chill environment, but this requires substantial effort to locate sites and testing partners. It is preferable to breed in a reduced-chill location, such as Pelotas, Brazil where 'Tupy' originated. Additionally, with the revolution in Central Mexico in blackberry production in a no-chill environment using a series of cultural manipulations, breeders have considered developing cultivars that are adaptable to the Mexican production system. Limited success has been achieved, although diversification of cultivars has occurred to a limited extent in Mexico. Another approach to reducing or eliminating chilling requirement is primocane fruiting. The first primocane-fruiting production was developed in Mexico and early indications were that this type of plant will be adapted in this no-chill environment.

Diseases

In some climates, like the Pacific Northwest, where the summers are dry and the seasons relatively mild year-round, blackberries have few disease problems that cannot be handled with a dormant fungicide spray, and therefore, there is little incentive to breed for tolerance to specific diseases.

In the southern USA or similar climates, diseases are important. The most limiting southern disease, double/blossom rosette, (caused by the fungus *Cercosporella rubi* (Wint.) Plakidas) (Gupton, 1999) has been reduced in occurrence by planting of thornless cultivars. Other fungal diseases that are seen at times include anthracnose (*Elsinoe veneta* (Burkholder) Jenk.), cane botrytis, botrytis fruit rot (*Botrytis cinerea* Pers.: Fr.), and cane blight (*Leptosphaeria coniothyrium* (Fuckel) Sacc.) (Martin et *al.*, 2016). Production for the shipping market usually includes routine fungicide sprays near and during the harvest period. A more recent disease development in the southern USA has been orange cane blotch, caused by the parasitic alga *Cephaleuros virescens* Kunze ex E.M. Fries (P.M. Brannen, pers. commun.). Two other diseases of international significance are downy mildew (*Peronospora sparsa* (Berk) = *Peronospora rubi* (Rabenh. Ex J. Schröt)), which is very common as a nursery disease. It may be severe, especially in tissue cultured plants soon after exflasking. Purple blotch (*Septocyta ruborum* (Lib.) Petrark) has had a significant role in causing winter dieback in susceptible blackberry cultivars (Hall, personal communication).

A number of new viruses and combinations of viruses have been identified in recent years in the South and other areas of the USA, and have resulted in reduced planting productivity (Martin et *al.*, 2016; Tzanetakis, personal communication). Fortunately, increased virus testing in initial stock plants has contributed to cleaner plants for growers, addressing the increased virus problems.

Breeding for disease resistance is primarily done by selecting healthy plants in progeny and for release, as no specific disease-resistance screening activities are being conducted currently (Finn and Clark, 2012). Breeders are constantly selecting genotypes that do not show a great deal of susceptibility to diseases, so that over time the breeding pool is improved for disease resistance.

Heat/cold tolerance

Tolerance of blackberry plants to heat is usually considered high, although heat or sunlight damage to fruit is a concern occasionally (see prior section) and high temperature conditions often limit fruit set, making the performance of primocane-fruiting blackberries significantly better in cooler conditions, like California and Oregon. However, winter hardiness has been a major limitation in successful blackberry production in cold regions, such as the Midwest to upper-Midwest to northeastern USA. Damage to canes is the most common winter injury, often being killed to the ground, and thus, all crop potential eliminated. Less severe damage can occur, such as full to partial bud kill, damage to cane laterals, or other partial cane damage. Breeding for winter hardiness is very limited currently in blackberries.

The potential for primocane fruiting to contribute to production in cold climates exists, but use of this type of plant has not occurred in commercial production. A major limitation in evaluation of the first primocane-fruiting cultivars in colder location was the failure of much of the fruit to ripen prior to the first freeze of autumn. A similar limitation was present in early primocane-fruiting red raspberries, and breeding for earlier bloom and ripening resulted in an expansion of cultivars for cooler locations. Primocane-fruiting blackberries can likely be bred for earlier bloom and ripening also.

Yield

High yields are critical to the viability of blackberry growers in the future. Yield is a complicated trait that at its simplest is a function of the number of fruit and their size. However, too many fruit on a plant and fruit that are too large can be a problem leading to poor-quality fruit or fruit that can't be effectively packed. As breeders make selections, selecting for high yields of medium-sized fruit with sufficient fruit quality will identify those with good yield potential. The other challenge faced by breeders is that they are using a standard set of horticultural management practices in one or a few locations for their trials. Once a cultivar is provided to commercial growers, they will test it in a broader range of environments and will optimize cultural practices for each genotype. Therefore, breeders must balance getting new selections into grower's hands as quickly as possible to determine their real commercial potential against sending the growers substantial numbers of inferior selections.

GENOMICS

Blackberry genomics and related molecular work lags behind most other fruits in 2016 due to the limited resources available for this crop. Most research has focused on pedigree analysis and identification of polymorphic regions. A genetic map was constructed by Castro et al. (2013), focusing on the primocane fruiting and thornless traits in a population of tetraploid blackberries. Blackberry transcriptome was analyzed, producing a cDNA library from which over

8000 simple sequence repeats (SSRs) and 67,000 single-nucleotide polymorphisms (SNPs) were detected (Garcia-Seco et al., 2015).

Earlier, using random amplified polymorphic DNA (RAPD) markers, Stafne et al. (2003) were able to calculate mean genetic contribution for 16 cultivars of blackberry and raspberry. Further, they reported that RAPD markers were able to determine percentage similarity between 1% and 5%, although ultimately RAPD and pedigree data were not completely correlated, with RAPD markers overestimating relatedness of individuals.

Stafne et al. (2005) used an F1 mapping population derived from a cross between 'APF-12' (Prime-Jim®) (a primocane-fruiting blackberry) and 'Arapaho.' To account for the hybridization within *Rubus*, SSR primer pairs were selected from 'Glen Moy' red raspberry, whose genetic background includes several *Rubus* species, the European species *R. alceifolius* Poir., and strawberry (*F. x ananassa* Duch. and *F. vesca* L.). The SSR screening of the 'APF-12' × 'Arapaho' population indicated that 19 of 45 primer pairs were able to detect base-pair differences between the two parents, and thus could be used for future mapping of blackberry (Stafne et al., 2005).

The identification and differentiation of blackberry cultivars has also been investigated. Bassil et al. (2010) evaluated 29 dinucleotide SSR primer pairs and identified 10 of these pairs were able to distinguish between 16 North American blackberry cultivars from DNA extracted from the torus. Lewter et al. (2015) found that as few as three SSR markers were required to correctly identify 30 University of Arkansas-developed genotypes.

While transgenic efforts have been made in blackberries at this time (2016), they have not been found to be necessary in the development of new cultivars (Meng et al., 2004).

REFERENCES

Ballington, J.R. and Moore, J.N. (1995) NC 194 Primocane-fruiting thorny erect tetraploid blackberry germplasm. *Fruit Varieties Journal* 49(2), 101–102.

Bassil, N.V., Muminova, M. and Njuguna, W. (2010) Microsatellite-based fingerprinting of western blackberries from plants, IQF berries and puree. *Acta Horticulturae* 859, 73–80.

Castro, P., Stafne, E.T., Clark, J.R. and Lewers, K.S. (2013) Genetic map of the primocane-fruiting and thornless traits of tetraploid blackberry. *Theoretical and Applied Genetics* 126(10), 2521–2532.

Clark, J.R. (2005) Changing times for eastern United States blackberries. *HortTechnology* 15(3), 491–494.

Clark, J.R. (2008) Primocane-fruiting blackberry breeding. *HortScience* 43(6), 1637–1639.

Clark, J.R. (2013) 'Osage' thornless blackberry. *HortScience* 48(7), 909–912.

Clark, J.R. (2014) 'Prime-Ark® Freedom' primocane-fruiting thornless blackberry. *HortScience* 49(8), 1097–1101.

Clark, J.R. (2016) Breeding southern blackberries: idea to industry. *Acta Horticulturae* 1133, 3–12.

Clark, J.R. and Finn, C.E. (2011) Blackberry breeding and genetics. In: Flachowsky, H. and Hanke, V.-M. (eds.) *Methods in Temperate Fruit Breeding: Fruit, Vegetable and Cereal Science and Biotechnology* 5 (Special Issue 1). Global Science Books, Ltd., pp. 27–43.

Clark, J.R. and Moore, J.N. (2005) 'Ouachita' thornless blackberry. *HortScience* 40(1), 258–260.

Clark, J.R. and Moore, J.N. (2008) 'Natchez' thornless blackberry. *HortScience* 43(6), 1897–1899.

Clark, J.R. and Perkins-Veazie, P. (2011) 'Prime-Ark® 45' primocane-fruiting blackberry. *HortScience* 46(4), 670–673.

Clark, J.R., Moore, J.N., Lopez-Medina, J., Perkins-Veazie, P. and Finn, C.E. (2005) 'Prime Jan' (APF-8) and 'Prime-Jim' (APF-12) primocane-fruiting blackberries. *HortScience* 40(3), 852–855.

Clark, J.R., Stafne, E.T., Hall, H.K. and Finn, C.E. (2007) Blackberry breeding and genetics. *Plant Breeding Reviews* 29, 19–144.

Connor, A.M., Finn, C.E. and Alspach, P.E. (2005a) Genotypic and environmental variation in antioxidant activity and total phenolic content among blackberry and hybridberry cultivars. *Journal of the American Society for Horticultural Science* 130(4), 527–533.

Connor, A.M., C.E. Finn, T.K. McGhie, and P.A. Alspach (2005b) Genetic and environmental variation in anthocyanins and their relationship to antioxidant activity in blackberry and hybridberry cultivars. *Journal of the American Society for Horticultural Science* 130(5), 680–687.

Darrow, G.M. (1937) Blackberry and raspberry improvement. *USDA Yearbook of Agriculture, Yearbook 1937.* United States Department of Agriculture, Washington, DC, pp. 496–533.

Daubeny, H. (1996) Brambles. In: Janick, J. and Moore, J.N. (eds.) *Fruit Breeding, Volume II Vine and Small Fruits.* John Wiley and Sons, New York, pp. 109–190.

Du, X.A., Finn, C. and Qian, M. (2010) Volatile composition and odour activity value of thornless 'Black Diamond' and 'Marion' blackberries. *Journal of Food Chemistry* 119(3), 1127–1134.

Fernandez-Salvador, J., Strik, B.C., Zhao, Y. and Finn, C.E. (2015) Trailing blackberry genotypes differ in yield and postharvest fruit quality during establishment in an organic production system. *HortScience* 50(2), 240–246.

Finn, C. (2001) Trailing blackberries: from clearcuts to your table. *HortScience* 36(2), 236–238.

Finn, C.E. and Clark, J.R. (2011) Emergence of blackberry as a world crop. *Chronica Horticulturae* 51(3), 13–18.

Finn, C.E. and J.R. Clark (2012) Blackberry. In: Badenes, M.L. and Byrne, D.H. (eds.) *Handbook of Plant Breeding: Volume 8: Fruit Breeding.* Springer, New York, pp. 151–190.

Finn, C.E. and Clark, J.R. (2014) Blackberry. In: Gasic, K. and Preece, J.E. (eds.) Register of new fruit and nut cultivars, list 47, pp. 399–400. *HortScience* 49(4), 396–421. Available at: http://hortsci.ashspublications.org/content/49/4/396.full (accessed May 11, 2017).

Finn, C.E., Strik, B. and Lawrence, F.J. (1997) 'Marion' trailing blackberry. *Fruit Varieties Journal* 51(3), 130–133.

Finn, C.E., Yorgey, B., Strik, B.C., Hall, H.K., Martin, R.R. and Qian, M.C. (2005a) 'Black Diamond' trailing thornless blackberry. *HortScience* 40(7), 2175–2178.

Finn, C.E., Yorgey, B., Strik, B.C. and Martin, R.R. (2005b) 'Metolius' trailing blackberry. *HortScience* 40(7), 2189–2191.

Finn, C.E., Yorgey, B., Strik, B.C., Martin, R.R. and Kempler, C. (2005c) 'Obsidian' trailing blackberry. *HortScience* 40(7), 2185–2188.

Finn, C.E., Strik, B.C., Yorgey, B.M., Qian, M., Martin, R.R. and Peterson, M. (2010) 'Wild Treasure' thornless trailing blackberry. *HortScience* 45(3), 434–436.

Finn, C.E., Strik, B.C., Yorgey, B.M., Peterson, M.E., Lee, J., Martin, R.R. and Hall, H.K. (2014) 'Columbia Star' thornless trailing blackberry. *HortScience* 49(8), 1108–1112.

Garcia-Seco, D., Zhang, Y., Gutierrez-Mañero, F.J., Martin, C. and Ramos-Solano, B. (2015) RNA-Seq analysis and transcriptome assembly for blackberry (*Rubus* sp. Var. Lochness) fruit. *BMC Genomics* 16, 1–11.

Gupton, C.L (1999) Breeding for rosette resistance in blackberry. *Acta Horticulturae* 505, 313–322.

Hall, H.K. (1990) Blackberry breeding. In: Janick, J. (ed.) *Plant Breeding Reviews*. Timber Press, Inc., Portland, Oregon, pp. 249–312.

Hall, H.K. and Stephens, J. (1999) Hybridberries and blackberries in New Zealand – breeding for spinelessness. *Acta Horticulturae* 505, 65–71.

Hall, H.K., Skirvin, R.M. and Braam, W.F. (1986a) Germplasm release of 'Lincoln Logan,' a tissue culture-derived genetic thornless 'Loganberry.' *Fruit Varieties Journal* 40(3), 134–135.

Hall, H.K., Quazi, M.H. and Skirvin, R.M. (1986b) Isolation of a pure thornless Loganberry by meristem tip culture. *Euphytica* 35, 1039–1044.

Hall, H.K., Stephens, M.J., Stanley, C.J., Finn, C. and Yorgey, B. (2002) Breeding new 'Boysen' and 'Marion' cultivars. *Acta Horticulturae* 585, 91–96.

Hedrick, U.P (1925) *The Small Fruits of New York*. J.B. Lyon, Albany, New York.

Heidt, L.J (2014) United States Patent Application: Willamette Thornless Marion, US 13/573,608. U.S. Patent and Trademark Office, Washington, DC.

Jennings, D.L (1988) *Raspberries and Blackberries: Their Breeding, Diseases and Growth*. Academic Press, London.

Lewter, J., Clark, J.R., Bassil, N., Nyberg, A. and Oliver, F.N. (2015) A breeder-friendly DNA fingerprinting protocol for blackberry fruit. *HortScience* 50, S327 (abstract).

Logan, M.E. (1955) *The Loganberry*. Mary E. Logan (Mrs. J.H. Logan), Oakland, California.

Lopez-Medina, J., Moore, J.N. and McNew, R.W (2000) A proposed model for inheritance of primocane fruiting in tetraploid erect blackberry. *Journal of the American Society for Horticultural Science* 125(2), 217–221.

Martin, R.R., Williamson, B., Williams, R. and Ellis, M. (2016) *Compendium of Raspberry and Blackberry Diseases and Insects*. APS Press, St. Paul, Minnesota.

Meng, R. and Finn, C. (1997) Determining ploidy level and nuclear DNA content in *Rubus* by flow cytometry. *Journal of the American Society for Horticultural Science* 127(5), 767–775.

Meng, R., Chen, T.H.H., Finn, C.E. and Li, Y. (2004) Improving in vitro plant regeneration from leaf and petiole explants of 'Marion' blackberry. *HortScience* 39(2), 316–320.

Salgado, A (2015) Applying molecular and phenotypic tools to characterize flesh texture and acidity traits in the Arkansas peach breeding program and understanding the crispy texture in the Arkansas blackberry breeding program. Unpublished PhD dissertation, University of Arkansas, Fayetteville, Arkansas.

Siriwoharn, T., Wrolstad, R.E., Finn, C.E. and Pereira, C.B. (2004) Influence of cultivar, maturity and sampling on blackberry (*Rubus* L. hybrids) anthocyanins, polyphenolics, and antioxidant properties. *Journal of Agricultural and Food Chemistry* 52(26), 8021–8030.

Stafne, E.T., Clark, J.R., Pelto, M.C. and Lindstrom, J.T. (2003) Discrimination of *Rubus* cultivars using RAPD markers and pedigree analysis. *Acta Horticulturae* 626, 119–124.

Stafne, E.T., Clark, J.R., Weber, C.A., Graham, J. and Lewers. K.S. (2005) Simple sequence repeat (SSR) markers for genetic mapping of raspberry and blackberry. *Journal of the American Society for Horticultural Science* 130(5), 722–728.

Takeda, F., Glenn, M.D. and Tworkoski, T. (2013) Rotating cross-arm trellis technology for blackberry production. *Journal of Berry Research* 3(1), 25–40.

Thompson, M.M. (1997) Survey of chromosome numbers in *Rubus* Rosaceae: Rosoideae. *Annals of the Missouri Botanical Garden* 84(1), 128–163.

Waldo, G.F. (1977) 'Thornless Evergreen' – Oregon's leading blackberry. *Fruit Varieties Journal* 31, 26–30.

Wang, Y., Finn, C. and Qian, M.C. (2005) Impact of growing environments on 'Chickasaw' blackberry (*Rubus* L.) aroma evaluated by gas chromatography olfactory dilution analysis. *Journal of Agricultural and Food Chemistry* 53(9), 3563–3571.

Weygandt, R.A. (1964) United States Patent: Machine for harvesting berries and similar produce from their plants, US3126692. U.S. Patent and Trademark Office, Washington, DC.

Yorgey, B. and Finn, C.E. (2005) Comparison of 'Marion' to thornless blackberry genotypes as individually quick frozen and puree products. *HortScience* 40(3), 513–515.

6

NURSERY PRODUCTION OF PLANTS

Ioannis E. Tzanetakis[1],* and Robert R. Martin[2]

[1]University of Arkansas, Fayetteville, Arkansas, USA;
[2]USDA-ARS, Corvallis, Oregon, USA

INTRODUCTION

Breeding, from the initial cross to the release of a named cultivar, involves a significant amount of time which, in some cases, may exceed a decade. During this period, seedlings, accessions, and advanced selections are propagated clonally and grown under diverse environmental conditions, quite often on different continents, to determine their suitability for commercial release. As with any other organism, plants, and blackberries in particular, are susceptible to viruses and other systemic pathogens that accumulate during the selection process. As noted later in this book (Chapter 13), viruses pose a major threat for producers because they can affect yield, fruit quality, and plant longevity. For this reason, during the establishment of blackberries as a horticultural crop, there have always been coordinated efforts between breeders and virologists to eliminate targeted systemic pathogens from selections prior to commercial release (Converse, 1987).

It is important to note that not all viruses can be eliminated from the germplasm, as some may integrate in the host genome and become part of the plant chromosome (see Rubus yellow net virus, Chapter 13). Others move through the plant via cell division, being virtually impossible to eliminate as they invade the gametes. These viruses are often prevalent in the germplasm because they can be transmitted by pollen and seed with 100% efficiency. In all cases, these viruses (cryptic and amalgaviruses) are not associated with disease symptoms in single or mixed infections; this is the reason that they have not been targeted for elimination from commercial genotypes. In addition to the aforementioned cases, there may be new viruses identified in plants (Martin et al., 2013), but if they do not cause symptoms on indicator plants, they often remain undetected, as was the case with several viruses involved in blackberry yellow vein disease complex (see Chapter 13). With new technologies involving large-scale

---

* Corresponding author: itzaneta@uark.edu

sequencing, this possibility is greatly reduced, but still exists. For this reason, the term 'free of targeted viruses/pathogens' rather than 'virus/pathogen-free' will be used throughout this chapter.

When a plant is free of targeted pathogens, it enters the propagation pipeline for increase to meet market demand. The propagation process per se will be discussed in Chapter 7, whereas this chapter will focus on the available processes for virus elimination as well as the best management practices in a nursery setting to safeguard plants and guarantee the best material possible for the end user, allowing for profitable and sustainable production.

STEP 1 – PATHOGEN DETECTION

As aforementioned, blackberries are subject to infection by systemic pathogens. Once a selection is designated to become a named cultivar it is tested for pathogens that have the potential to cause disease. Worldwide, the industry standard for detection is grafting onto susceptible accessions/cultivars with the most widely used being black raspberry cv. Munger. Not all viruses infecting blackberries cause symptoms when grafted onto 'Munger' or other indicators (Martin et al., 2013). Thus, in most laboratories, pathogen-specific tests: immunological (e.g. ELISA) or molecular (e.g. PCR/RT-PCR) are used in addition to biological indexing (Fig. 6.1) (MacFarlane et al., 2015).

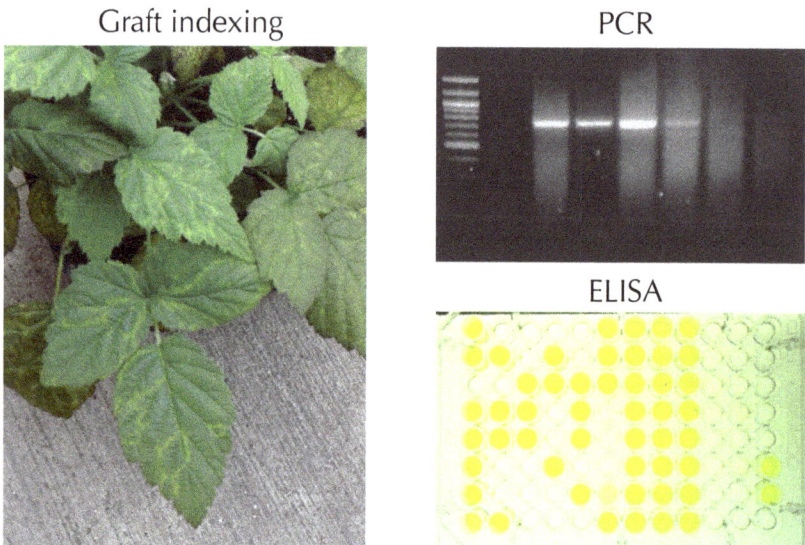

Fig. 6.1. Methods regularly used in clean plant centers for virus indexing and testing (R.R. Martin).

The possibility of a new virus that remains asymptomatic on indicator plants still exists. To minimize the possibility of such an occurrence new technologies involving large-scale sequencing (LSS, aka next generation sequencing) are being applied in several facilities around the globe. During this process, nucleic acids depleted of plant genomic DNA and ribosomal RNA are reverse transcribed, amplified, and sequenced using one or more of the several available LSS platforms (Illumina, pyrosequencing, etc). The output sequences are analyzed using bioinformatics tools that could detect known viruses and identify new ones based on similarity with existing ones or by having signature viral enzymatic and/or structural motifs (Fig. 6.2) (Ho et *al.*, 2015; Ho and Tzanetakis, 2014).

Once the presence of a virus is verified via any of the aforementioned processes, the elimination process is initiated.

STEP 2 – ELIMINATION

There are different approaches to eliminate systemic pathogens from an advanced selection and they may all be applied on a genotype to provide for the best chance of pathogen elimination and minimize the timeframe between elimination and commercial release. Here we will discuss thermo-, chemo-, and cryotherapy.

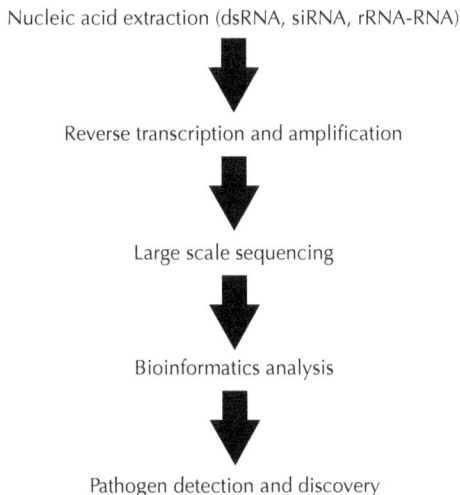

Nucleic acid extraction (dsRNA, siRNA, rRNA-RNA)

⬇

Reverse transcription and amplification

⬇

Large scale sequencing

⬇

Bioinformatics analysis

⬇

Pathogen detection and discovery

Fig. 6.2. Steps taken to insure the health status of advanced selections before release (I. Tzanetakis).

Thermotherapy

Thermotherapy is the most widely used method for virus elimination. Early in the 20th century, it was observed that during summer time plants tend to display ameliorated symptoms, whereas new growth may be asymptomatic altogether. In some cases, mechanical or graft transmissions from asymptomatic tissue was unsuccessful, indicating that viruses may have been eliminated or be in low enough titer that could not reinitiate infection onto a new host. As a side note, this is also why testing for some viruses during the hottest part of the year is not advisable, as it can result in false negatives.

It is understood that ameliorated symptoms are due to gene silencing. The enzymes participating in the pathway function at accelerated rates at higher temperatures minimizing or even eliminating virus replication. This innate mechanism to fight virus infections and reduced virus replication at elevated temperatures has been exploited to obtain plants free of targeted viruses.

Rooted material is placed in a growth chamber with fluctuating temperatures of 37°C (99°F) and 28°C (82°F), or at a constant 37°C (99°F), with 12- to 16-hour day length, for a period of 4–6 weeks. Alternatively, the plants are established in tissue culture and then subjected to thermal therapy in a growth chamber using the fluctuating temperature regime. During this period of heat treatment, gene-silencing affects virus replication and allows for new growth of the meristematic tissues to be infection-free. Viruses that infect multiple types of tissue and move readily from cell to cell are more difficult to eliminate than those that are phloem-restricted, since the meristematic dome lacks differentiated phloem tissue.

After the heat treatment is completed, the meristematic dome (usually <0.5 mm (0.02 in)) is excised and put in tissue culture to regenerate plants (Fig. 6.3).

The tissue culture microenvironment and the presence of several hormones/growth factors essential for growth of young plantlets make pathogen detection unreliable. Once plantlets are rooted and can survive transplant to potting media, they are grown for a few months in protected environment to eliminate the possibility of reinfection before being tested for the previously detected virus(es).

Chemotherapy

Chemotherapy is based on the premise that virus replication may be stalled or slowed down in the presence of replication inhibitors. Those inhibitors may target the virus polymerase or host factors that are essential for replication, or are nucleotide analogs that interfere with RNA synthesis or capping. Chemotherapy can be used by itself or in conjunction with thermotherapy, making the process more robust.

Fig. 6.3. Meristeming material to insure they are free of targeted pathogens. *Top left:* Meristematic dome (circled); *top right:* The size of the excised material vs. the scalpel used for excision; *bottom:* young plantlet after tissue regeneration (R.R. Martin).

Cryotherapy

Cryotherapy is a relative new pathogen elimination method, first applied at the end of the 20th century. Unlike thermotherapy, there is no need for excision of a thin layer of cells from the meristematic dome, but it can accommodate a larger size of shoot tips. Tissue that is used for cryotherapy is partially dehydrated through a series of osmotic treatments before being submerged briefly in liquid nitrogen. As apical meristem cells are smaller and contain small vacuoles, they are better suited to withstand dehydration and the $-196°C$ ($-321°F$) liquid nitrogen treatment. Given that most pathogens could not achieve meristem invasion and differentiated cells cannot survive the cold shock, the only tissue that regenerates to full plants are the pathogen-free meristematic cells (Wang and Valkonen, 2008). Still this technique is not 100% successful, especially with viruses that are able to invade the meristematic dome, such as raspberry bushy dwarf virus. In such cases, a combination of traditional thermotherapy followed by cryotherapy could provide exceptional results as seen in raspberry infected with the virus (Wang *et al.*, 2009). As in the case of thermo-therapy, no protocol is successful with all genotypes and modifications to already developed protocols are necessary to accommodate the new genotypes that are going through the process.

STEP 3 – NURSERY MULTIPLICATION

Given the complex nomenclature that may be used in different countries this chapter will use the generation (G) designation (Fig. 6.4). Designations and standards are part of the draft national certification guidelines for *Rubus* species in the United States (www.ncpnberries.org).

A plant that is free of all targeted pathogens, a status achieved after two consecutive seasons of testing by any verified detection method for a particular pathogen is designated as G1. The material is 'fingerprinted' using molecular techniques to insure trueness-to-type and minimize the possibility of germplasm mix-up or spontaneous mutations in tissue culture during regeneration. Plants are grown under protected culture (concrete-paved screenhouses or screened greenhouses in soil-less media, eliminating flowers and with a significant distance from *Rubus* plants or hosts of blackberry viruses) or in tissue culture. They are tested regularly (if grown in open air), for all targeted blackberry pathogens, even those that are not known to be present in the country of origin. Additionally, they are tested for new virus pathogens of *Rubus* that are discovered and may not be on the targeted pathogen list. This insures that the material that provides all tissue for subsequent generations is of the highest quality possible.

The first propagation step yields G2 plants. G2 plants are normally an intermediate step in the process of achieving appropriate plant numbers for commercial use. Depending on the nursery location, they may be grown in tissue culture, screenhouses, screened greenhouses, or the field, given that all appropriate measures are taken to minimize systemic pathogen infection

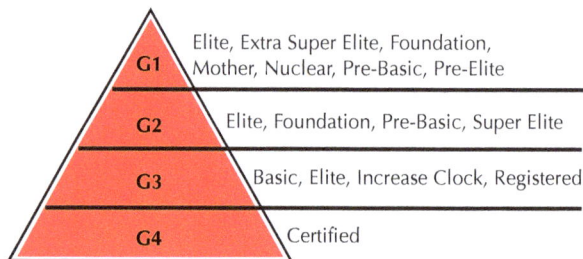

G1 Elite, Extra Super Elite, Foundation, Mother, Nuclear, Pre-Basic, Pre-Elite

G2 Elite, Foundation, Pre-Basic, Super Elite

G3 Basic, Elite, Increase Clock, Registered

G4 Certified

Fig. 6.4. Visualization of the generation (G) scheme and the commonly used terms for each of the tiers (I. Tzanetakis)

(e.g. testing for nematode vectors, elimination of weeds that may harbor black-berry viruses, appropriate control measures for vectors that are indigenous to the area). Individual plants are tested regularly for all viruses that are indige-nous to the area.

The next propagation step yields G3 plants which are normally grown in containers in canyards or in the ground. Guidelines and best management practices allow for field growth but focus on approaches that minimize reinfec-tion. Those include measures that are similar or identical to those listed for G2 plants, but with more relaxed criteria. For example, it may be that pathogen testing is done at the block level vs. individual plants, allowable distances between nursery stock and wild *Rubus* may be shorter for G3 plants, and there may be a tolerance on infection rate in the field, normally this being <1%.

G4 presents the final step in nursery propagation and those are plants that are marked for sale to producers. All the guidelines and best management prac-tices that apply to the previous propagation generations also apply here, but, again, with more relaxed criteria. A major difference with the previous genera-tions is that monitoring for infection is based on visual symptoms. In the USA, there is a coordinated effort to abolish this practice, as it is well established that the majority of blackberry viruses are asymptomatic in single infections, and therefore, symptoms at this level suggest the presence of multiple viruses. With this being the case, a significant number of symptomatic plants in a G4 block is a sign of heavy infection with the potential to initiate an epidemic once the plants are moved to a commercial field setting. The proposed procedure will include laboratory tests for the viruses indigenous to the region, with the prem-ise that in such a setting an infection rate of 5% that is detected with 95% confidence will be the signal of a major breakdown in the guidelines put in place to allow for the production of high-quality blackberry plants.

SUMMARY

In today's (2017) globalized market, blackberry propagation material moves between continents and is grown in the field or protected culture from the sub-tropics to the subarctic areas around the world. Those practices, in addition to climate change, that have allowed for the expansion of pathogen vectors to new areas, are the reasons that there are several new pathogens, new to black-berries or science altogether, associated with emerging diseases infecting the crop (Martin et *al.*, 2013).

Whereas it takes significant effort and time investment to understand the biology and implement control strategies for new pathogens and diseases, it is well understood how systemic pathogens can be eliminated from the germlines. Even though starting with plants free of targeted pathogens does not guaran-tee a successful crop, especially in areas with significant disease pressure, this practice does eliminate a major factor in disease development and control.

Starting with clean plant material is essential for establishing a healthy planting and producing a productive crop.

This chapter introduced the reader to available technologies for systemic pathogen elimination and new approaches evaluated to safeguard propagation material and keep it at the highest health status possible. It is unquestionable that once a plant is in the field for a significant length of time it will get infected by one or more pathogens, but if the guidelines presented here are followed, production fields will provide better-quality fruit for a longer period of time, the ultimate goal for sustainability and producer profitability.

REFERENCES

Converse, R.H. (ed.) (1987) *Virus Diseases of Small Fruits.* U.S. Dept. of Agriculture, Agriculture Handbook No. 631. U.S. Government Printing Office, Washington, DC.

Ho, T. and Tzanetakis, I.E. (2014) Development of a virus detection and discovery pipeline using next generation sequencing. *Virology* 471–473, 54–60.

Ho, T., Martin., R.R. and Tzanetakis, I.E. (2015) Next generation sequencing of elite berry germplasm and data analysis using a bioinformatics pipeline for virus detection and discovery. *Methods in Molecular Biology – Plant Pathology* 1302, 301–313.

MacFarlane, S., McGavin W. and Tzanetakis I.E. (2015) Virus-testing by PCR and RT-PCR amplification in berry fruit. *Methods in Molecular Biology – Plant Pathology* 1302, 227–248.

Martin, R.R., MacFarlane, S., Sabanadzovic, S., Quito-Avila, D.F., Poudel, B. and Tzanetakis, I.E. (2013) Viruses and virus diseases of *Rubus. Plant Disease* 97(2), 168–182.

Wang, Q. and Valkonen, J.P.T. (2008) Combined thermotherapy and cryotherapy for efficient virus eradication: relation of virus distribution, subcellular changes, cell survival and viral RNA degradation in shoot tips. *Molecular Plant Pathology* 9(2), 237–250.

Wang, Q., Cuellar, W.J., Rajamaki, M.-L., Hirata Y. and Valkonen J.P.T. (2009) Cryotherapy of shoot tips: novel pathogen eradication method. *Trends in Plant Science* 14(3), 119–122.

7

PROPAGATION OF BLACKBERRIES AND RELATED *RUBUS* SPECIES

Barbara Reed,[1,]* Sukalya Poothong,[2] and Harvey K. Hall[3]

[1]*USDA-ARS National Clonal Germplasm Repository, Corvallis, Oregon, USA; [2]University of Phayao, Phayao, Thailand; [3]Shekinah Berries, Ltd, Tauranga, New Zealand*

INTRODUCTION

Blackberries and hybrid blackberries are clonally propagated crops with each commercial cultivar being produced from an original mother plant. Propagation by seed is not used because seedlings vary widely from each other and from both the seed and pollen parents. Traditional propagation methods for blackberries and hybrid blackberries have varied according to the type of blackberry plants being propagated. This has included digging of canes, division of the crown, rooting of dormant hardwood cuttings, softwood cuttings and leaf-bud cuttings, taking of suckers, the use of roots for root cuttings, and planting of root mass for establishing plants in the field. In 'Brambles' by Ourecky (1975), the use of tissue culture was not mentioned. However, this propagation technique became important during the latter part of the 20th century, and now is the most common way to propagate blackberries and hybrid berries (Broome and Zimmerman, 1978; Harper, 1978; Zimmerman, 1979; Zimmerman and Broome, 1980; Skirvin et al., 1981; Swartz et al., 1983; Slivinski et al., 1984; Hall, 1990; Reed, 1990; Fernandez and Clark, 1991; Bobrowski et al., 1996; Finne, 1997; Velchev et al., 1997; Gonzalez et al., 2000; Dziedzic and Jagla, 2012).

TRADITIONAL METHODS OF PROPAGATION

Thorny (spiny) and thornless (spineless) upright blackberries

The growth habit of the wild upright blackberry species in the northeast USA and Canada, and similar species native to Europe, is to form a loose crown of

---

* Corresponding author: reedba@onid.oregonstate.edu

upright canes surrounded by new canes that grow as suckers from buds formed on the roots of these plants. Traditionally, these were easily propagated by division of the crown and by digging the suckers. These same propagation methods were also adopted for propagating cultivated types derived from these species, including 'Hedrick' and 'Darrow' (Caldwell, 1984). The upright type 'Brazos' could also be propagated by similar methods. Both the upright growth habit and the ability to sucker in 'Brazos' were derived from the 'Brilliant Red' raspberry; it may be that the origin of upright blackberry species is from the introgression of uprightness from raspberries. Experimentally, as with 'Brazos,' controlled crosses of tetraploid semi-erect blackberry cultivars with tetraploid raspberries has produced hybrids similar to upright blackberry types that propagate by division, suckers, and from root buds (Moore *et al.*, 1978). These are similar to the wild and cultivated upright blackberry types (Hall, personal observation).

Propagation by digging of canes or division of the plants was ineffective in generating large numbers of plants for commercial plantings or sales. The practice of propagation by root cuttings and digging of the root mass was found to be more effective for obtaining a large number of plants (Busby and Himelrick, 1999). When thornlessness from the crown-forming species *R. ulmifolius* was introgressed into upright blackberry types from 'Merton Thornless,' it was found that many of the hybrids had many of the characteristics of 'Merton Thornless.' These hybrids had little ability to produce buds on the roots or any canes or suckers that were not attached to the crown. Further breeding has given rise to upright thornless types that sucker more freely and can be propagated like the thorny upright types.

Semi-erect blackberries

With semi-erect blackberry types, the traditional propagation method was to propagate by tip layering, or to collect the rooted tips of canes during the dormant season. More recent propagation practices of semi-erect blackberries have been to take dormant hardwood cuttings and place them in peat, sand, or pumice in a shaded cloche or mini greenhouse (glasshouse) or propagate by softwood cuttings, rooted under mist (Bray *et al.*, 2003).

Hybrid blackberries and other hybrid berries

These hybrids, including 'Boysenberry' and 'Loganberry,' have been propagated in a manner similar to semi-erect blackberries since the beginning of commercial production (Garner and Hammond, 1939; Wood, 1989). Many production fields were planted using tip-rooted plants collected from other fields, but the spread of pests (insects and nematodes) and diseases (fungi, bacteria, and viruses) from one field to another was increased by this practice, affecting the plant health of the entire industry (Hansen, 1985; Donnelly and Daubeny, 1986; Hall, personal observation). Production of plants by hard-wood cuttings as an alternative to using tip-rooted plants was limited due to low numbers of planting stock. Attempts to get a higher multiplication of desirable clones of semi-erect blackberries and hybrid types included pruning to increase branching, layering and digging of roots, and the rooting of soft-wood cuttings under mist. These methods are rarely used for commercial production of plants.

Micropropagation

Initial attempts to micropropagate clonally propagated plants were focused on rapid propagation, elimination of disease and improvement of asexually prop-agated varieties (Skirvin and Chu, 1977). However, the first attempts at doing this in the blackberry 'Bedford Giant' and the hybrid berry 'Tayberry' were not useful due to the poor rate of multiplication (Harper, 1978). Nevertheless, the methods initially pioneered in micropropagation of blackberries and hybrid berries by tissue culture, as initiated by Broome and Zimmermann (1978) and Harper (1978), have been refined. As of 2016, this technology is the preferred technique for producing large numbers of plants within a short time as well as a means of creating and conserving healthy plant stocks.

Micropropagation protocols of various blackberry species and cultivars are well represented in the literature. Breeders and germplasm facilities com-monly use shoot tip culture for rapid propagation and virus elimination. Micro-propagation is also important in nursery crop production. Propagation using apical or axillary explants is commonly used for uniform mass production of plants for the field (Trigiano and Gray, 2000; George *et al.*, 2008; Debnath, 2011). *Rubus* is a very diverse genus and the response of 256 species, hybrids, and cultivars to *in vitro* culture, was quite variable (Reed, 1990). Some general protocols can be used to initiate and propagate the various *Rubus* types, and optional protocols are also available for those that do not respond well to the more general methods (Dziedzic and Jagla, 2012; Poothong and Reed, 2014, 2015, 2016). Some blackberries are easy to propagate in standard growth media and plant growth regulators (PGR), while others grow poorly or do not grow or multiply. Recommendations for growth in each of the stages of *in vitro* culture are suggested in the following paragraphs.

Initiation

Explants for shoot cultures should be obtained from mother plants that are free of bacteria, fungi, and viruses. Sheltered growth of mother plants provides the best conditions for obtaining clean explants. Virus testing of mother plants is recommended. If viruses are present, the growth of shoot cultures and the health of the resulting plants will be at risk (Tsao *et al.*, 2000). General surface disinfestation procedures are used to remove epiphytic organisms from the surface of the explants. These include various steps with alcohol, bleach, and other steriliants (Trigiano and Gray, 2000; Wu *et al.*, 2009; Dziedzic and Jagla, 2012). All initiated cultures should be tested for fungi and bacteria at the beginning of the culture process and during the multiplication cycle (Reed *et al.*, 1995). At the USA National Clonal Germplasm Repository (NCGR) laboratory, 5 cm (2 in) shoots from virus-tested, pot-grown screenhouse plants were stripped of all leaves and disinfected for 10 minutes in 10% household bleach (Clorox bleach, 5.25% sodium hypochlorite diluted tenfold) with 0.1 ml/l Tween 20 on a rotary shaker, followed immediately by two rinses in sterile water. Shoots were trimmed to single node sections and placed in 16 mm (0.6 in) tubes with 5 ml liquid detection medium (Reed and DeNoma, 2014) for 1 week before being transferred to solid Murashige and Skoog (MS; 1962) medium. Blackberry shoots can be forced from field-grown canes, but these canes should be treated with dormant oil or chemical sprays to kill insects (thrips, mites). Shoots for forcing should be placed in water with florist mix to inhibit bacterial growth and the water changed weekly (Reed and DeNoma, 2014). Shoots produced from field-grown canes should be tested for viruses before any distribution or field planting of the propagules as they may have become infected in the field. Wu *et al.* (2009) pre-treated 32 *Rubus* clones by chilling stock plants at 4°C (39°F) for 6 weeks before shoots were forced to increase initiation in half-strength Murashige and Tucker (MT; 1969) medium. Chilled mother plants produced 62–75% successfully initiated shoots as compared to 32–50% for non-chilled mother plants. Less bacterial and fungal contamination was detected compared to non-chilled plants. Initiation medium normally contained N^6-benzyladenine (BA) (4.44–13.31 µM). The use of BA, indole-3-butyric acid (IBA), and charcoal promoted multiple shoot development.

Contaminant detection

Detection medium used at NCGR is half-strength liquid (MS; 1962) medium with 256 mg peptone and 88 mg yeast extract and at pH 6.9. The neutral pH is more suitable for bacterial and fungal growth. Contamination appears as cloudiness or flocculent growth in the medium. Shoots in the detection medium are held in standard growing conditions, and if no contamination is evident at one week, then shoot bases are removed and tested on an agar-based detection medium (Viss *et al.*, 1991) or nutrient agar when the explants are moved to tubes of multiplication medium. If contamination appears at either step, the shoots are discarded. If a second group is contaminated, the growth conditions of the mother plant should be improved or an antibiotic treatment should be considered (Reed et *al.*, 1995; Tanprasert and Reed, 1997). Some plants do not tolerate submersion in liquid medium for a week, so they should be only partially submerged and shaken occasionally to rinse contaminants from the surface into the medium. Shoots should not be grouped until all testing is completed.

Growth medium

Growth media for *Rubus* are mostly based on Murashige and Skoog (MS) derived growth medium (Murashige and Skoog, 1962; Murashige and Tucker, 1969), but some laboratories use Woody Plant Medium (Lloyd and McCown, 1980) or Anderson's medium (Anderson, 1980). Plant growth regulators (PGR) vary, but are often similar to those of Reed (1990), with 0.29 mM gibberellic acid (GA_3), 0.49 mM indole-3-butyric acid (IBA), and 4.45 mM N^6-benzyladenine (BA). Iron (Fe) in the medium varies from standard to double concentrations of MS Fe-EDTA to Fe-EDDHA (Tsao and Reed, 2002; Zawadzka and Orlikowska, 2006; Wu et *al.*, 2009; Poothong and Reed, 2015). Sucrose (30 g/l) and agar or agar and gellan gum combinations equivalent to 0.75% are commonly used and the medium pH adjusted to 5.7 or 5.8. Trimmed explants may be placed in petri dishes, tubs, or glass tubes for propagation. Growth room conditions vary in lighting from a 16-h photoperiod (cool white fluorescent illumination, 17–30 Mmol m^{-2} s^{-1}) (Wu et *al.*, 2009) to a mixture of cool and warm fluorescent bulbs (70–90 Mmol m^{-2} s^{-1}) (Poothong and Reed, 2015). Growth room temperatures of 25–26°C (77–79°F) are most common.

Rubus species and cultivars are quite variable in their response to micropropagation. While raspberries are often considered the most difficult, some blackberry species or cultivars can also grow slowly or callus extensively on standard growth medium. PGRs for shoot proliferation range from 4.45 to 8.9 µM BA, 0.5–5.0 µM IBA, and 0.29 µM GA (Anderson, 1980; Donnelly et *al.*, 1980; Reed, 1990; Poothong and Reed, 2014). Wu et *al.* (2009) found that three patterns of growth were common for initiation of the 32 genotypes tested: a rosette of microshoots, a single shoot with many nodes, or multiple shoots from a single-node segment. To manage these diverse responses, the shoots were grown with 0.49 µM IBA, and with alternating high and low BA concentrations. A separate wide-ranging study of diverse *Rubus* germplasm found that the various cultivars required different amounts and types of PGRs and growth media for successful micropropagation (Reed, 1990). Generally, blackberries grew better on MS medium while raspberries propagated best on Anderson's medium and increased growth regulators were required for some hard to propagate selections. Preece (1995) noted that increased PGRs can sometimes compensate for sub-optimum mineral nutrients. Broome and Zimmerman (1978), in their study of blackberry shoot culture, also indicated that cultivars varied in their responses on the same growth medium.

Elongation of shoots was noted as a major problem with *Rubus* cultures (Welander, 1985). In some cases, lower light intensity may be used to elongate shoots (Zawadzka and Orlikowska, 2006; Wu et *al.,* 2009) or changes in mineral nutrients may be used to improve shoot length (Poothong and Reed, 2014). Recent studies on mineral nutrition for *Rubus* cultivars and species indicated that the mineral composition of MS medium is not suitable for diverse *Rubus* germplasm (Poothong and Reed, 2014, 2015, 2016). Increased calcium chloride ($CaCl_2$) and potassium phosphate (KH_2PO_4) improved shoot length and higher concentrations of magnesium sulfate ($MgSO_4$) improved leaf color and decreased leaf-edge necrosis for most of the genotypes, while iron at concentrations higher than MS decreased shoot length (Poothong and Reed, 2015). Zawadzka and Orlikowska (2006) also determined that a 2× concentration of $MgSO_4$ was best for growth of raspberry cultures.

Shoots of some *Rubus* genotypes grown on MS medium exhibited abnormal growth, stunting, hyperhydricity, leaf-edge necrosis, and red or purple leaf coloration (Fig. 7.1) (Poothong and Reed, 2014). Most or all of these growth defects were alleviated by increasing the mesos ($MgSO_4$, $CaCl_2$, and KH_2PO_4) content of the growth medium. Increasing iron concentrations over the normal MS levels resulted in greener, but much shorter shoots than those on MS iron. Increased shoot length was promoted by increased mesos (2.5× MS) and low iron (≤MS), while shoot multiplication was improved with moderate nitrogen (N) concentrations and increased mesos. Overall, improved shoot quality was controlled mostly by the mesos concentrations and somewhat by N and Fe (Poothong and Reed, 2014, 2015).

Most of the 12 *Rubus* genotypes tested for response to nitrogen compounds had good multiplication, shoot length, and green color with increased (2.5× MS concentration) mesos nutrients ($MgSO_4$, $CaCl_2$, and KH_2PO_4) and the standard MS nitrogen (Table 7.1) (Trt 2: 20 mM ammonium (NH_4^+) and 40 mM nitrate (NO_3^-) (Poothong and Reed, 2016). Three other nitrogen treatments (Table 7.1) also improved shoot quality for many of the tested genotypes: lower total N with lower NH_4^+ (Trt 1: 10 mM NH_4^+ and 40 mM NO_3^-), higher total N and intermediate (Trt 3: 30 mM NH_4^+ and 60 mM NO_3^-) or high (Trt 4: 45 mM NH_4^+ and 60 mM NO_3^-) concentrations of both components. Most of the diverse genotypes responded well to one or more of these treatments as compared to standard MS medium (Fig. 7.2).

Rooting and acclimation

Generally *Rubus* shoots are not difficult to root or acclimate. For most genotypes, standard growth medium with 0.5–1 μM IBA or with 10 μM IBA for one week then on no PGR medium for three weeks is effective. Two weeks in a mist chamber with decreasing mist cycles was effective for acclimation in the greenhouse (Reed and DeNoma, 2014). Cloudberry shoots cultured without growth

Fig. 7.1. *Rubus* shoots grown on Murashige and Skoog medium showing common physiological disorders: A. callus and leaf vein chlorosis, B. leaf chlorosis and necrosis, C. hyperhydricity, D. Leaf discoloration and spotting (S. Poothong, unpublished).

Table 7.1. The four best nitrogen treatments for *Rubus* genotypes grown on MS medium nutrients with 2.5× MS mesos.

	Mixture components		Numeric factor		Total N (mM)
	NH_4^+ (mM)	K^+ (mM)	NO_3^- (mM)	NH_4/NO_3	
1	0.25 (10)	0.75 (30)	40.00	0.25	50.00
2	0.50 (20)	0.50 (20)	40.00	0.50	60.00
3	0.50 (30)	0.50 (30)	60.00	0.50	90.00
4	0.75 (45)	0.25 (15)	60.00	0.75	105.00

(Adapted from Poothong and Reed, 2016)

Fig. 7.2. Growth response of shoots of four diverse *Rubus* genotypes cultured on Murashige and Skoog (1962) medium and three of the four nitrogen treatments with 2.5× MS mesos shown in Table 7.1. Control MS medium. Trt 1: ($NH_4^+/NO_3^- = 0.25$ (50 mM total N)) and 2.5× MS mesos. Trt 2: MS nitrogen ($NH_4^+/NO_3^- = 0.5$ (60 mM total N)) and 2.5× MS mesos. Trt 4: ($NH_4^+/NO_3^- = 0.75$ (105 mM total N)) and 2.5× MS mesos (S. Poothong, unpublished)

regulators rooted in 6–8 weeks *in vitro* under standard growth conditions. In another study, rooted shoots were acclimated in plug trays with 2 peat: 1 perlite (v/v) in a humidity chamber for 2–3 weeks (20°C (68°F), 95% relative humidity, with a 16-h photoperiod at 55 Mmol m^{-2} s^{-1}) (Debnath, 2004). A decrease in basal salts (1/4 or 1/2) or the addition of activated charcoal often produced a good rooting response, and *ex-vitro* rooting by dipping in commercial rooting compound often resulted in high rooting percentages (Pelto and Clark, 2000). In some cases, lower light intensity and the addition of charcoal to the medium were required for rooting recalcitrant blackberries (Wu *et al.*, 2009).

Although there are many culture media and PGR combinations available for the wide range of blackberry and other *Rubus* species, the general protocol used at NCGR as noted below provides good growth for most if not all the diverse genotypes.

THE GENERAL NCGR *IN VITRO* GROWTH PROTOCOL

Initiation: Explant shoot tips and axillary buds from dormant or spring growth of mother plants in protected growth conditions following cold treatment. Surface sterilize, 10 minutes in 10% household bleach with surfactant, rinse in sterile water. Test for contaminants and grow on MS medium with 2.5× mesos nutrients, 0.29 μM GA3, 0.49 μM IBA, and 4.45 μM BA (NCGR Rubus 2015 medium).

Multiplication: Grow on NCGR Rubus 2015 medium (Poothong and Reed, 2016) under 16-h photoperiod with a mixture of cool and warm fluorescent bulbs (70–90 Mmol m^{-2} s^{-1}) at 25°C (77°F). Divide and pass to new growth medium at 3-week intervals. Multiplication can be increased as needed with slight increases in N or BA. These will also decrease shoot length.

1. Rooting: Transfer to 1 μM IBA medium for rooting. Dipping shoots directly in rooting compound is also successful for *ex-vitro* rooting.

2. Acclimation: Rinse growth medium from the roots and plant in plug trays in a light potting mix. Place under mist with decreasing intervals for about 2 weeks.

REFERENCES

Anderson, W.C. (1980) Tissue culture propagation of red raspberries and black raspberries, *Rubus idaeus* and *R. occidentalis*. *Acta Horticulturae* 11, 13–20.

Bobrowski, V.L., Mello-Farias, P.C. and Peters, J.A. (1996) Micropropagation of blackberries (*Rubus* sp.) cultivars. *Revista Brasileira de Agrociencia* 2(1), 17–20.

Bray, M.M., Rom, C.R. and Clark, J.R. (2003) Propagation of thornless Arkansas blackberries by hardwood cuttings. *Horticultural Studies – 2003. Arkansas Agricultural Experiment Station Research Series* 520, 11–13.

Broome, O.C. and Zimmerman R.H. (1978) *In vitro* propagation of blackberry. *HortScience* 13(2), 151–153.

Busby, A.L. and Himelrick, D.G. (1999) Propagation of blackberries (*Rubus* spp.) by stem cuttings using various IBA formulations. *Acta Horticulturae* 505, 327–332.

Caldwell, J.D. (1984) Blackberry propagation. *HortScience* 19(2), 193–195.

Debnath, S.C. (2004) Clonal propagation of dwarf raspberry (*Rubus pubescens* Raf.) through *in vitro* axillary shoot proliferation. *Plant Growth Regulation* 43(2), 179–186.

Debnath, S.C. (2011) Bioreactors and molecular analysis in berry crop micropropagation – a review. *Canadian Journal of Plant Science* 91(1), 147–157.

Donnelly, D.J. and Daubeny, H.A. (1986) Tissue culture of *Rubus* species. *Acta Horticulturae* 183, 305–314.

Donnelly, D.J., Stace-Smith, R. and Mellor, F.C. (1980) *In vitro* culture of three *Rubus* species. *Acta Horticulturae* 112, 69–75.

Dziedzic, E. and Jagla, J. (2012) Micropropagation of *Rubus* and *Ribes* spp. In: Lambardi, M., Ozudogru, E.A, and Jain, S.M. (eds.) *Protocols for Micropropagation of*

Selected Economically-important Horticultural Plants: Methods in Molecular Biology, pp. 149–160. Available at: www.springerprotocols.com/Abstract/doi/10.1007/978-1-62703-074-8_11 (accessed May 1, 2017).

Fernandez, G.E. and Clark, J.R. (1991) In vitro propagation of thornless 'Navaho' blackberry. *HortScience* 26(9), 1219.

Finne, A. (1997) Micropropagation of *Rubus* spp. *Journal of Agricultural Science in Finland* 58(3), 193–196.

Garner, R.J. and Hammond, D.H. (1939) Leaf-bud propagation of loganberry, youngberry and blackberries. *Report of the East Malling Research Station for 1938*. East Malling Research Station, East Malling, UK, pp. 218–222.

George, E.F., Hall, M.A. and De Klerk, G J. (2008) The components of plant tissue culture media. I: Macro- and micro-nutrients. In: George, E.F., Hall, M.A. and De Klerk, G.J. (eds.) *Plant Propagation by Tissue Culture*, 3rd edn. Springer, New York, pp. 65–113.

Gonzalez, M.V., Lopez, M., Valdes, A.E. and Ordas, R.J. (2000) Micropropagation of three berry fruit species using nodal segments from field-grown plants. *Annals of Applied Biology* 137(1), 73–78.

Hall, H.K. (1990) Blackberry breeding. In: Janick, J. (ed.) *Plant Breeding Reviews*, Vol. 8. Timber Press, Inc., Portland, Oregon, pp. 249–312.

Hansen, A.J. (1985) An end to the dilemma – virus free all the way. *HortScience* 20(6), 852–860.

Harper, P.C. (1978) Tissue culture propagation of blackberry and tayberry. *Horticultural Research Institute* 18, 141–143.

Lloyd, G. and McCown, B. (1980) Commercially-feasible micropropagation of mountain laurel, *Kalmia latifolia*, by use of shoot-tip culture. *Combined Proceedings of the International Plant Propagators Society* 30, 421–427.

Moore, J.N., Pavlis, G C., Brown, G.R. and Lundergan, C.A. (1978) Establishing blackberry plantings with root cuttings. *Arkansas Farm Research* 27(2), 4.

Murashige T. and Skoog, F. (1962) A revised medium for rapid growth and bio-assays with tobacco tissue culture. *Physiologia Plantarum* 15(3), 473–497.

Murashige, T. and Tucker, D.H.P. (1969) Growth factor requirements of *Citrus* tissue culture. In: *Proceedings of the First International Citrus Symposium*, vol. 3. International Society of Citriculture (ISC), Riverside, California, pp. 1155–1161.

Ourecky, D.K. (1975) Brambles. In: Janick, J. and Moore, J.N. (eds.) *Advances in Fruit Breeding*. Purdue University Press, West Lafayette, Indiana, pp. 98–129.

Pelto, M.C. and Clark, J.R. (2000) *In vitro* shoot tip culture of *Rubus*, Part 1. *Small Fruits Review*, 1, 69–82.

Poothong, S. and Reed, B.M. (2014) Modeling the effects of mineral nutrition for improving growth and development of micropropagated red raspberries. *Scientia Horticulturae* 165, 132–141.

Poothong, S. and Reed, B.M. (2015) Increased CaCl2, MgSO4, and KH2PO4 improve the growth of micropropagated red raspberries. *In Vitro Cellular and Developmental Biology – Plant* 51(6), 648–658.

Poothong, S. and Reed, B.M. (2016) Optimizing shoot culture media for *Rubus* germplasm: the effects of NH_4^+, NO_3^- and total nitrogen. *In Vitro Cellular and Developmental Biology – Plant* 52(3), 265–275.

Preece, J. (1995) Can nutrient salts partially substitute for plant growth regulators? *Plant Tissue Culture Biotechnology* 1, 26–37.

Reed, B.M. (1990) Multiplication of *Rubus* germplasm in vitro: a screen of 256 accessions. *Fruit Varieties Journal* 44(3), 141–148.

Reed, B.M. and DeNoma, J.S. (2014) *Rubus* tissue culture procedures. *National Clonal Germplasm Repository – Corvallis Operations* Manual. USDA-ARS National Clonal Germplasm Repository, Corvallis, Oregon.

Reed, B.M., Buckley, P.M. and DeWilde, T.N. (1995) Detection and eradication of endophytic bacteria from micropropagated mint plants. *In Vitro Cellular and Developmental Biology – Plant* 31(1), 53–57.

Skirvin, R.M. and Chu, M.C. (1977) Tissue culture may revolutionize the production of peach shoots. *Illinois Research* 19(4), 18.

Skirvin, R.M., Chu, M.C. and Gomez, E. (1981) In vitro propagation of thornless trailing blackberries. *HortScience* 16(3), 310–312.

Slivinski, J.A., Preece, J.E. and Myers, O. (1984) In vitro micropropagation of thornless blackberries utilizing single-node explants. *Plant Propagation* 30, 4–5.

Swartz, H.J., Galletta, G.J. and Zimmerman, R.H. (1983) Field performance and phenotypic stability of tissue culture-propagated thornless blackberries. *Journal of the American Society for Horticultural Science* 108(2), 285–290.

Tanprasert, P. and Reed, B.M. (1997) Determination of minimal bactericidal and effective antibiotic treatment concentrations for bacterial contaminants from micropropagated strawberries. *In Vitro Cellular and Developmental Biology – Plant* 33(3), 227–230.

Trigiano, R.N. and Gray, D.J. (eds.) (2000) *Plant Tissue Culture Concepts and Laboratory Exercises*, 2nd edn. CRC Press, Boca Raton, Florida.

Tsao, C.W.V. and Reed, B.M. (2002) Gelling agents, silver nitrate, and sequestrene iron influence adventitious shoot and callus formation from *Rubus* leaves. *In Vitro Cellular and Developmental Biology – Plant* 38(1), 29–32.

Tsao, C.W.V., Postman, J.D. and Reed, B.M. (2000) Virus infections reduce *in vitro* multiplication of 'Malling Landmark' raspberry. *In Vitro Cellular and Developmental Biology – Plant* 36(1), 65–68.

Velchev, V., Toshkova, A. and Mladenova, O. (1997) In vitro propagation of thornless blackberry. II. Rooting micropropagated shoots of cv Thornfree. *Horticulture and Viticultural Science* 20(7), 16–23.

Viss, P.R., Brooks, E.M. and Driver, J.A. (1991) A simplified method for the control of bacterial contamination in woody plant tissue culture. *In Vitro Cellular and Developmental Biology* 27(1), 42.

Welander, M. (1985) In vitro culture of raspberry (*Rubus ideaus*) for mass propagation. *Journal of Horticultural Science* 60(4), 493–499.

Wood, G.A. (1989) Propagation, virus-screening and heat therapy of Northern Hemisphere imports of *Ribes, Rubus,* and *Vaccinium. New Zealand Journal of Crop and Horticultural Science* 17(3), 271–274.

Wu, J.-H., Miller, S.A., Hall, H.K. and Mooney, P.A. (2009) Factors affecting the efficiency of micropropagation from lateral buds and shoot tips of *Rubus. Plant Cell Tissue and Organ Culture* 99(1), 17–25.

Zawadzka, M. and Orlikowska, T. (2006) Factors modifying regeneration *in vitro* of adventitious shoots in five red raspberry cultivars. *Journal of Fruit and Ornamental Plant Research* 14(2), 105.

Zimmerman, R.H. (1979) Fruit plants micropropagation at Beltsville Fruit Laboratory and in North America. *Ortoflorofrutticolutura* 64, 241–256.

Zimmerman, R.H. and Broome, O.C. (1980) Micropropagation of thornless blackberries. In: *Proceedings of the Conference on Nursery Production of Fruit Plants through Tissue Culture – Applications and Feasibility.* USDA-ARS, Beltsville, Martland.

8

SITE SELECTION

Ellen Thompson*

Pacific Berry Breeding, LLC, Salinas, California, USA

INTRODUCTION

Amid the recent and continued global expansion of blackberry production, site selection is undergoing a shift in perspective. The introduction of primocane-fruiting cultivars, along with trellis, tunnel, and substrate-growing innovations have further pushed the boundaries where prospective sites are geographically chosen. The majority of commercial blackberries are currently grown in latitudes ranging from 19–45°N to 27–41°S, in climates ranging from subtropical to Mediterranean coastal, and elevations of 0–2000 m (0–6500 ft). In the coming years, blackberry cultivation is expected to expand into locations previously not thought possible (Strik et al., 2007; John R. Clark, personal communication).

SOIL

Blackberries can tolerate a wide spectrum of soils, provided the drainage is moderate to good. If drainage is unknown, a soil percolation test should be conducted. In the event that drainage is poor, as can be exhibited in heavier soils, planting on raised beds is recommended. From pure sand to heavy clay loam, soil testing is critical prior to planting. Representative sampling is typically done using a soil probe to a depth of 20–30 cm (8–12 in), in an 'X' pattern across the field or blocks. In planting sites that vary in elevation or other features, it is important to extract representative samples of each distinct feature in the landscape. Deep ripping may be necessary if a subsurface hardpan is present (Bolda et al., 2012).

Determining the pH, electrical conductivity (EC), elemental content of macro- and micronutrients, sodium absorption ratio (SAR), and percent

---

* Corresponding author: Ellen@pacificberrybreeding.com

organic matter of the proposed site is necessary for building subsequent fertility and irrigation programs. The ideal soil pH range is slightly acid to neutral, in the range of 5.6–6.8 (see Chapter 11; Bolda et *al.*, 2012; Bushway et *al.*, 2008; Strik, 2008). In more arid regions, like southern coastal California or the Middle East, particularly in sites that have been previously covered with tunnels, salt accumulation may be of concern. Ideally, the range for sodium (Na) and chloride (Cl) values should not exceed 3 meq/l; however, blackberries, once established, can tolerate more than double this rate for short periods (personal observation).

Organic matter promotes soil tilth, aeration, and nutrient and water holding capacity. Applying well-composted manure (free of weed seeds) and/or the use of cover crops are common methods to incorporate an increase in organic matter. Cool- and warm-season cover crops may be established annually in inter-row pathways and later incorporated to increase organic matter. Annual cane prunings are also commonly turned into mulch within the inter-row path using a flail mower, and in some locations this mulch is wind-rowed under the plants. In blackberry-growing regions that do not use tunnels, particularly those that reliably receive summer rain, it is common to establish grass in the inter-row alleys to reduce dust and prevent erosion. Any use of cover, whether grass, legume, or landscape fabric, will help prevent soil losses from wind and rain, and further reduce dust on the crop.

Knowledge of previous crops can be extremely helpful in avoiding or ameliorating problem weeds and disease, aiding understanding in how best to prepare the future planting. All soils harbor weeds and their seeds, therefore it is imperative to control perennial weeds prior to establishment. Fumigation, though effective in killing weeds, dormant seeds, soil-borne diseases, ground-burrowing insects and mammals, is increasingly becoming restricted. Pre-emergent herbicides, if timed correctly, can work well to prevent broadleaf weeds from establishing. Clean-cultivation, along with spot weeding by hand or implement, is effective particularly after weed seeds are allowed to germinate but aren't well established. Organic and low-input systems commonly control weeds using a combination of crop rotation, landscape fabric or plastic mulch, flaming, clean-cultivation, and hand weeding. Wild *Rubus* and other weedy species in the immediate vicinity (within 100 m (330 ft)) may harbor pests and disease, and therefore should be eliminated prior to planting.

WATER

A secure source of good-quality water is important for any agricultural operation. Municipal sources may be available and most secure, yet cost may be prohibitive. Other sources of water include wells and surfaces (e.g. streams, ponds, lakes). Despite the source, water samples should be tested for pH, salt levels, heavy metals, and biological organisms of concern for food safety, such as *E. coli*. Local regulations may restrict water for agricultural use, or require on-site ability to chlorinate surface water (Bolda et *al.*, 2012). Capture, treatment, and reuse of irrigation water is likely to increase in the future. Recycling of water is most easily done in above-ground, or substrate, growing systems. Re-use of irrigation water requires well-planned engineering, proper water storage, and the capacity to filter out biological contaminants. The EC of remaining nutrients in the recycled irrigation water must be monitored and adjusted prior to re-use within the planting.

TOPOGRAPHY, EXPOSURE, AND LAYOUT

A surface that is flat or moderately sloped (<3% grade) is ideal for movement of equipment and employees alike. Erosion is more likely to occur along steeply sloped areas (>3% grade), thus under such conditions, preparing and maintaining ditches and drains for proper water slowing and infiltration, respectively, is critical to avoid loss of soil, plants, and productivity. It is important to note that hillsides may naturally be less fertile, due to the erosive nature of the topography. Downward flow of water on slopes can be mitigated by arranging rows perpendicular to the rise, a design similar to terracing. Blackberries planted on steeper hills are likely to enjoy better air movement and avoid cold air settling, of particular importance in colder and continental climates. Improved air movement can further help reduce the likelihood of certain fungal diseases that thrive under still conditions. In contrast, air drainage can be poorer in flat areas. Low-lying sites and valley floors tend to attract settling air and may even be frost-pockets in certain climates. Flat areas are commonly laser-leveled to allow uniform water penetration, with the production zones surrounded by drains or ditches to channel excesses away.

Exposure toward the equator is generally preferred over the contrary. North–south row orientation will optimize uniform sunlight exposure through the blackberry canopy. Site-dependent, ideal row orientation is not always possible, nor desired. In some climates, particularly at higher elevations, afternoon radiation on a western or equator-facing exposure can promote sunburn on the berries. With the use of a Rotating Cross Arm (RCA) trellis system, increasingly used in the upper Midwestern USA, row orientation is purposefully done on an east–west axis. This is done to ripen fruit on the shaded side of the trellis, away from the equator, protecting both berry and hand harvester from sun exposure.

Typical field layouts are arranged into evenly sized blocks, each with an access road along the periphery. A balance in block size and design promotes organization and aids in overall management of labor, irrigation, and supplies required to operate successfully.

A field office is most commonly located near the entry of the operation, with proper signage to signal prospective employees and supplier deliveries alike. Nearby, it is common to find temporary or permanent storage structures to house safety bulletins, legal disclosures, equipment, chemicals, and packing supplies. A designated area for bee hives should be situated in an area that receives early morning sun exposure, allows easy access by beekeepers, and with an appropriate distance from areas commonly utilized by employees.

ADAPTATION AND CLIMATE

Blackberries are widely adapted, though limitations exist on where they can be cultivated optimally. Mediterranean-like climates, those with mild wet winters and cool dry summers, offer ideal conditions for many temperate-zone fruiting crops, including blackberries. Inland, or continental climate zones, can be more challenging with extreme temperatures experienced in both summer and winter. Intense summer heat can adversely affect pollen viability (Stanton et al., 2007) and damage berry quality of ripe fruit. Winter injury, or even death, can occur in cold inland climates. Damaging winds can be mitigated with the use of windbreaks or shelter belts (see Appendix 1). Blackberries that are fruited on their floricanes require environmental cues to satisfy the chilling requirement. Chilling hours, or units, are commonly calculated as 1 hour below 7.2°C (45°F). Floricane blackberries tend to require 500–1200 chilling units (CU), while primocane types are thought to require only 100–300 CU (Clark et al., 2005). Blackberries grown in zero-chill areas, such as Mexico, rely on chemical treatments to stimulate budburst and flowering. Reliable weather data can often be accessed through local climatological stations. If the location is distinct or remote, it is advisable to install one on site.

LOCATION AND LABOR

Along with the aforementioned considerations, location is frequently concomitant with labor availability. As the labor force in many countries continues to shift, ease of access and convenience of location can play an instrumental role in attracting employees. In areas where employees drive individual automobiles to work, ample parking spaces must be available. If parking is limited, providing transport for your employees may be necessary.

Distance to cold storage or processing facilities is an important factor to consider when choosing a location, including the roads on which trucks laden with blackberries must drive. The delicate nature of blackberry fruits can easily be compromised if the ability to carefully deliver and cool flats is challenging, or requires several hours to accomplish. Convenient locations will likely promote improved fruit quality for these reasons.

SUBSTRATE

Choosing a site for substrate (e.g. soil-less, containerized) blackberry production may, in some ways, be easier than finding a proper spot for in-ground operations. With soil removed from the equation, the focus shifts to the other considerations described previously.

Adequate area in which to receive and store bulk quantities of substrate and pots is necessary. A structure, even a temporary one, under which to prepare pots prior to planting can be useful in higher rainfall climates. Adequate space to comfortably use pot-filling and moving equipment, if used, should also be taken under consideration.

SUMMARY

Production and consumption of blackberries is expected to increase globally in the coming decades (Strik et al., 2007). Under current industry trends observed in Europe and California, expansion of blackberry production may increase most rapidly through utilization of low-/no-chill primocane varieties grown in substrate. Therefore, site selection for fresh market production may shift away from fertile soils and toward urban centers; away from valleys and toward building rooftops.

REFERENCES

Bolda, M., Gaskell, M., Mitcham, E. and Cahn, M. (2012) *Fresh Market Caneberry Production Manual*. University of California Agriculture and Natural Resources Publication 3525. The Regents of the University of California Agriculture and Natural Resources, Richmond, California.

Bushway, L., Pritts, M. and Handley, D. (eds.) (2008) *Raspberry and Blackberry Production Guide for the Northeast, Midwest, and Eastern Canada*. Natural Resource, Agriculture, and Engineering Service Cooperative Extension, NRAES-35, Ithaca, New York.

Clark, J.R., Moore, J.N., Lopez-Medina, J. and Perkins-Veazie, P. (2005) 'Prime-Jan' ('APF-8') and 'Prime-Jim' ('APF-12') primocane-fruiting blackberries. *HortScience* 40(3), 852–855.

Stanton, M.A., Scheerens, J.C., Funt, R.C. and Clark, J.R. (2007) Floral competence of primocane-fruiting blackberries Prime-Jim and Prime-Jan grown at three temperature regimines. *HortScience* 42(3), 508–513.

Strik, B.C. (2008) *Growing Blackberries in Your Home Garden. Growing Small Fruits*. Oregon State University Extension Publication, EC 1303. Oregon State University, Corvallis, Oregon.

Strik, B.C., Clark, J.R., Finn, C.E. and Bañados, M.P. (2007) Worldwide blackberry production. *HortTechnology* 17, 205–213.

SITE PREPARATION, SOIL MANAGEMENT, AND PLANTING

Marvin Pritts[1,*] and Eric Hanson[2]

[1]Cornell University, Ithaca, New York, USA; [2]Michigan State University, East Lansing, Michigan, USA

INTRODUCTION

The treatment and management of the site and soil before planting can impact growth and productivity of blackberries for years to come. Many consider the year prior to planting to be the most important because of the substantial effects that soil factors have on plant growth. Modifications to soil are much easier to make without blackberries present on the site, hence the emphasis on modifications prior to planting. Furthermore, changes to soil properties take time, and waiting until just before planting to attempt modifications may not be sufficient.

Fortunately, blackberries are tolerant of a wide range of soil types. They have root systems that can penetrate through heavier soils, and they are more tolerant than their raspberry cousins to the pathogens that thrive under wet soil conditions. Nonetheless, optimal performance requires loamy soils with good drainage. In addition, soils almost always can be improved before a particular crop is planted, whether it be enhancement of drainage, modification of soil pH, addition of organic matter, or reduction of the weed seed bank. These improvements may take months to implement, so it is recommended that the year before planting be spent on making these improvements. Pritts *et al.* (2015) describe in detail soil and nutrient management practices for berry crops.

PHYSICAL PROPERTIES

Soil physical properties are difficult to modify, so it is best to select a site with an appropriate soil type. Blackberries (thorny, thornless, erect, semi-trailing, or trailing types; primocane or floricane; summer- or autumn-bearing cultivars)

---

* Corresponding author: mpp3@cornell.edu

grow best in well-drained, loamy soils but will tolerate a range of soil types. Restrictions on growth and productivity are often associated with compacted soils or those with a perched water table. If the water table is shallow, blackberries may benefit from planting on raised beds.

The Cation Exchange Capacity (CEC) is made up of contributions from both the mineral and organic components of the soil. In general, a high CEC (15–25) indicates that the soil is relatively fertile and has the ability to hold on to potassium (K), magnesium (Mg), and calcium (Ca) ions. Although the mineral contribution to CEC cannot be modified significantly, the organic component can be increased through the addition of organic matter, such as manure, compost, or incorporated cover crops. Increasing the organic matter generally increases the CEC and the water-holding capacity of the soil. It also improves soil structure, the aggregation of soil particles, and soil tilth. Increasing organic matter and water-holding capacity are nearly always beneficial.

Increased aggregation of soil particles, facilitated by organic matter, allows for better soil structure and root development. Without structure, soil particles collapse on each other, resulting in compaction and small pore spaces between particles. Soils without pore space have a difficult time holding on to water; water tends not to drain through the soil profile, and roots have a difficult time growing and penetrating the soil. Soil structure is also damaged by repeated cultivation. Although a field that has been rototilled several times may look good prior to planting, the structure may be degraded so that when it rains, the soil becomes compacted with little pore space for air and water. Also, repeated tillage and the subsequent exposure of organic matter to oxygen will accelerate decomposition of the organic matter. Driving heavy equipment onto a field when soil is wet also causes compaction. Each of these practices should be avoided. If a field does have a plow layer or is otherwise compacted, then deep subsoiling may be required to break up the compacted zone. Repeated shallow plowing or tilling will only increase compaction and soil degradation.

Green manure cover crops can be incorporated to increase soil organic matter and provide nitrogen (N) if the cover crop is a legume. They also tend to suppress weeds while growing. Obviously cover crops require time to grow, so plan for site preparation and soil modification well in advance. Popular pre-plant cover crops for blackberries include peas, lupines, alfalfa, buckwheat, rye, wheat, oats, clover, sudangrass, vetch, and canola.

BIOLOGICAL PROPERTIES

Living organisms in the soil can have a significant impact on plant growth. These include bacteria, viruses, fungi, nematodes, protozoa, algae, insects and earthworms. The vast majority of these are beneficial or benign. These organisms recycle nutrients and bind soil particles together, improving soil structure. Although the role of each species of microorganism is not well understood, some generalizations can be made. Soils with a large amount of organic matter to sustain these organisms and fuel biological activity tend to be healthier and more stable than those without much activity (Hargreaves et al., 2008). This biological activity tends to suppress harmful organisms that attempt to become established and enhances nutrient uptake. Organic matter that is 'aged' tends to have a more stable complex of organisms than fresh material. Some composts have been shown to suppress plant pathogens in the soil and reduce plant disease.

The incorporation of beneficial microorganisms, such as mycorrhizae or bacteria, into soil before planting, or inoculating plant roots prior to planting, has been promoted as a way to enhance growth and productivity. There is only scattered evidence that this is an effective practice under field conditions. Providing a growing medium for beneficial organisms (e.g. compost and organic matter) seems to be critical for obtaining consistent suppression of soil pathogens.

Certain cover crops also can help suppress disease-causing organisms in berry crops. Marigolds, mustards, and certain cultivars of oats and sudangrass, for example, will suppress nematodes when grown prior to planting blackberries. Nematodes, in particular, can cause problems in blackberries (Wehunt et al., 1991). Testing for harmful nematodes is recommended prior to planting, particularly since they can harbor and transmit virus diseases. If levels are very high, fumigation may be recommended as opposed to natural means of nematode suppression. One risk of fumigation is that it kills all soil organisms, even those that are beneficial and benign. Beneficial organisms may not re-establish well after fumigation, leading to an increased risk of pathogen infection in later years. Cover crops are usually a better choice if sufficient time is available to grow them prior to planting. Fumigation will suppress these pathogens quickly, but it is very expensive and difficult to apply properly.

Among the most detrimental organisms present before planting are weeds. Cover crops can help suppress weeds, especially those that grow fast and tall. Suppression is greatest when a broad-spectrum, non-selective, non-residual herbicide is used to kill perennial weeds prior to cover cropping. Repeated cultivation also can reduce weed pressure, but is detrimental to soil structure.

A sequential mixture of cover crops appears to work best for suppressing a range of harmful organisms while improving organic matter. Many growers use a combination of summer and winter cover crops in sequence. For example, summer cover crops, such as sudangrass or buckwheat, may be followed with vetch or rye for the fall/winter. Plants will almost always perform better following one or two years of cover crop rotations prior to planting.

CHEMICAL PROPERTIES

Soils contain many chemicals that are naturally occurring or added by human activity. For the most part, the soil is a large reservoir of minerals and organic material that slowly releases chemical elements, some of which are taken up and used by plants as essential nutrients. Most evaluation methods seek to estimate the amount of these nutrients available to the plant for growth and development – these are generally grouped into those that are required only in small amounts: boron (B), iron (Fe), zinc (Zn), manganese (Mn), copper, (Cu), molybdenum (Mo), chlorine (Cl) and those required in larger amounts: N, phosphorus (P), K, Ca, Mg, sulfur (S). Soil amendments usually consist of fertilizers containing nutrients from this latter group. Certain other elements may be beneficial for certain plants, but are not essential for growth. These are silicon (Si), cobalt (Co), nickel (Ni), selenium (Se), sodium (Na), and aluminum (Al). The role of these particular nutrients in berry crops is not well understood.

Soil samples for testing should be taken randomly (or systematically to assure thorough representation) from a field to be planted. A minimum of ten sub-samples or cores should be taken from a field that has the same soil type and cropping history. A sampled section should be 4 ha (10 acres) or less. If the field is larger than 4 ha (10 acres), then a second set of sub-samples should be collected. The sub-samples should come from the future root zone, and then combined and mixed in a bucket. The composite sample should be sent to a soil testing laboratory where it will be extracted and analyzed. Different labs use different extractants, so the estimate of the amount of essential plant nutrients will differ from lab to lab. Recommendations should be based on the specific extractant used by the lab. Do not use results from one lab to generate recommendations from another.

The first test to consider is the soil pH. The optimal pH for blackberries is 6.0–6.5 (Fig. 9.1), although good performance is realized within a range of 5.6–6.8. If the soil pH is outside this range, then nutrient uptake can be compromised and an amendment will be recommended. Sulfur is used to lower pH and lime is used to raise pH to the target of 6.5. The amount of S or lime required to change pH to the optimum is dependent on the ability of the soil to hold onto alkaline-forming cations, such as K, Mg, and Ca. Typically, heavier soils and those with high organic matter (a high CEC) require more amendment to change pH than sandier soils with low organic matter (a low CEC). The soil test result should provide a recommendation for the amount of S or lime to add, and will indicate if the lime should have a high proportion of Ca (calcitic) or Mg (dolomitic). A common myth is that certain organic materials, such as pine needles, can lower soil pH. Studies have shown that incorporating most organic materials, while perhaps increasing the CEC slightly, have little effect on overall soil pH.

Lime is not 100% pure, so it will come with a percentage that represents its effective neutralizing value (ENV). For example, if the lime has a 90% ENV rating, then it will take 110% of the recommended rate to bring about the desired change in soil pH. Both lime and S are available in pelletized form to facilitate spreading and to reduce dust and blowing. However, pelletized forms require a longer period to break down and react with soil to change pH. Most soil testing

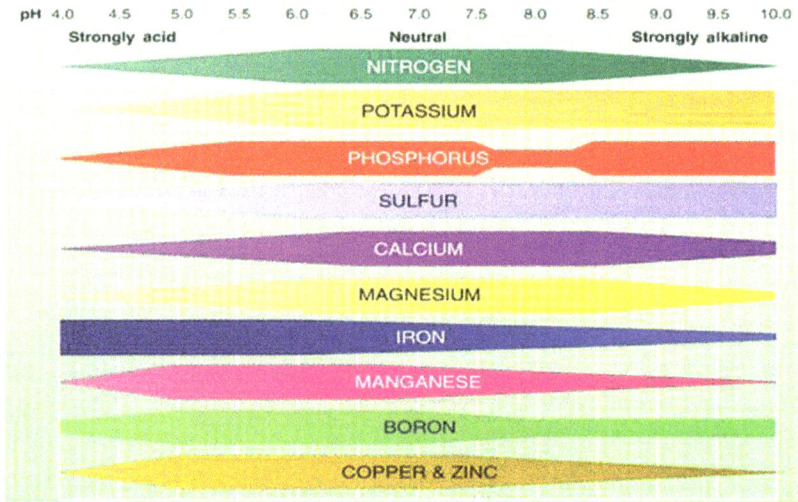

Fig. 9.1.   Availability of soil nutrients varies with pH (M. Pritts).

labs assume a 15 cm (6 in) slice of soil for making recommendations. For black-berries that have a slightly deeper rooting depth (i.e. 20 cm (8 in)), amounts should be increased by 25% to bring about a change in pH to a depth of 20 cm (8 in).

The lime or S should be broadcast over the entire area and incorporated to the depth of the rooting zone. Correcting the soil pH with lime will usually provide all of the required Ca and Mg as well. Some soils may require supplemental Mg even when the pH is 6.5 and no lime is required. In these situations, magnesium sulfate can provide additional Mg with little change in soil pH. Similarly, if additional Ca is required but the pH is already high, then calcium sulfate (gypsum) can be added. Evidence also exists that Ca ions can suppress certain soil-borne diseases (Maloney *et al.*, 2005).

Soil pH may also be affected by the source of N, but this comes into play only after the plants are established. Nitrogen fertilizers should not be used to modify pH prior to planting. Other nutrient sources have only a small impact on soil pH. Therefore, lime and sulfur are the basic tools for managing soil pH.

A standard soil test analysis also reports the estimated amount of plant available P and K, which will vary depending on the extractant used. If additions are recommended, then follow the recommendation of a lab or consultant with blackberry experience. Generalizations about soil tests and sufficiency ranges cannot be made, due to variation among lab procedures and use of extractants (see Chapter 11).

As is the case with lime and S, the more finely ground the source of P and K, the faster it will become available to plants. In addition, sources that are readily soluble in water will be available faster than those less soluble. Organic sources of P and K, whether occurring naturally in the soil or provided through organic fertilizers, tend to have nutrients tied up in complex mineral structures that are only slowly released to the plant. Conventional inorganic fertilizers provide a more readily available form of nutrient, but they contribute saltiness to the soil (e.g. chlorides, sulfates) that can negatively affect plant growth if their levels are too high. A combination of organic and conventional sources is often desirable.

Many soils are sufficiently high in P for blackberries, especially if they have received manure applications in the past or have had annual field crops grown on them. Excessively high P can interfere with uptake of certain other essential nutrients, so P should not be applied without first testing the soil to determine whether it is needed.

In areas that typically have low B, it is prudent to test for B even though this usually incurs a separate charge from the laboratory. Boron affects root development, so low B can reduce uptake of all other nutrients even though adequate levels of these nutrients may be present in soil. Boron also has profound effects on bud break, and in B deficient soils, bud break and production may be severely reduced.

Soil test values for other nutrients, especially micronutrients and N, do not provide a reliable basis for making recommendations. Adjustments to these nutrients are best made once plants are established and leaves can be used as an indicator of nutritional status.

All amendments should be worked into the root zone prior to planting. Applying nutrients to the soil surface after plants are established is not very effective because most fertilizers move slowly through the soil profile. Incorporating lime, S, and nutrients several months prior to planting will allow sufficient time for appropriate changes to take place in soil chemistry.

Nitrogen is an essential nutrient for plant growth and is generally considered desirable in soils (see Chapter 11). Higher soil N usually means that the soil is fertile. Nitrogen is available in the soil in three basic forms: ammonium N, nitrate N, and organic N. Blackberries preferentially use nitrate N. Typically, nitrate forms of N are low in soil, so they are usually added through fertilizers. Newly planted blackberries do not require large amounts of nitrate N, so it is best if it can be supplied by converting from the other sources already present in the soil. Ammonium N is converted to available nitrate N when soils are warm and the pH is not too low. Organic N is converted to available nitrate N when biological activity is high and there is not too much carbon in the soil. The best method of providing a pool of nitrate N to plants prior to planting is by incorporating a source of organic N that is low in carbon. Examples include manure or a legume cover crop. These materials break down slowly and provide a source of nitrate N to the plants without significant leaching. In situations where organic N is low, supplemental nitrate fertilizer may need to be provided, but this rapidly available form is subject to leaching. Too much nitrate or ammonium N fertilizer can be toxic to plants. It would be rare for organic matter to provide excessive N to the plant, except in situations where it is released late in the season as plants attempt to harden off for winter. In most situations, increasing organic matter prior to planting is a valuable step in the soil management process.

Other chemical elements can be toxic to plants and people, so soils high in these elements should be avoided. Arsenic (As) Se, Na and Al are naturally part of certain soils, but when they are taken up by plants in large amounts, detrimental responses result. Uptake of these elements is usually increased as acidity increases, so pH should be maintained above 6.0 if these elements are present at high levels. Excessive amounts of essential nutrients can be toxic as well, so it is critical not to over-fertilize. Certain fertilizers may contain other elements that are toxic in high amounts. For example, muriate of potash (potassium chloride) is a useful fertilizer for blackberries when rates are modest. If applied amounts are high, the excessive chloride in the fertilizer can be harmful if there is not sufficient rainfall to leach the chlorides. In such cases, potassium sulfate is recommended, even though it is more expensive.

A second potential source of toxic chemicals in soils is irrigation water. Often the toxicity is caused by too much salt in the water. This is especially problematic where drip irrigation is used under dry conditions. Test the irrigation water for salt content before using. It should be less than 2.0 dS/m and preferably less than 1.0.

A third potential source of toxicity is herbicides that have been used for weed control in previously planted crops. Laboratory testing for herbicide residue is inexact and expensive, so a bioassay is recommended. A bioassay is simply a comparison of the growth of several types of seeds (e.g. radish, bean, rye, cucumber) planted into a sample of soil from the intended field and in a similar soil that has never been treated with herbicide. If growth differences occur or injury symptoms are expressed, then it might be attributed to herbicide residue. If a difference exists, waiting another year before planting may be prudent.

SOIL QUALITY EVALUATION

Recently soil scientists have been developing soil health tests that consist of a representative set of chemical, physical, and biological variables. Traditionally, only chemical assessments have been made because they are easy to obtain. However, physical and biological properties of soils play large roles in plant performance as well. Chemical assessments are provided by standard soil tests. Among the many physical properties of soils, the following are good indicators of soil health: aggregate stability, available water-holding capacity, surface and subsurface hardness, and soil texture. Good biological indicators are soil organic matter content, active carbon content (the quality and type of organic matter), potentially mineralizable N (ability of organic matter to supply N), and root health (presence of plant disease and nematodes). One assumes that there is cause-and-effect between these soil quality variables and plant performance; therefore, improving any of these should eventually result in improved plant performance (Pritts *et al.*, 2015).

SITE PREPARATION

Preparing a site for planting requires a thoughtful plan to create an environment in which plants perform to their maximum potential. Select a site with an appropriate soil type – in particular, the soil should have adequate internal drainage. If the site has standing water, it should be drained. If drainage is not possible, then plan on planting on raised beds of at least 25 cm (10 in) in height (see Chapter 10).

Perennial weeds should be controlled at least one year prior to the intended planting date. This may require two or more applications of a broad-spectrum, post-emergent, non-selective, non-residual herbicide, followed by plowing and disking after each herbicide application. The purpose of disking is to promote the germination of weeds in the seed bank and the sprouting of root pieces so they can subsequently be killed with the second herbicide application or disking event. Planting blackberries into a field that has previously been planted to field crops where perennial weeds have mostly been eliminated is an approach that works well for many growers. Herbicides used in field crops often are active against difficult-to-control weeds in blackberries, leaving fields relatively free of problem weeds. Care should be taken, however, to insure that herbicide carry over does not affect the young blackberry transplants (see previous text on testing for herbicide toxicity).

Conduct soil and nematode tests the summer before planting and incorporate recommended soil amendments to a depth of 20 cm (8 in), if possible. Grow a sequence of cover crops one year or more before planting, including some legumes, and use deep tillage to incorporate them. Certain cover crops will suppress nematode levels if they are high.

Preparing a site for blackberries does not require special equipment. Applying a broad-spectrum herbicide requires a boom sprayer. Plowing, disking, applying lime or fertilizer, and tilling require a tractor with a plow, disc, spreader, and tiller attachment. These are all commonly used for site preparation for most crops.

Some growers fumigate prior to planting blackberries. Fumigation suppresses levels of root lesion nematodes, certain pathogens, and weeds that negatively impact plant growth. The benefits of fumigation have been well documented for strawberries, less so for raspberries, and little for blackberries. In most situations, pathogens and nematodes can be reduced prior to planting through good crop rotations and planting specific nematode-suppressive cover crops. If blackberries will be replanted into a site used for berry crops within the last few years, then fumigation might be warranted. However, the benefits of fumigation dissipate after the first few years. Growers considering fumigation need to assess the costs and benefits; fumigation is very expensive, and legal options and regulations change each year.

Other strategies have been proposed to eliminate pathogens from sites prior to planting. These include solarization, anaerobic soil disinfection, and adding antagonistic or beneficial organisms (Blok et al., 2000; Butler et al., 2012; Pinkerton et al., 2009; Shennan et al., 2014). While these strategies often work for shallow-rooted annual crops, they have been less reliable for deeply rooted perennial crops such as blackberries.

By following the recommended steps, the site should be ready for planting the following spring. The soil pH and nutrient levels should be adequate for the life of the planting. Drainage, if necessary, has been installed. Perennial weeds should have been eliminated and nematodes suppressed. Organic matter levels should have increased, along with biological activity and water-holding capacity. Soil tilth and aggregation are preserved as much as possible. Irrigation water has been tested and is low in salts with a neutral or slightly acid pH.

FIELD LAYOUT

On flat ground, it is generally better to plant rows oriented north–south rather than east–west as light interception by plant rows will be greater. Since the sun rises in the east and sets in the west, the entire canopy receives exposure to direct sunlight at some point during the day (as long as rows are not too close together). The only exception is if autumn-bearing blackberries are being grown in a high tunnel for late-season production. In this case, east–west rows intercept more sunlight when the sun is low in the southern sky (in the northern hemisphere) after the autumn equinox.

The aspect of the slope is also an important consideration when the goal is to maximize light interception of the canopy. Steep, north-facing slopes at high latitudes may never receive direct sunlight, and hence, plants may not grow well under these conditions. In contrast, in the northern hemisphere, plants on a south-facing slope will receive more direct sunlight during summer than those facing other directions. Differences in light interception are small when the angle of the slope is small, but become significant when the angle of the slope increases, especially at higher latitudes.

Blackberries are particularly sensitive to wind damage and low pockets where cold air settles. Even if direct damage and visible symptoms are not apparent, blackberries grown under these conditions generally have shorter canes and lower yields. Two strategies are used to mitigate wind damage. The first is to use natural or artificial shelter belts to break the wind (Appendix 1). Typically, if wind comes from the west, a north–south windbreak on the west end of the field will help reduce wind damage. Windbreaks need to be spaced more closely on windier sites. Blackberry rows themselves may be less suscep-tible to wind damage if planted in the direction of the prevailing wind, but light interception must be considered as well. Windbreaks can provide protection for a distance equal to about ten times their height.

A second strategy to mitigate wind damage is to use a trellis. All brambles benefit from a trellis, even those grown for a fall-crop only. A trellis not only helps prevent canes from moving in wind and damaging their vascular con-nections with the root, but also holds canes erect, helps improve exposure of fruit, and improves efficiency of harvest and fruit quality. Trellises for autumn-fruiting types need not be elaborate, but simply need to hold canes erect. Trellises for floricane-fruiting plants need to be more rugged to hold primocanes and fruit-bearing canes erect throughout the year.

Materials for constructing trellises have traditionally consisted of 5 by 10 cm (2 by 4 in), or 10 cm^2 (4 by 4 in) treated posts, with metal wires strung between posts. Posts are usually 8–10 m (25–30 ft) apart down the row, some-times with metal stakes in-between. Erect floricane-fruiting types benefit from having the canes spread into a V-shape so that light interception and penetration, especially into the lower canopy, is improved. Floricane-fruiting, trailing blackberries are most commonly grown on two-wire trellises with the top wire at about 2 m (6 ft) and the lower wire at about 1.5 m (4.5) ft from the ground.

Because blackberries are vigorous, tall trellises may be required – perhaps 2 m (6 ft) above the ground. Newer, lighter-weight materials have been devel-oped for trellises. Further, a special trellis and cane training system using metal arms, called the Rotating Cross-Arm (RCA), has long cross-arms which rotate towards the ground so that the entire plant is covered in winter to protect the plants from low winter temperatures and high winds. Monofilament plastic wire is as strong as steel wire, and is easier to work with as it does not conduct electricity in the event of a lightning strike. Fiberglass posts are easy to adjust and just as strong as steel.

If blackberries are to be planted in a high tunnel, additional steps are recommended prior to planting. Supplemental compost (a 5-cm (2-in) layer) should be added to the soil and incorporated. If blackberry tunnels are covered all year, rainfall will not leach out salts. Therefore, using compost to supply the long-term nutrient and organic matter needs of blackberry plants is preferable to salty mineral fertilizers. Overuse of mineral fertilizers can compromise plant growth. A water extraction of soil that exhibits an electrical conductivity read-ing of greater than 2 dS/m is potentially damaging. Some composts, particu-larly those derived from animal manure, can also exceed this salt limit, and should be applied at low rates if at all.

It is also very important to have a level site if installing a high tunnel for blackberries. Sites that are not level will be difficult to cover since the compo-nents may not fit together on uneven ground. There are many sources of infor-mation on how to construct high tunnels (Heidenreich *et al.*, 2012).

PLANTING

Several choices of planting stock are available, depending on the nursery. The traditional type of blackberry is the bare-root sucker taken from mother plants in a nursery. These have a short piece of cane attached to a rather large root and are dug dormant in autumn. Bare-rooted plants are kept in a cooler until shipping in spring. They can be planted in spring as soon as the soil can be worked. Some blackberries can be propagated by root tips. They, too, are dug in the autumn with roots attached to a shoot, and are best transplanted in early spring.

Tissue-cultured green plants are much smaller than bare-rooted plants; similar in size to pepper or tomato transplants. The plugs are easily handled by most mechanical transplanters. They should be planted only after the danger of heavy frost is over because the leaves are green and sensitive to extreme cold. These are becoming the propagule type of choice among most growers.

Green tissue-cultured plugs can be chilled in the nursery or subjected to conditions that encourage leaf drop and dormancy. These propagules are very small – just a root ball with a leafless shoot, a few centimeters tall. They can be planted while there is still a risk of frost. These dormant tissue-cultured plugs ship well because they have no leaves.

Studies have compared the performance of these three types of propagules. By the time these plants reach maturity, the performance is about equal among the propagule types. Therefore, the choice of which to purchase is based primarily on characteristics other than yield. Green plugs might be the best choice if the planting date is after frost and irrigation water is readily available. Dormant, bare-root plants might be the best choice if no irrigation is available.

Typically, blackberries are set in rows that vary in spacing with the distance between rows depending on the plant type and their tendency to produce laterals. Between row spacing is 3.0–4.0 m (9–12 ft), while the intra-row spacing is from 1 to 2 m (3 to 6 ft), depending on the type of blackberry grown and the training system. Primocane-fruiting blackberries are planted at the closest spacing (3 m or 10 ft) between rows and 0.5 m (1.5 ft) between plants within rows. In high tunnels and greenhouses where tractors are not used, between row spacing can be as close as 1.8–2 m (5.5–6 ft).

Planting through plastic mulch or weed mat (porous polyethylene 'landscape fabric') or mulching with straw or newspaper after planting helps suppress weeds while retaining soil moisture. Plants mulched their first year tend to outperform unmulched plants because herbicide use and cultivation are avoided and moisture levels remain consistently high. Plants can also be set into killed rye or other cover crop, mimicking a no-till or strip-till situation. This strategy helps suppress weeds while the plants become established without a lot of soil disturbance.

Once set, plants should be irrigated and mulched (Dixon *et al.*, 2015). The shallow root system of plug plants cannot access much soil moisture;

therefore, sufficient water should be applied to maintain some moisture in the root zone of the plants. Plantings of tissue-cultured plants have been success-fully established using only drip irrigation. Once plants are set, mulched, and irrigated, they are well on their way to becoming established. If recommended procedures are followed, non-nitrogenous fertilizer should not be required the first year, because nutrients should have been incorporated into the soil during the pre-plant step. Nitrogen may be required shortly after planting if organic matter reserves are not high. The most important function for the grower after planting is to keep weeds under control. If the planting establishes well and weeds are kept under control the first year, then it will be difficult for weeds to invade the planting later (Harkins et al., 2013).

In many cases, a perennial grass is planted between rows of berries to help suppress weeds and provide a suitable surface for equipment and foot traffic, especially after rains. Fescues and dwarf perennial ryes are good choices. Grass seed germination is best when soil is moist and temperatures are cool, so autumn planting is often the best time to seed row middles.

An initial investment in the pre-plant site preparation and modification can pay major dividends for growers in the ensuing years, but planning ahead is the key to success.

REFERENCES

Blok, W.J., Lamers, J.G., Termorshizen, A.J. and Bollen, G.J. (2000) Control of soilborne plant pathogens by incorporating fresh organic amendments followed by tarping. *Phytopathology* 90(3), 253–259.

Butler, D.M., Kokalis-Burelle, N., Muramoto, J., Shennan, C., McCollum, T.G. and Rosskopf, E.N. (2012) Impact of anaerobic soil disinfestation combined with soil solarization on plant-parasitic nematodes and introduced inoculum of soilborne plant pathogens in raised-bed vegetable production. *Crop Protection* 39, 33–40.

Dixon, E.K., Strik, B.C., Valenzuela-Estrada, L.R. and Bryla, D.R. (2015) Weed management, training, and irrigation practices for organic production of trailing blackberry. I. Mature plant growth and fruit production. *HortScience* 50(8), 1165–1177.

Hargreaves, J., Adl, M.S., Warman, P.R. and Vasantha Rupasinghe, H.P. (2008) The effects of organic amendments on mineral element uptake and fruit quality of blackberries. *Plant and Soil* 308(1/2), 213–226.

Harkins, R.H., Strik, B.C. and Bryla, D.R. (2013) Weed management practices for organic production of trailing blackberry. I. Plant growth and early fruit produc-tion. *HortScience* 48(9), 1139–1144.

Heidenreich, C., Pritts, M., Demchak, K., Hanson, E., Weber, C. and Kelly, M.J. (2012) High tunnel raspberries and blackberries. Available at: http://blogs.cornell.edu/newfruit/files/2016/12/hightunnelsrasp2012-vegdsq.pdf (accessed May 4, 2017).

Maloney, K., Pritts, M., Wilcox, W. and Kelly, M. (2005) Suppression of Phytophthora root rot in red raspberries with cultural practices and soil amendments. *HortScience* 40(6), 1790–1795.

Pinkerton, J.N., Bristow, P.R., Windom, G.E. and Walters, T.W. (2009) Soil solarization as a component of an integrated program for control of raspberry root rot. *Plant Disease* 93(5), 452–458.

Pritts, M., Heidenreich, C., McDermott, L. and Miller, J. (2015) *Berry Soil and Nutrient Management: A Guide for Educators and Growers.* Available at: http://www.google.co.uk/url?sa=t&rct=j&q=&esrc=s&source=web&cd=1&ved=0ahUKEwi3ooOK6avUAhVMLVAKHQbyDbcQFggkMAA&url=http%3A%2F%2Fwww.sare.org%2F-content%2Fdownload%2F74320%2F1253195%2Ffile%2FBerrySoilandNutrientManagementGuide.pdf&usg=AFQjCNE7cSEopup-XFlRrmIwtb8WZIrnIQ&sig2=4rq03k2kXpcFlZR2nrI07g (accessed May 4, 2017).

Shennan, C., Muramoto, J., Lamers, J., Mazzola, M., Rosskopf, E.N., Kokalis-Burelle, N., Momma, N., Butler, D.M. and Kobara, Y. (2014) Anaerobic soil disinfestation for soil borne disease control in strawberry and vegetable systems: current knowledge and future directions. *Acta Horticulturae* 1044, 165–175.

Wehunt, E.J., Baker, E.C., Brown, M.A., Kirkpatrick, T.L., Golden, A.M. and Clark, J.R. (1991) Nematodes associated with blackberry in Arkansas. *Journal of Nematology* 23(4S), 620–623.

10

SOIL AND WATER MANAGEMENT

Richard C. Funt[1,]* and David S. Ross[2]

[1]The Ohio State University, Columbus, Ohio, USA; [2]University of Maryland, College Park, Maryland, USA

INTRODUCTION

Each grower must intelligently weigh certain fundamentals in site and soil selection that are important for the success of a blackberry production unit. A farm may grow different crops for different markets, and therefore requires a management plan for the location of the infrastructure for the water supply and how water is to be applied to different soil types and different crops. A site should be chosen with the soil characteristics needed for the best long-term blackberry production (Slate et al., 1949) (see Chapters 8 and 9).

An effective management plan should include the future expansion of the business in regards to water usage, movement of vehicles across main irrigation lines, size of irrigation equipment, and the rotation and/or expansion of crops. These need to be oriented to the type of production and marketing system that the grower selects. In general, the grower needs to have both long-term and short-term economic planning horizons to be effective and efficient with an operation. Soil and water management over the short and long term is very important.

The primary goal of soil management is to maintain fertility, minimize erosion of the topsoil, and maintain good soil structure with minimum tillage. The primary goal of water management is to maintain adequate soil moisture for producing an optimal canopy for maximum fruit production (Ross and Auchter, 1930). Water also needs to be supplied from rainfall and/or irrigation from bloom to harvest for maximum yield and fruit size, but without leaching nutrients into the water table. Also, the management of the soil fertility and soil moisture throughout the growing season will determine the effective production of large canes for the current and subsequent years' crops. In this chapter, 'blackberry plants' or 'blackberries,' including hybrid blackberries,

---

* Corresponding author: richardfunt@sbcglobal.net

Loganberry and Boysenberry types, refer to erect or trailing types; floricane or
primocane types; and/or summer or autumn types.

SOIL MANAGEMENT

Blackberries require open, porous, and slightly acidic soil having a pH range of
5.6–6.8 with a preferred range of 6.0–6.5 (see Chapter 11) for good growth,
movement of water, and good nutrition availability. Water from rainfall and
supplemental irrigation is necessary to allow the transfer of nutrients from the
soil to the plant roots. The ideal soil is well drained (good percolation below the
root zone), a sandy or silt loam soil with good moisture retention and an
organic matter content of 2–4%. Wet soil stimulates late-autumn growth and
the plant fails to harden off properly (see Chapters 2 and 11). While blackberry
root systems can penetrate through heavier soils and can tolerate soil pathogens more than raspberries, they will perform best under loamy well-drained
soils (see Chapter 9). Blackberry roots are generally in the top 0.6 m (2 ft) of the
soil but can be as low as 2 m (6 ft) beneath the soil surface (see Chapter 2). In
the Midwest and eastern parts of the USA, the water level may come within
1 m (3 ft) of the soil surface. Subsurface water levels in the soil should not reach
within 1 m (3 ft) of the soil water surface (see Chapter 9). Where soil water
percolation may be marginal, soils can be modified slightly to correct some
problems. Raised beds are beneficial in an area with poorly drained soils. Raised
beds that are 20–25 cm high (8–10 in) can reduce root rot problems (Funt and
Bierman, 2000). Raised beds are drier than flat surfaces and will require more
frequent irrigation (Fig. 10.1), particularly during harvest time. Blackberry
plants on raised beds suffer less injury than plants on flat areas because there is
less chance of roots being in saturated soil.

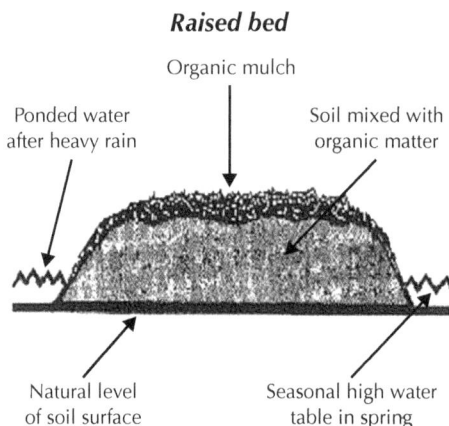

Raised bed

Organic mulch

Ponded water
after heavy rain

Soil mixed with
organic matter

Natural level
of soil surface

Seasonal high water
table in spring

Fig. 10.1.   Raised bed for water
management of blackberries.

Soils can be modified with organic matter (humus), using green manure, animal manure, and plant- or animal-based composts prior to planting (see Chapter 9). This will improve the moisture-holding capacity, the fertility, and the cation exchange capacity of the soil. A silty clay loam soil or a sandy clay soil, well supplied with organic matter, is desired and hardens less than soils with a lower level of organic matter. Adding organic matter, such as leaf compost, to a raised bed and incorporating it into the top 10 cm (4 in) creates an improved rooting environment for plant growth. In Ohio, where 60% of the soil requires improvements in internal drainage, raised beds have been recommended for blackberries. Generally, the top layer of soil (10–15 cm (4–6 in)) is a silt loam, but going deeper it can be a clay loam, which is slow in water percolation. By placing an additional layer (10–15 cm (4–6 in)) of top soil in the planting row, with a one- or two-bottom plow, the grower can create an environment of 12–25 cm (5–10 in) of well-drained soil. Loosening the soil, with deep rototilling prior to making the beds in dry soil, can provide greater success in heavy soils. Generally, beds are made in autumn and prepared for planting in spring, although in milder climates autumn planting with dormant plants may be favored for facilitating herbicide control of weeds. A soil test should be completed and based on the soil test results; nutrients, such as calcium (Ca), potassium (K), zinc (Zn), and phosphorus (P), should then be incorporated into the top 10–13 cm (4–5 in) with a rototiller (see Chapter 11). However, deep rototilling is not advised because it will destroy or flatten the raised bed.

Soils should be tested prior to planting. Any major or minor nutrients/minerals that are deficient should be incorporated, down to the till layer, into the upper 30 cm (12 in) of soil before planting (see Chapter 11). Soil fertility, gas exchange (oxygen), pH, and cation exchange capacity (CEC) may be improved with organic matter. Organic matter can be incorporated into the soil prior to planting (see Chapter 9; Glossary 1).

The CEC is generally related to the soil texture; clay soils have a higher CEC than sandy soils because they have a higher total particle surface area than sandy soils. One to two years prior to planting, blackberry growers should take a soil sample for determining the amount of sand, silt and clay (soil texture), organic matter, and CEC. Materials should be applied to improve the soil if required (see Chapter 9).

Grass planted in the sod drive row can reduce soil compaction when farm equipment is used under wet soil conditions. In areas of slow soil water percolation with clay-type soils and during periods of heavy rainfall, grasses, such as tall fescue, can aid equipment travel between the rows for applying pesticides and/or the use of mechanical harvesters. Also, sod is mowed and creates a solid pathway for pickers. Overall, this reduces soil compaction, allows water infiltration under rapid rainfall, reduces erosion on hillsides, and reduces weeds (biological weed control) between rows.

Many blackberry plantings are grown in soils where peach and cherry trees can be grown. These soils tend to be fertile, well drained, and have a porous structure, with ample oxygen for roots in the upper 20–31 cm (8–12 in). In Washington County, Maryland, black raspberries are grown on gravelly loams; the rocks in these soils include sand, slaty, and shaly types (Gourley and Howlett, 1941) (see Chapter 9). Further, there are soils as in Oregon receiving adequate rainfall for blackberries where irrigation may not be necessary to increase or enhance yields (Dixon *et al.*, 2015).

Root health is increased with improved soil structure from the application of organic matter; and therefore, providing increased pore space, which increases oxygen and gas exchange and internal water drainage. In Ohio, on Crosby silt loam, composted yard waste decreased soil bulk density; decreased water-filled pore space; and therefore, increased porosity of the soil by 10–40%. When compost was incorporated into the soil and applied to the surface of the soil, soil water-holding capacity and the rate of infiltration were increased by 36–40%, respectively and were seven to 21 times faster, in the first 2.5 cm (1 in) depth as compared to the non-treated soils (Funt and Bierman, 2000).

Raised beds allow better soil water drainage under heavy and prolonged rainfall conditions. Accordingly, the reduction in disease severity provided by the raised bed was due to the provision of a rooting zone in which the soil water tension generally exceeded that which supported zoospore activity. With all these factors, plants are healthier and can resist more insect and disease pressure than plants grown with lower amounts of nutrients, organic matter, or porosity soil. However, raised beds dry out much faster and to a greater depth than flat (non-raised) beds (Fig. 10.1). Thus, irrigation (Fig. 10.2) becomes a part of the raised bed culture system. Most blackberry plantings throughout other regions of the USA and world are not grown on raised beds.

Irrigation and water management

The quantity of water available for blackberry production is important throughout the life of the planting. Adequate soil moisture is the goal. Too much water over a long period of time on poorly drained soil will cause plant root disease and plant loss. Improving internal soil water drainage and porosity in the upper 30–50 cm (12–18 inches) will greatly enhance the performance of the life of the blackberry planting. Too little water causes plant water stress and poor plant growth regardless of the soil fertility and internal drainage (Ames and Byers, 2006). Reduced amounts of water (low soil-moisture content) over the short and long term may decrease shoot growth and berry size for the current season and decrease fruit bud development for the next season's crop. Blackberries are deeper rooting than raspberries so are more tolerant to drought conditions (Fernandez and Ballington, 1999). The need for irrigation depends upon the annual rainfall and its seasonal distribution, particularly during the growing season. Soil types and their water-holding capacities are major factors. If a growing season drought occurs, then moisture to promote

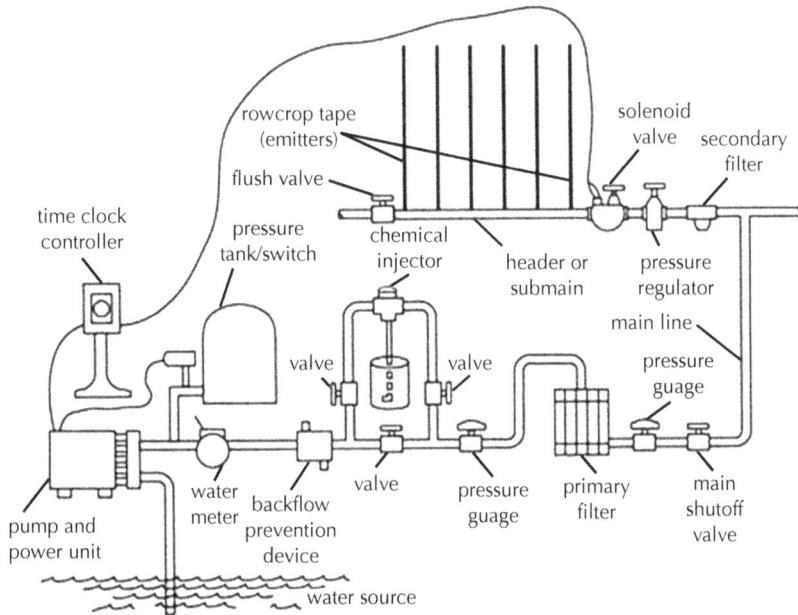

Fig. 10.2. Components of a typical drip irrigation system for blackberries. (Source: Ross, 2004.)

young plant establishment and later sufficient moisture for good berry production is affected. Some crops may need little irrigation water after they are well established and deep rooted in good soil, but the irrigation system is insurance against a drought.

Soil water-holding capacity

Irrigation requirements and timing depend a lot on the soil in the growing area. A sandy soil (larger coarse particles) holds little water and crops grown on sandy soil require frequent irrigation.

The amount of water held by the soil (water-holding capacity) is a factor of the amount of clay and organic matter present. An understanding of the hydrological cycle is necessary for a grower to make soil water management decisions (Fig. 10.3). When rainfall or irrigation exceeds the infiltration and percolation rates of a soil (i.e. 2.5 cm (1 in) per hour for well-drained soil), water runs off the surface and can cause erosion and loss of nutrients (Ross and Wolf, 2008). Grasses between rows can reduce erosion as explained above. Contour planting can help reduce runoff and erosion.

Twenty-eight liters or $0.03\,m^3$ (one cubic ft) of nearly ideal soil contains 50% soil particles, 25% moisture (about $7.5\,cm$ ($3\,in$) or about $7\,l$ ($2\,gal$) of water) and 25% air space, if one envisioned these as parts of a cube. Different

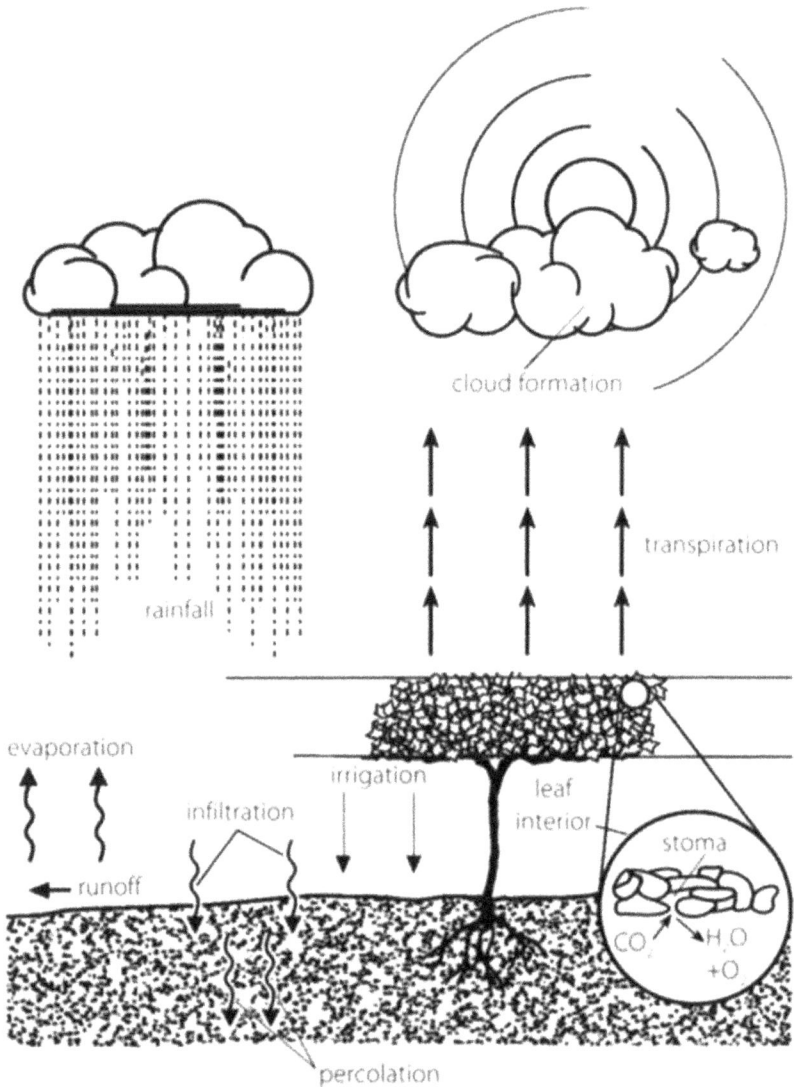

Fig. 10.3. The blackberry hydrologic cycle. Water enters the blackberry planting as rainfall or irrigation and is removed through gravity, runoff, evaporation, and transpiration through plant leaves. (Source: Ross and Wolf, 2008.)

soil types hold different amounts of water (Table 10.1). At 0.5 cm (0.2 in) of water loss by evapotranspiration per day, the plant could be near water stress in about 4–5 days, when more than 50% of the available soil water has been used by the plant or evaporated, depending on the soil type and blackberry grown. This amount is also referred to as the plant available water (PAW) and is the amount of water between field capacity and the permanent wilting point. Field capacity is the amount of water the soil can hold after it is saturated and allowed to drain away the water not held by soil particles. The permanent wilting point is the amount of water held so tightly by the soil that plants cannot extract it. The goal is to maintain the PAW at 50% or higher, so plants are not stressed for lack of water.

Further, the presence of a grass (sod) alley and/or raised bed will affect the amount of water needed to produce a crop of blackberries if overhead irrigation is used. In warm climates, the amount of water needed could range from 2.5–3 cm (1.0–1.3 in) per week for the crop and when an additional 30–40% air evaporative loss is calculated during midsummer, the total water loss could be 3–5 cm (1.3–2.0 in) of water per week. Drip irrigation, however, puts the water only in the root zone within the row of the blackberries, so there is much less influence by the grass alley or air evaporation and only the crop requirements must be met. The water placed by the drip irrigation system in the crop row will not be accessed very much by the grass or blackberry roots out in the row middle.

CROP WATER REQUIREMENTS

When calculating the amount of water needed for blackberries using drip irrigation, only the root zone area needs to be irrigated. For example, assuming a

Table 10.1. Water-holding capacity of different soil textures.

Soil texture	Available water-holding capacity (cm of water/cm of soil = inch of water/inch of soil)
Coarse sand	0.02–0.06
Fine sand	0.04–0.09
Loamy sand	0.06–0.12
Sandy loam	0.11–0.15
Fine sandy loam	0.14–0.18
Loam and silt loam	0.17–0.23
Clay loam and silty clay loam	0.14–0.21
Silty clay and clay	0.13–0.18

Source: Pritts and Handley (2008).

crop root zone width of 1 m (3.25 ft) times 100 m (325 ft) length of row, the equivalent area to wet is 0.01 ha (1076 ft^2 or 0.025 acres). To apply 2.5 cm (1 in) of water requires 2.5 m^3 of water (660 gal). One must estimate how many times in a season this amount of water will be applied in order to estimate the seasonal requirement for the crop. Typically, an evapotranspiration rate, ET, of 0.25–0.76 cm (0.10–0.3 in) per day of water in the root zone is required by the crop over the season as the plant grows (canopy development/berry production). Some allowance must be added to account for the efficiency of the application of irrigation water due to evaporation.

The water requirements can also be determined using a crop coefficient, Kc, which reflects the specific crop conditions over a growing season. The seasonal Kc values are available for specific crops and will vary with seasonal canopy development, type and amount of ground cover, and other factors. For sizing an irrigation system the maximum seasonal daily/weekly water requirement should be considered so that demand can be met. For sizing the required water supply capacity the seasonal Kc and the local evapotranspiration rate can be used. If the local potential evapotranspiration rate, ET_o, is 0.5 cm (0.20 in) per day and the Kc value is 0.60 at a time during the growing season, then one can expect the crop water consumption rate, ET_c, to be ET_c (or ET) = 0.60 × 0.5 cm (0.20 in) = 0.30 cm (0.12 in) per day at that time.

The total volume of required water might be estimated over a growing season based on an estimated water use curve (average daily water use in inches per day × number of days) and area to be irrigated, based on the Kc value. Generally, it is most critical to understand that the water supply demand will be greatest during the peak demand periods of the growing season (high Kc) and to understand the effects of a long drought on the water supply. Some streams and rivers can be very low during dry periods. Ground water table levels can drop during a drought. Learn about the potential water supplies of your growing area before making plans for crop production.

IRRIGATION SYSTEM CAPACITY

A general recommendation is to design an irrigation system to replace the water loss during the months of greatest water loss (July and August in the USA) plus 25% for extreme hot and windy conditions (Fig. 10.4). The average daily evapotranspiration, ET (plant use and evaporation loss) in the humid eastern USA ranges from 0.45 to 0.62 cm (0.2–0.25 in) per day. In California, Oregon, and Washington State, USA, that rate is as much as 0.8–0.95 cm (0.3–0.35 in) per day. In the Willamette Valley in Oregon, the average peak evapotranspiration rate is 0.62 cm (0.25 in) per day in July, while precipitation is less than 2.5 cm (1 in) per day. Other areas may vary and local data should be considered. A general recommendation for drip irrigation of blackberries is to apply water to the plant root zone according to the evaporative loss plus

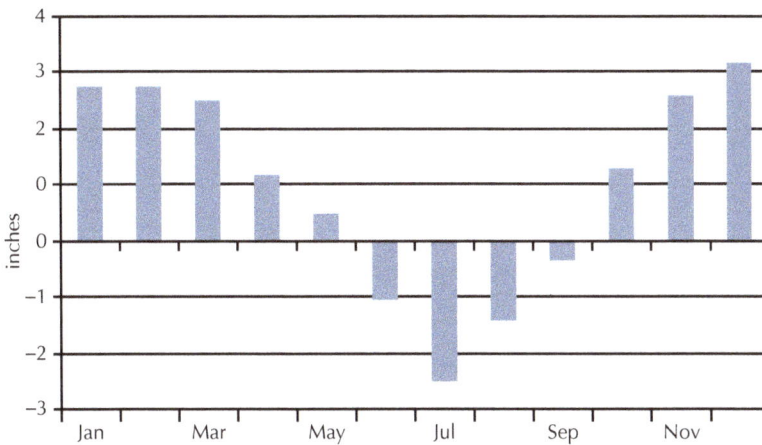

Fig. 10.4. The imbalance between precipitation and potential evapotranspiration (PET) is illustrated for Washington Dulles International Airport in northern Virginia. (Source: http://climate.virginia.edu.)

whatever weather conditions occurred over the past 24 hours. The local crop coefficient, Kc, should be considered to account for seasonal variation. In Oregon, the Kc is about 0.15 when the first leaf opens and 1.0 for the mature plant in full bloom (Hess *et al.*, 1997). Water should be applied to mature plants unless there has been 2.5 cm (1 in) of rainfall recently. Applying water in the morning can be effective in reducing water or leaf stress and help to cool the plant (Fig. 10.5). However, blackberry growers should invest in soil moisture monitoring equipment (see below) to understand the actual soil moisture conditions before judging the quantity of water needed to maintain peak plant performance.

WATER SUPPLY

The irrigation water supply must be large enough to meet the needs of the total enterprise. Large ponds, lakes, streams, springs, groundwater, municipal water, and waste water are all potential water sources. These may be supplemented by manmade reservoirs, which are built to collect runoff or to store water received steadily over time to allow an irrigation application (Ross and Wolf, 2008). In some areas, permits are required before surface water or high-capacity wells can be developed and/or used for irrigation. In general, when rainfall does not occur for several weeks the planting is under drought conditions. Water sources need to hold or be able to supply sufficient water for all crops and must be functional in droughty seasons. A storage reservoir can be charged (filled) from an alternate source, such as a well or stream.

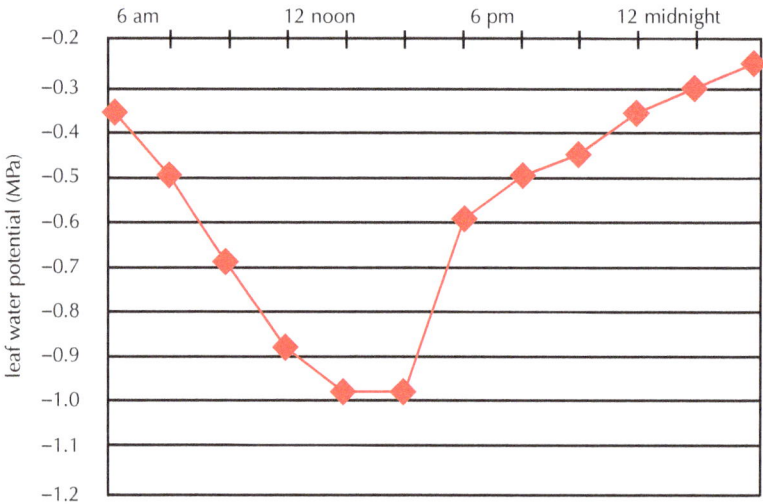

Fig. 10.5.   Changes in leaf water potential throughout the course of a day. Degree of leaf stress is indicated by increasingly negative values (Source: Ross and Wolf, 2008.)

A 2.5 cm (1 in) depth of water, over the surface of 1 ha (2.47 acres), contains 250 m^3 (66,043 gal) of water. This is equivalent to about 27,154 gal per acre for a 1-in depth. Under average conditions, this amount of water may be required by the crop each week. The water supply must be able to satisfy the most demanding criteria. Overhead, flood, or drip irrigation systems are widely used in various parts of the world. Further, climatic conditions (normal weather, heavy rains, or drought) determine water needs.

In many farming areas, the amount of water available for irrigation is limited to less than is required to operate an overhead or furrow system. Drip (trickle) irrigation becomes the only option. Drip irrigation can work with water supplies as low as 0.5 m^3 (500 l) per hour per hectare (53 gal per hour per acre) or less than 10 l per minute per hectare (1 gal per minute per acre), but that is not much water. In some cases water must be pumped or received 24 hours a day into storage tanks or holding ponds to irrigate a small area during the day. The water supply must be evaluated to determine whether it can meet irrigation requirements.

Water quality

Good-quality water is required for drip irrigation systems. Good water quality is defined as having a suitable pH, from 5.5 to 7.0 and being free from dissolved salts (low electrical conductivity (EC)); soil particles, as sand, silt, or clay; and biological materials, such as plants (algae); waste products containing *E. coli*; heavy metals (cadmium, lead); and pesticides (Ross, 2004). Drip irrigation emitters can be clogged by physical, biological, or chemical means. Physical clogging results from solid materials, such as sand, organic matter, or seeds. The control is filtration with sand filters, screens, or disc filters. Biological clogging is usually from surface water where algae, algal residues, or bacterial slimes may not pass through the screen filter. Control may require chemical treatment of the water; for example, copper sulfate to kill algae or chlorination to kill iron or sulfur bacteria followed by a sand filter to remove the materials. Chemical clogging results from precipitation of various chemical compounds in the water. Bicarbonates and fertilizer nutrients, such as iron (Fe), manganese (Mn) and P can form precipitates which can be reduced by acidification of the water.

A commercial water test should be obtained initially and then every year or two to establish the stability of the water quality. Testing avoids finding trouble later that might require costly treatment.

Generally, the highest water quality comes from wells and/or springs which are protected against contaminants. Well water is used for drip irrigation systems in fields, greenhouses, and high tunnels. Water may also be obtained from streams (when authorized), irrigation canals or ditches, or reservoirs, ponds, or lakes, and applied with overhead or furrow types of irrigation, including permanent pipes or center pivot or traveling guns. These sources offer larger amounts of water per minute but may contain soil, algae, and other materials. After application, monitoring of soil water quantity and quality is also beneficial in order to schedule irrigations.

Moisture monitoring

Irrigation management is best done by monitoring the soil moisture in the root zone under the irrigation system on a frequent basis and providing water to maintain a reasonable plant-available amount of water. Trying to estimate water usage by daily evaporation bookkeeping methods or manually reading tensiometers and other sensors is rapidly being replaced by newer technologies (Belayneh *et al.*, 2013). Newer capacitance sensors can be calibrated to specific soils or growing substrates to give accurate volumetric moisture content and relationship to moisture tension, the measure of water availability to plants (Lea-Cox *et al.*, 2008). Real-time data can now be obtained for scheduling irrigations to maintain uniform soil-moisture conditions for plant growth.

Since late 2010, moisture sensors attached to wireless radio nodes (battery powered) send the moisture data (volumetric moisture content) on a frequent basis to a wireless radio node attached to a computer in the office. Software in the computer stores the data and graphically displays it for the grower to make irrigation scheduling decisions in real time. The wireless radio nodes in the field

create a network to relay the data via line of sight, so node-to-node transmission pathways can change if something blocks the pathway. Battery life
of the radio nodes lasts for the full growing season (Lea-Cox et *al.*, 2013). Also,
some of the current moisture sensors can also measure soil/substrate temperature and electrical conductivity. Weather station sensors (i.e., rainfall, air temperature, photosynthetically active radiation (PAR) light, leaf moisture, and air
velocity and direction) are also connected to wireless radio nodes to send real-
time data to be graphically displayed (Lea-Cox et *al.*, 2015).

Wireless switching radio nodes with microprocessors collect and evaluate
local data. Based on set points, these radio nodes turn irrigation solenoid valves
on and off in an irrigation zone (Ross, personal communication, 2015). This
technology gives the grower real-time information about his growing conditions to allow for better management. It is true that one must learn where to
place the sensors to get the most meaningful data, but this can easily be learned
in a short time. The systems are portable for trial-and-error movement. Growers who have experience with the sensor networks consider them essential for
efficient moisture management, saving water and nutrients, and for understanding what is happening in the crop environment. Regardless of the method
used to monitor soil water content, it is always advised to use a soil core to
visually check the soil immediately after an irrigation and before the next
scheduled irrigation, to insure that the irrigation system is properly wetting the
root-zone area.

Irrigation systems must also be well maintained, checking for proper rotation of sprinkler heads, and emitter performance. Drip irrigation systems will
require proper maintenance including flushing and cleaning to insure optimal
performance. Irrigation systems should be properly designed and installed by
competent persons to deliver the water uniformly across a planting to insure
good soil-moisture conditions (Funt et *al.*, 1980). Monitoring the water pressure serves as a check on proper operation. Pressure is a good indicator of
proper operation and water delivery to sprinklers and drip emitters.

Make an irrigation business plan

Blackberry growers need to have a business/management plan for designing
(building) the irrigation system for present and future needs (Fig. 10.6). An
overall plan and design can reduce system and installation costs, energy costs,
and water usage over a long period of time (Funt et *al.*, 1980). Main supply
lines should be designed for future expanded water usage but need to be based
on the available water supply capacity. Also, growers may wish to install equipment to apply chemicals and fertilizer through the drip system. Fertilizer and
chemical injectors are available to add nutrients or other chemicals into the
irrigation water for soluble delivery. A dealer can suggest the appropriate
equipment to meet these needs. Fertigation (see Appendix 2) is an easy way to

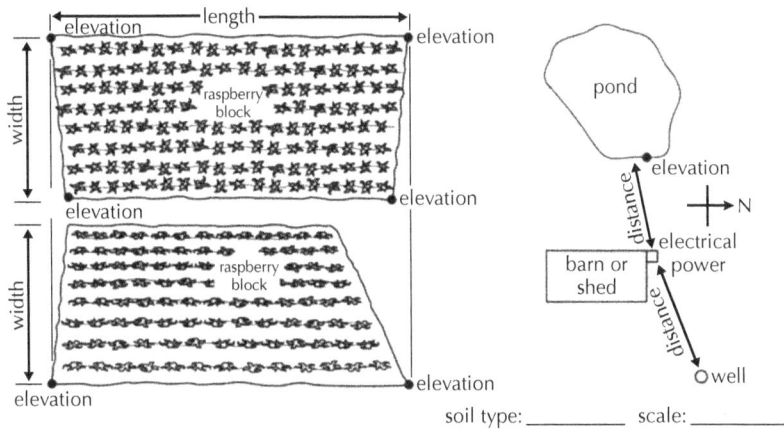

Fig. 10.6. Planning information for irrigation designer. (Adapted from Ross, 2004.)

apply a uniform application of nutrients to the crops in an irrigation zone as required and when needed. Chemicals can be applied per their label.

Water management and weeds

Blackberry growers need to eliminate weeds in order to obtain effective plant growth, to reduce moisture and fertilizer loss and to reduce shading and competition among blackberry plants (Dixon *et al.*, 2015). Weeds that grow tall, have thorns (thistle or horse nettle), or produce vines climbing over plants, impede hand harvesting because people do not want to be injured or they cannot find the fruit. Growers also do not want weeds to restrict mechanical harvest or have pieces of these plants mixed in with harvested fruit.

The use of herbicides (see Chapter 13) and cultivation by plowing or rototilling can greatly reduce the weed pressure before, during, and after planting. When planting into fields that have been in a monoculture (pasture, hay, or woody plants) for many years, it is wise to use a combination of herbicides and cultivation one to two years (years −1 to 0) prior to planting to reduce perennial weeds and their roots. Using a safe herbicide in the row just after setting the plants can allow much less pressure of weeds in the second or third year. A system of weed management benefiting many growers is the use of a herbicide in the row, and either a seasonal cover crops or permanent sod in the drive row (alley). As described above, using certain grasses to establish a permanent sod can reduce insects in harsh environments. Further, if a thick sod is immediately formed after planting, fewer weeds will emerge, and thus, sod becomes a biological weed control method. Research in Oregon showed that the presence of weeds in the row reduced blackberry yield by more than half compared to bare soil (hoed) or weed mat used in the row (Dixon et *al.*, 2015). Fertilizer and irrigation are more efficiently used when the root zone is maintained free of weeds. Growers may want to apply mulches, such as composted yard waste or sawdust, particularly during the first year. An application of pea straw can be very effective. These can improve water-holding capacity but may contain weed seeds. Root systems tend to be shallower under these systems, and therefore, additional applications in year 2 or 3 may not be advisable.

REFERENCES

Ames, G.K. and Byers, P.L. (2006) *Growing Blackberries in Missouri.* Bulletin 39. Missouri State University, Springfield, Missouri.

Belayneh, B.E., Lea-Cox, J.D. and Lichtenberg, E. (2013) Benefits and costs of implementing sensor-controlled irrigation in a commercial pot-in-pot container nursery. *HortTechnology* 23(6), 760–769.

Dixon, E.K., Strik, B.C., Valenzuela-Estrada, L.R. and Bryla, D.R. (2015) Weed management, training, and irrigation practices for organic production of trailing blackberry. I. Mature plant growth and fruit production. *HortScience* 50(8), 1165–1177.

Fernandez, G. and Ballington, J.R. (1999) *Growing Blackberries in North Carolina.* AG-401. North Carolina Cooperative Extension Service. North Carolina State University, Raleigh, North Carolina.

Funt, R.C. and Bierman, P. (2000) Composted yard waster improves strawberry soil quality and soil water relations. Proceedings of the XXV International Congress. *Acta Horticulturae* 517, 235–240.

Funt, R.C., Ross, D.S. and Brodie, H.L. (1980) Economic comparison of trickle and sprinkler irrigation of six fruit crops in Maryland, 1978. *Maryland Agricultural Experiment Station Bulletin M950*, University of Maryland, College Park, Maryland, pp. 1–16.

Gourley, J.H. (1946) *Modern Fruit Production.* Macmillan, New York.

Hess, M., Strik, B., Smesrud, J. and Selker, J. (1997) *Caneberry Irrigation Guide.* Oregon State Univeristy Extension Service EM8713.

Lea-Cox, J.D., Ristvey, A.G., Arguedas-Rodriguez, F., Ross, D.S., Anhalt, J. and Kantor, G. (2008) A low-cost multihop wireless sensor network enabling real-time management of environmental data for the greenhouse and nursery industry. *Acta Horticulturae* 801, 523–529.

Lea-Cox, J.D., Bauerle, W.L., van Iersel, M.W., Kantor, G.F., Bauerle, T.L., Lichtenberg, E., King, D.M. and Crawford, L. (2013) Advancing wireless sensor networks for irrigation management of ornamental crops: an overview. *HortTechnology* 23(6), 717–724.

Lea-Cox, J.D., van Iersel, M.W. and Burnett, S. (2015) Precision irrigation – how and why? *Greenhouse Grower* 33(1), 60, 62, 64, 66.

Pritts, M. and Handley, D. (eds.) (2008) Water management. In: Pritts and Handley (eds.) *Raspberry and Blackberry Production Guide.* Northeast Regional Agricultural Engineering Service-35, pp. 51–63.

Ross, D.S. (2004) Drip irrigation and water management. In: Lamont, W.J. (ed.) *Production of Vegetables, Strawberries, and Cut Flowers Using Plasticulture*, NRAES-133, Ch. 3. Natural Resource, Agriculture, and Engineering Services (NRAES), Ithaca, New York, pp. 15–35.

Ross, D.S. and Wolf, T.K. (2008) Grapevine water relations and irrigation. In: Wolf, T.K. (ed.) *Wine Grape Production Guide for Eastern North America*, NRAES-145. Natural Resource, Agriculture and Engineering Service (NRAES), Ithaca, New York, pp. 169–195.

Ross, H. and Auchter, E.C. (1930) *A Production and Economic Survey of the Blackberry Industry in Washington County, Maryland*. University of Maryland Agricultural Experiment Station, College Park, Maryland, pp. 207–245.

Slate, G.L., Braun, A.J. and Mundinger, F.G. (1949) *Raspberry Growing: Culture, Disease and Insects*. Cornell Extension Bulletin 719. Cornell University, Ithaca, New York, pp. 1–68.

PLANT NUTRIENT MANAGEMENT

Bernadine C. Strik*

Oregon State University, Corvallis, Oregon, USA

INTRODUCTION

The unique growth habit of blackberry plants, where nutrients are accumulated in the primocanes, crowns, and roots and are lost in the fruit, floricanes, and leaves (in autumn) (see Chapter 2), makes nutrient management somewhat complex. In the spring, growth of fruiting laterals and fruit is very dependent upon nutrient reserves stored in the floricanes, crowns, and roots and on additional nutrients available from soil or new fertilizer (Strik and Bryla, 2015b). However, new primocane growth in the spring is most dependent on nutrients available in the soil or from fertilization (Malik *et al.*, 1991; Mohadjer *et al.*, 2001; Naraguma *et al.*, 1999). Good nutrient-management programs are important for sustained growth and production. Strik and Bryla (2015b) further discuss the uptake and partitioning of nutrients in blackberry and raspberry.

In order to gain benefits from fertilization, crop management (from selecting certified plants to good irrigation and pest management) must be appropriate and timely. Proper fertilization or excess fertilizer will not compensate for poor growth that is caused by soil properties that are not ideal, or disease, weed, nematode, or insect problems.

The goal of fertilizing any high-value crop is to supply the plant with ample nutrition in advance of demand, thereby improving yield and quality (Hart *et al.*, 2006; Strik, 2015b; Strik and Bryla, 2015a). Important considerations include the economic return from the investment of fertilizer, while being a good environmental steward, and complying with government regulations. A fertilizer application should produce measurable changes in plant growth or nutrient status, or otherwise benefit the crop in a measurable way. The increased fruit yield and/or quality can aid in an increased return on the total investment.

---

* Corresponding author: Bernadine.Strik@oregonstate.edu

Growers, with the assistance of local extension agents and field represent-atives, should consider the nutrient needs of each field or type of blackberry. Key questions that need to be answered with regard to any nutrient manage-ment program are: How much nutrient should be applied? When is the best time to apply the nutrient? What is the best source of the nutrient for the plant? and What is the best method to apply the nutrient?

In this chapter, various components of successful nutrient management in commercial blackberry production will be discussed. Soil and tissue sampling, interpreting soil and tissue results, and making timely nutrient applications are all inclusive in producing a successful crop. Soil and tissue sample analyses help in determining appropriate fertilizer nutrient applications. Keeping records of weather, yield, disease and insect problems, and nutrient application rates and timing will help in interpreting soil and tissue analysis data over time. Observations of annual growth (visual assessments of cane number, diameter, and height, and fruiting lateral length), leaf color, and fruit quality (amount of rot, drupelet set, and firmness), in addition to yield, will also help in adjusting nutrient-management programs as needed (Hart *et al.*, 2006; Strik, 2015a). The terms 'blackberry plants' or 'blackberry' refer to trailing, erect, or semi-erect types, and either floricane- or primocane-fruiting cultivars, unless specifi-cally referred to as such.

SOIL SAMPLING

Soil testing and interpreting soil test results is important to adjust nutrients prior to planting, if needed (see Chapter 9). This not only gets the plants off to a good start, but incorporation of nutrients or amendments is very effective for those that are immobile or do not move readily into the rooting zone with a surface application (e.g. lime). Take soil samples well in advance of planting so that pH can be adjusted if needed (e.g. sample in early fall for spring planting) – it takes time for incorporated, pulverized, or ground agricultural lime (strictly calcium oxide CaO; commonly used in agriculture as calcium carbonate ($CaCO_3$) with or without magnesium) or sulfur (S) to react and change soil pH. Blackberry plantings have performed very well within a range in soil pH of 5.6–6.8. Soil that is outside this range should generally be modified prior to planting. As soil pH decreases, the solubility of iron (Fe), zinc (Zn), manganese (Mn), and aluminum (Al) increases. The concentration of Mn and Al can reach levels that are toxic or at least inhibit root growth. Crop sensitivity to Mn and Al varies, and blackberries are moderately sensitive. As soil pH increases, the solubility of Fe, Zn, and Mn decreases. The concentration of Mn and Fe can reach levels that are deficient, causing yellowing of leaves; in such situations the problem is best fixed by correcting soil pH. Primocane leaf tissue analysis (see 'Tissue testing') for Mn concentration is a good indicator of declining soil pH. As soil pH declines, Mn availability increases and leaf Mn concentrations rise. If primocane leaf Mn during late July or early August is above 300 ppm, or if levels are increasing over the years, check the soil pH.

If the soil analysis indicates pH is below 5.6, lime is required. However, you may wish to add lime (pulverized or ground agricultural limestone) before planting even if the pH is at 5.8, because:

1. 5.6 is at the lower end of the sufficiency range;
2. soil pH declines over time with fertilization (see 'Nitrogen'); and
3. blackberries are a long-lived crop and application of lime after planting is not very effective (not mobile in the soil).

A second soil test, the lime requirement or buffer test (sometimes called SMP), provides an estimate of the amount of lime needed to achieve a target soil pH. For detailed information on interpreting SMP values and adjusting pH with lime, see the References section for key publications (e.g. Anderson et *al.*, 2013).

Sandy soils to which fertilizers have not been recently applied sometimes record low pH and high SMP buffer values. In such cases, a light, broadcast application of 2–4 tonnes/ha of lime (1–2 tons/acre) should suffice to neutralize soil acidity. Agricultural lime is suggested when the soil pH is 5.6 or below, or when calcium (Ca) levels are below recommended levels (Table 11.1) or (1000 ppm = 5 meq Ca/100 g of soil). For acidic soils low in magnesium (Mg; Table 11.1) (60 ppm = 0.5 meq Mg/100 g of soil), 2 tonnes/ha of dolomitic lime ($CaMg(CO_3)_2$) can be used as a source of Mg. Dolomite and ground agricultural limestone have about the same ability to neutralize soil acidity. Lime applications are most effective when the lime is mixed with the soil and are effective over several years. Lime can be applied to the soil surface and using a mold board plow, plowing deep to a 20–23 cm (8–9 in) depth and then deep rototilling can move lime to lower soil depths.

If lime is needed to increase soil pH in established plantings, top dressing is the common method of application (Fig. 11.1). Lime can be top dressed in

Table 11.1.   Suggested nutrient levels for soil in blackberry plantings grown in western Oregon. (Adapted from Hart et *al.*, 2006.)

Nutrient	Deficient at less than (ppm)
Phosphorus (P; Bray)	20–40
Phosphorus (Olsen)	10–20
Potassium (K)	150–350
Calcium (Ca)	1000
Magnesium (Mg)	120
Manganese (Mn)	20–60
Boron (B)	0.5–1.0

Fig. 11.1. Broadcast lime application in a 'Silvan' trailing blackberry field in Oregon in spring (B. Strik).

plantings that have an in-row porous polyethylene mulch such as weed mat (Fig. 11.2). Soil amendments, such as lime, are typically applied as broadcast applications (to the whole field). A top-dress lime application should not exceed $4 \, t \cdot ha^{-1}$. Top-dressed lime moves downward 1–3 cm per year until reaching a depth of 5–8 cm. Low soil pH below 8 cm will not be corrected by top dressing lime. There is no evidence that other methods of lime application (e.g. liquid products) are more effective at this time.

The deficiency levels of nutrients in the soil presented in Table 11.1 are used as a guide for blackberry growers in western Oregon, USA. Similar levels may be used in other regions, but check with your local extension faculty or crop advisors. If a nutrient is not listed, no standards are available and plant tissue nutrient status should be used to assess nutrient needs after planting. A range in deficiency levels is provided, as there is no absolute value and ideal levels may depend on growing region or soil type. Soil-testing laboratories use different methods to extract plant-available nutrients, making it difficult to provide standards that apply to all soil-test results or production regions. Work with a reliable laboratory to interpret soil-test results as they apply to blackberry and ideally stay with the same testing lab to be able to compare results over time. Any needed nutrients should be applied as a broadcast application to the entire field and then incorporated into the top 10–20 cm (4–8 in) of soil.

Fig. 11.2. Broadcast lime application in a 'Black Diamond' trailing blackberry field with a weed mat mulch in the row (B. Strik).

After planting, periodic soil analyses can be helpful in diagnosing problems, such as low or high soil pH or the presence of excessive salts. Collect soil samples every 2–3 years to monitor changes in soil nutrient status and request a 'salts' test if it is otherwise not included. In established fields, sample soil at the same time of year, so that years can be more easily compared. Soil pH fluctuates over the season. Do not collect soil samples in spring, right after fertilization has occurred. The irrigation wetting front, fertigation, and band applications of fertilizer affect soil sample results. Collect soil samples in the plant row (where the fertilizer is applied) and, in drip irrigated fields, sample within a few inches of a drip emitter in all sub-sample locations. If mulch is present, remove the mulch layer before taking the soil sample.

Sample from the soil surface to about 50 cm (20 in) deep. The fact that nutrient concentrations vary horizontally between the aisles (alleys) and crop rows when fertilizers are banded is relatively easy to understand. Less obvious is vertical stratification of nutrient concentrations. Lime, P, and K are relatively immobile in the soil and tend to remain where they are placed or to migrate slowly downward. This immobility can result in soils with decreasing pH, and P and K concentrations as depth increases. This should be kept in mind when interpreting soil analysis results and planning nutrient management programs. You can take soil samples at two depths occasionally (e.g. 0–8 cm (0–3 inches) and 8–45 cm (3–18 in)) to better understand the level of stratification occurring in your farm. The soil-test report may have a measurement of Cation Exchange Capacity (CEC). A high reading will indicate a good level of soil fertility and nutrient uptake (see Chapter 9).

Testing soil to predict nitrogen (N) application rates is not advised. Testing the soil for high N concentration in late summer/fall can be used as an indicator of excessive N fertilization.

TISSUE TESTING

Leaf tissue analysis provides information on the nutrient content of the plant. The results of tissue analysis, when compared with published standards, indicate which elements the tissues contain in adequate, deficient, or excessive amounts. Routine tissue analysis can help in detecting low nutrient concentrations before visible symptoms or yield reduction occur (Hart *et al.*, 2006). Tissue analysis is a valuable tool to help diagnose visible plant problems and to evaluate fertilizer programs. Sometimes, even when the soil nutrient content is adequate, the plant is not able to take up the nutrients required (e.g. when soil pH is incorrect; in dry or waterlogged soils; during cool weather; and under certain cultural issues such as with too much or insufficient irrigation) (Strik and Bryla, 2015a). However, using tissue test results to anticipate current-season fertilizer needs does not work well for perennial crops such as blackberries. In part, this is due to the minimal short-term effects of fertilizer on yield. Changes in tissue nutrient concentrations may not be observed for 1–2 years after fertilization. In addition, primocanes, which respond to new fertilizer nutrients, do not fruit until the following year in floricane-fruiting cultivars. Delays in plant uptake are common, particularly when relatively immobile materials, such as P, K, and lime, are top dressed. The only exception is for correction of micronutrient deficiencies (e.g. boron) and N deficiency, where corrections can be made quite quickly. However, in general, leaf testing is more of a tool to assess how the nutrient management program may need to be changed for sustainable growth and production.

If problems such as poor cane growth or foliar discoloration appear during the growing season, a comparative tissue test can be used to check for possible nutrient deficiencies (see 'When to sample' below). Foliar symptoms of nutrient deficiency or toxicity may be on the older or the younger leaves on the canes, depending on the mobility of the nutrient in the plant. Nutrients in the plant move in either the xylem (with water – this tissue is dead) or the phloem (with 'food' – this tissue is alive). Nutrients that move in the xylem are not mobile within the plant, as they simply move with water to the leaves (not from leaves to the fruit or from old leaves to new leaves, for example). Nutrients that move in the phloem are mobile within the plant. The nutrients that are mobile in the plant are: N, P, K, Mg, and Cl. The ones that are immobile are: S, Fe, Mn, Cu, Zn, Ca, and B (Strik and Bryla, 2015a).

Before using tissue testing to predict or evaluate fertilizer needs, the following information is needed: when to sample; how to sample (plant part, number of leaves); and how to interpret the laboratory analysis results. These topics are discussed below.

When to sample

Primocane leaf tissue nutrient levels have been shown to vary over the growing season in floricane-fruiting blackberry (Clark et *al.*, 1988; Mohadjer et *al.*, 2001; Strik and Vance, 2017) and primocane-fruiting blackberry (Strik, 2015b). The recommended time of sampling leaves for tissue analysis is related to a period of time when the leaf nutrient concentration is most stable. In floricane-fruiting blackberry, collect leaves from the primocanes in mid- to late August in the western or northeastern USA and after fruit harvest in the southeastern USA. In primocane-fruiting blackberry, collect primocane leaf samples during the bloom to green fruit stage of development (for the primocane crop) regardless of the production region.

Tissue testing to diagnose possible nutrient deficiencies or toxicities (based on visual symptoms in the field) may be done at any time during the season. However, when outside the normal, recommended mid- to late August sampling period in the northern hemisphere, also collect a sample from an unaffected plant for comparison. If leaves are sampled outside of the recommended sample time, you will need the results from these 'comparator leaves' collected from normal-looking plants to try to deduce the cause of the symptoms. The recommended tissue standards are not applicable for leaves sampled at any other time than the recommended period (see subtopic 'How to interpret laboratory results').

Annual sampling of all fields is ideal for gathering information on plant nutrient status and to optimize nutrient management. Annual sampling, however, is not always necessary or financially feasible. Regardless of whether you collect samples every year, develop a plan for regular sampling.

How to sample

In addition to changing over the growing season, tissue nutrient levels will also change with location or age of the leaf and what type of leaf it is. For example, results from floricane leaves (Pereira et *al.*, 2015; Strik and Vance, 2017) will be different than primocane leaves and older primocane leaves will have different levels of many nutrients than younger leaves. Collect leaves that are free of disease or other damage if possible and a sample that represents the entire block/field.

Collect 50 leaves per block or field. Each sample should represent a single cultivar because blackberry cultivars have been shown to differ in leaf nutrient levels (Fernandez-Salvador et *al.*, 2015a,b,c; Harkins et *al.*, 2014; Strik, 2015b; Strik and Vance, 2017). A single sample should not represent more than 2 hectares (5 acres) nor contain leaves from more than 50 primocanes.

In floricane-fruiting cultivars, sample the most recent, fully expanded leaves from primocanes – about 30 cm from the tip of the cane (Fig. 11.3). In

Fig. 11.3. Tissue sampling in trailing 'Marion' blackberry. Selecting the most recent, fully expanded leaf on the primocane (B. Strik).

primocane-fruiting cultivars, sample fully expanded leaves from primocane branches.

Collect and ship samples in paper bags. Recommendations on whether to include the leaf petiole (stem) and whether to wash the leaves varies by production region (Table 11.2). If you do wash leaves, rinse them briefly to remove any dust and let them air dry before preparing for shipping; washing or soaking leaves may lead to leaching or leaking of certain nutrients leading to erroneous values. Ship fresh samples early in the week to insure delivery before the weekend and to reduce the likelihood of spoilage. Do not use plastic bags or containers, as the plant materials may mold or spoil. A list of analytical laboratories that perform tissue analyses is available from local university extension offices.

How to interpret laboratory results

Tissue standards have been developed using results from research experiments and estimated from large databases that relate tissue nutrient levels to good

Table 11.2. Recommended primocane leaf nutrient sufficiency levels for blackberry when sampled in late July to early August in Oregon, May to August in California, and the first week of August in the eastern, midwestern, and northeastern USA.

Nutrient	Oregon[a]	California[b]	Eastern, Midwestern, and Northeastern USA[c]
Nitrogen (%)	2.3–3.0	2.0–3.0	2.0–3.0
Phosphorus (%)	0.19–0.45	0.25–0.40	0.25–0.40
Potassium (%)	1.3–2.0	1.5–2.5	1.5–2.5
Calcium (%)	0.6–2.0	0.6–2.5	0.6–2.0
Magnesium (%)	0.3–0.6	0.3–0.9	0.6–0.9
Sulfur (%)	0.1–0.2	–	0.4–0.6
Manganese (ppm)	50–300	50–200	50–200
Boron (ppm)	30–70	30–50	30–70
Iron (ppm)	60–250	50–200	60–250
Zinc (ppm)	15–50	20–50	20–50
Copper (ppm)	6–20	7–50	6–20

[a]In Oregon, the recommendations are to use whole leaves – petioles included – and to leave them unwashed (Hart *et al.*, 2006).
[b]In California, there are no specifications for leaf petioles or washing (Bolda *et al.*, 2012).
[c]In the Northeast, recommendations include petiole removal and leaf washing (Bushway *et al.*, 2008).

yielding fields for each crop. Compare the results from a primocane leaf tissue analysis, when taken at the correct time of the year, to the sufficiency levels presented in Table 11.2. If a nutrient is deficient and observations of growth or plant performance indicate fertilizer is needed, apply the correct product (source of nutrient) at the right time to make the nutrient available for plant uptake. In perennial crops, such as blackberry, tissue analysis and observations of plant growth are best used to plan for and adjust nutrient management programs for the following year. See specific comments under each of the nutrients in Table 11.2.

Tissue analysis results outside the normal range cannot always be attributed to improper fertilization. Deficient mineral nutrient concentrations in tissues can be caused by saturated or dry soils; high temperatures; frost; shade; weed, insect, or disease pressure; or herbicide injury. In addition, several fungicides contain plant nutrients. Because tissue samples may not be washed before analysis, high copper (Cu), Mn, or Zn may be the result of fungicide residue. High B and Zn also may occur if liquid or foliar fertilizer was used.

MACRONUTRIENTS

Nitrogen (N)

Reviews of nitrogen (N) research in blackberry and raspberry were done by Dale (1989) and Strik (2008). Nitrogen requirements vary with cultivar, yield, cane growth, plant age, soil type, irrigation, and rainfall. The measured above-ground gain in N over a year in a mature planting has ranged from 37 to 62 kg/ ha of N (33–55 lb/acre), depending on the blackberry type grown and the research study (Dixon et al., 2016; Harkins et al., 2014; Mohadjer et al., 2001; Naraguma and Clark, 1998).

Cane growth is an initial indicator of N sufficiency. Some blackberry cultivars are more vigorous than others and may require less N to give the desired amount of cane growth. Less N is also required in the planting year than in subsequent years.

Excess N adversely affects yield and fruit quality, particularly fruit firmness (Nelson and Martin, 1986). It also can promote vigorous vegetative growth. Excessive vegetative growth leads to longer, thinner primocanes with longer-than-normal internodes (distance between buds), thereby reducing yield per cane. Excess N also can increase fruiting lateral length on floricanes. Long laterals increase the risk of breakage during machine harvest as well as the risk of disease, particularly botrytis.

Nitrogen fertilization should be based on tissue N concentration (Table 11.2), cane vigor, yield, and irrigation practices. Adjust fertilization programs based on the following (Hart et al., 2006).

Low tissue N concentration and abundant cane growth

Excessive cane growth usually is caused by an overabundance of N. Lower-than-normal tissue nutrient concentrations are common with excessive cane growth. In this situation, low tissue nutrient concentrations occur because the nutrient content of the tissue is diluted by the intensive growth. This condition should correct itself when growth returns to normal. Therefore, do not apply extra fertilizer N in this situation. Below-normal N and high vigor also can occur on canes with little or no crop.

Low tissue N concentration and weak cane growth

If canes are weak, discolored, or stunted, apply fertilizer N at the recommended rate (see below).

Normal tissue N concentration and cane growth

If tissue analyses and growth are within the normal range, continue with the current fertilizer program.

Above-normal tissue N concentration and weak cane growth
If the canes are weak, discolored, or stunted, and tissue analyses are above normal, look for stress from pests, poor drainage, drought, frost, or other factors.

Above-normal tissue N concentration and cane growth
If tissue N is above normal and cane growth is adequate or above normal, reduce the amount of fertilizer N applied.

Nitrogen source and rate

Blackberry plants have a low requirement for nitrogen compared to many other perennial crops. Annual above-ground plant gain in nitrogen has ranged from 37 to 58 kg/ha of N (33–52 lb/acre) depending on the blackberry type grown (Dixon et al., 2016; Harkins et al., 2014; Mohadjer et al., 2001). Research has shown that blackberry fruit contain about 0.9–1.3% N and 0.75–1% K (per unit of dry weight); these are by far the main nutrients in the fruit as the next-largest proportion is accounted for by P (0.17–0.2%) (Dixon et al., 2016; Strik, unpublished). Blackberry fruit contain about 16% dry weight (84% water). Thus one metric ton of fresh harvested blackberry fruit contains 1.4 kg (3 lb) of nitrogen (1000 kg/ton × 0.16 (percent dry weight) × 0.009 (percent N in fruit)) to 2.1 kg (4.6 lb) of N, depending on fruit N concentration. A field that yields 10 tonnes/ha (4.5 tons/acre) would thus have from 14 to 21 kg/ha (12.5–18.7 lb/acre) of N removed in the harvested fruit, depending on the cultivar.

Floricane removal or pruning in a commercial blackberry field typically occurs any time after harvest throughout the fall (see Chapter 12). In trailing blackberry, we (Dixon et al., 2016; Mohadjer et al., 2001) found from 32 to 51 kg/ha (28.5–45.4 lb/acre) of N was contained in the prunings. Delaying pruning until the fall allowed for more recovery of N from the dying canes. However, if these prunings are placed in the row middles and are flail mowed or chopped (a common commercial practice in many regions), the nutrients in these prunings will be returned to the soil and be available for the plants (Strik et al., 2006). While the nutrients in the floricanes at pruning time may be considered a loss from the plant, only the fruit is a true loss from the field. Of course, in addition to replacing true losses of N from the field, N would also be needed for above- and below-ground plant growth.

Nitrogen fertilizer rate recommendations for blackberry depend on planting age and the type grown, with 28–56 kg/ha of N (25–50 lb/acre) in the establishment year and 56–90 kg/ha of N (50–80 lb/acre) being common ranges (Bushway et al., 2008; Hart et al., 2006; Fernandez and Ballington, 1999; Krewer et al., 1999; Kuepper et al., 2003) and good starting points for growers. In the US Midwest, recommendations of 28–45 kg/ha of N (25–40 lb/acre) per year on silt loam soils having 2–4% organic matter have been suggested. An additional 22–34 kg/ha of N (20–30 lb/acre) is recommended at bloom to pri-mocane-fruiting types. The impact of N fertilization rate on yield varies among blackberry types and cultivars, growing regions, and soil type (Archbold et *al.*, 1989; Mohadjer et *al.*, 2001; Naraguma and Clark, 1998; Nelson and Martin, 1986; Rincon and Salas, 1987). Fertilizer rate recommendations provide the rate per planted acre – the given rate should be applied to the in-row area (concentrated into the row area). Rates are thus the same whether a granular product is applied or the fertilizer is applied through the drip irrigation system.

Blackberry plants use the nitrate (NO_3^-) form of nitrogen more readily than the ammonium (NH_4^+) form. Nitrate-N is soluble in water and moves into the soil or plant rapidly, but it also leaches easily from soil. Ammonium-N is less easily leached because it binds to soil particles. Because nitrate N generally is more expensive than ammonium forms, many growers apply urea or other ammoniacal sources of N. These will convert to the nitrate form at the recom-mended soil pH for blackberry (see the section 'Nitrogen and soil pH').

The most common N fertilizers applied to blackberry fields are granular or liquid/soluble forms (when applying through the drip irrigation system – 'ferti-gation') of calcium nitrate, urea, and ammonium sulfate in conventional sys-tems and OMRI-listed (Organic Materials Review Institute) granular fish emulsion, pelletized chicken litter, soybean meal, or feather meal or liquid fish or plant-based products in organic systems. Various organic sources of N have been applied to fresh market and processed blackberry using granular or liquid, fertigated products with no impact of source of N on blackberry plant growth or yield (Fernandez-Salvador et *al.*, 2015a,b). Organic fertilizer sources contain predominantly ammonium N and often differ in N release rate (Gutser et *al.*, 2005), which is an important factor to consider when targeting the crop needs. In addition, the cost (per kg of N applied) of the fertilizer can vary considerably among products. The rate of N applied may need to be increased by as much as 33% if the organic fertilizer used does not have 100% of the total N in the prod-uct available in the current year. For example, approximately 75% of the N in feather meal is available to plants in the current season (Collins et *al.*, 2013).

Many nutrients other than N are present in organic fertilizers and are thus applied to the planting, whether required or not (Dixon et *al.*, 2016; Fernandez-Salvador et *al.*, 2015a,b; Harkins et *al.*, 2014). Fertilization with different sources of organic fertilizers, or various rates of N when using inorganic sources, may affect the concentration of other macro- and micronutrients in the plants. All of these things need to be considered when developing or adjust-ing nutrient programs.

Nitrogen and soil pH

Ammonium-N is converted into the nitrate form through a process called nitri-fication. Soil pH is one factor controlling nitrification. Ammoniacal-N is rapidly converted to nitrate-N in warm, moist soil with a pH above 6.0. All nitrogen sources nitrify faster at pH 6.0 than 5.5. Maintaining soil pH at the upper end of the recommended range (6.0–6.8) would allow for use of all sources of N, as nitrification would occur relatively rapidly.

Standard fertilization programs lead to acidification of the soil in the row – the pH will thus decrease over time in all blackberry fields over time. Experience has shown that for every 112 kg/ha of N applied (100 lb/acre) in a typical clay loam soil in the Willamette Valley, Oregon, soil pH will decrease about 0.1 unit per year with urea and 0.2 units per year with ammonium sulfate (Hart *et al.* 2006; Strik and Bryla, 2015a). For example, if urea is applied at the rate of 112 kg/ha of N, the soil pH will decrease by approximately 0.1 pH unit (e.g. from 5.9 to 5.8). If 112 kg/ha of N is used for 3 years, soil pH will decline approximately 0.3 pH unit (e.g. from 5.9 to 5.6). It is thus best to monitor soil pH every few years and apply lime regularly to mitigate this decline.

Applying excessive N fertilizers has several costs:

1. the cost of the fertilizer itself, when not offset by increased yields or economic returns;
2. a reduction in fruit quality or disease (decreasing returns) and a potential increase in risk of winter cold injury; and
3. greater acidification of the soil, which then requires additional lime to raise the soil pH.

Applying 55 kg/ha of N above a crop's need will require an additional 300–700 kg/ha (267–623 lb/acre) of lime in 3 years.

Timing and method of application

In floricane-fruiting cultivars, N application in one season may affect both the current yield of the floricanes, mainly through increased fruit size, and the next season's yield, through its impact on primocane growth and flower bud development. In primocane-fruiting cultivars, N application may increase primocane number and advance growth and flowering, leading to increased yield. Fruit firmness may be reduced if excess N is applied from late winter through early spring, as a considerable portion of early-season N fertilizer goes to the fruit (Strik and Bryla, 2015b).

In blackberry, fertilizer N that is applied early (before new primocane emergence or when primocanes are less than 15 cm (6 in) tall) is taken up by the new primocanes and the fruiting laterals and fruit on the floricanes (Mohadjer *et al.*, 2001; Naraguma *et al.*, 1999). Nitrogen fertilizer that is taken up later, during or after fruiting, goes to the primocanes. This N will be stored in primocanes, crowns, and roots, and is important for sustaining yields from year to year. Floricane-fruiting red raspberry and blackberry plants have been shown to use about 30–40% of their stored N each year (Mohadjer *et al.*, 2001;

Rempel *et al.*, 2004). Excess fertilizer N that is applied early in the growing season or late fertilization with N may increase late growth of primocanes and increase the risk of cold injury.

Granular forms of N are commonly used in blackberry in late winter and spring, even in drip irrigated fields – fertigation is not commonly used when there is sufficient rainfall and fields do not need to be irrigated. Nitrogen can be efficiently applied with P and K, if these nutrients are also required. Apply granular fertilizer in bands about 60 cm (24 inches) wide and centered on the rows; for example, if the recommendation is to apply 34 kg/ha of N (30 lb/ acre) then apply this amount of actual N per planted acre, but concentrate the fertilizer into the row area. Nitrogen can be lost from surface-banded applications if the fertilizer is not washed into the soil by rain or irrigation within a few days following application.

Research has suggested that a split application of N fertilizer is best for maintaining current-season yields and acceptable primocane growth for next season's crop (Strik and Bryla, 2015b). If conventional granular fertilizers (e.g. urea) are going to be used for all of the needed N, apply one-half of the N fertilizer about 1–2 weeks before primocane emergence and one-half about 1 month (at first bloom) before the first harvest. As mentioned above, an additional application may be needed in primocane-fruiting cultivars. If organic sources of N are going to be used (e.g. feather meal), apply the first half in early March in the western USA or as soon as possible thereafter in colder regions and the second half about 1.5 months later (in late April in the western USA) to insure that the N is available when needed (Strik and Bryla, 2015a).

In drip irrigated plantings (Fig. 11.4), the first half of the required N is often applied using granular sources if irrigation is not yet required and there is an expectation of adequate rainfall after application. Once irrigation starts, soluble forms of fertilizer may be applied through the irrigation system (see Chapter 10) with an appropriate inorganic or organic source. Apply N from early spring through bloom or early fruit development, depending on the blackberry type grown.

While foliar sprays have proven effective for applying Zn, B, and other micronutrients (see below), they are not very effective for meeting blackberry plant N needs. Foliar nutrient sprays at bud break and/or before bud formation (August, USA) which include Zn and B, or other micronutrients appear to be beneficial in Midwest USA under cool, wet growing conditions.

Phosphorus (P)

Most soils in the USA contain adequate available P for blackberry production. However, on acidic, volcanic ash-influenced soils (e.g. near the Cascade Mountains and in northern Idaho and New Zealand) available P can be limiting for some crops. Still, there is no definitive research showing yield or growth

Fig. 11.4. A drip irrigated and fertigated 'Black Diamond' trailing blackberry field in Oregon (B. Strik).

responses from P applications to blackberry. There is often no relationship between soil P and tissue P.

Blackberry plants have relatively low needs for P with only 0.3 kg of P per metric ton (0.6 lb/ton) in the harvested fruit and 5 kg/ha (5.6 lb/acre) of P in the prunings (returned to soil if left in field).

When tissue P concentrations are below normal (Table 11.2), trial applications can help determine P needs for individual fields. Apply from 67 to 90 kg/ha (60–80 lb/acre) of phosphate (P_2O_5) under these circumstances. If P fertilization is used to correct a deficiency, tissue testing for 3–5 years may be necessary before differences are observed.

Surface applications of P are less effective than subsurface banding due to lack of mobility in soil (Hart *et al.*, 2006). For the fastest and most efficient movement of P to blackberry roots, place bands on each side of the row and 5–15 cm (2–6 in) deep. Double or triple the rate of phosphate given above for a pre-plant, broadcast, incorporated application. Lower rates of P can be applied through fertigation in deficient soils or plants.

Rock phosphate can be used by growers practicing organic production. Not all rock phosphates react or release P at the same rate. Finely ground rock phosphate mined in North Carolina, when applied to a P-deficient soil at double the rate of super-phosphates, produced wheat yields comparable to those produced by super-phosphate. Rock phosphate material has approximately 30% P_2O_5.

Potassium (K)

Potassium is essential for blackberry production. There is considerable K removed in harvested fruit (1.6 kg K/metric tonne (3.2 lb/ton)) and prunings (40 kg/ha of K (35 lb/acre)). While the importance of K nutrition in plant cold hardiness often is cited, no documentation exists to support the idea that higher than adequate tissue K concentrations increase cold hardiness.

Tissue analysis is the best indicator of K needs after crop establishment. Generally, little or no correlation exists between soil and tissue K levels. High surface soil K and low tissue K may indicate a gravelly subsoil low in K, inadequate irrigation, diseases, or other production problems. Generally low K can be seen under dry soil conditions particularly in July and August in the eastern and midwestern regions of the USA (Hart *et al.*, 2006).

In new plantings, one-half to two-thirds of the K requirement (Table 11.1) can be broadcast and incorporated before planting. The remaining one-half to one-third can be banded with N and P (if needed) after planting. No more than 45–67 kg/ha (40–60 lb/acre) of potash, K_2O, should be included in N–P–K fertilizers banded after planting. Excessive amounts of banded K can cause burning of new roots, particularly in sandy soils.

In fields two years old or older, K can be banded or broadcast, alone or in combination with N, P, and possibly other fertilizers. If tissue K is deficient (Table 11.2) apply 45–67 kg/ha (40–60 lb/acre) of K_2O if soil K is from 150–350 ppm and 67–112 kg/ha (60–100 lb/acre) if soil K is below 150 ppm. Avoid excessive applications of muriate of potash (KCl) as this may lead to excessive salt and injury to blackberry (Fig. 11.5).

Sulfur (S)

Sulfur deficiencies in blackberry crops are not common, but have been reported in some areas with very sandy soils. Soil S concentrations usually are adequate in blackberry fields because S often is added with other nutrients in fertilizers. Fertilizer materials such as ammonium sulfate (21–0–0), potassium sulfate (0–0–50), and gypsum contain sulfur.

For plantation establishment, apply 34 kg/ha of S (30 lb/acre) if the pre-plant soil test is below 10 ppm. No S should be needed if the soil test is above 10 ppm, assuming that pH is between about 5.6 and 7.0.

Like nitrogen, S is a key component of proteins, but tissue concentrations are rarely below sufficiency levels in most production regions (Table 11.2). When S applications are needed in established plantings, 34–45 kg/ha of S

(30–40 lb/acre) is adequate. Gypsum is a common source of S and has little impact on soil pH. When soil pH values are higher than desired and S is needed, elemental S can be used to lower soil pH while providing the needed sulfur.

Calcium (Ca)

Calcium deficiency may occur in blackberry primocanes that are rapidly growing (Fig. 11.6). If calcium is needed based on soil (Table 11.1) and tissue levels (Table 11.2), but the soil pH is in the desired range, then apply gypsum in the fall or late winter to promote availability (rainfall). Refer to the sections on soil pH above.

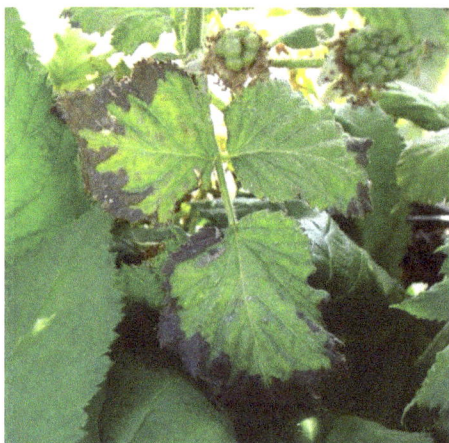

Fig. 11.5. Salt damage (from excessive application of muriate of potash, KCl) in 'Metolius' trailing blackberry (B. Strik).

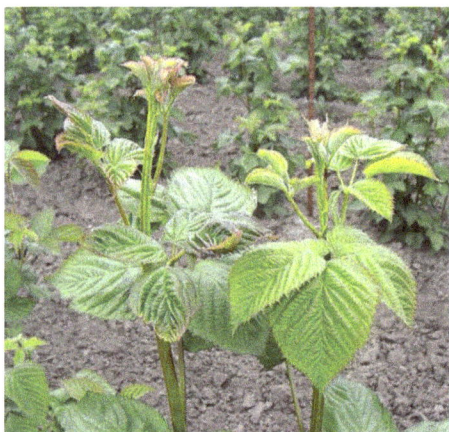

Fig. 11.6. Symptoms of calcium deficiency in leaves at the tips of primocanes in erect blackberry 'Apache' (B. Strik).

Less than 0.5 kg Ca/tonne (1 lb/ton) is removed in harvested fruit. While 35 kg/ha (30 lb/acre) of Ca is in the prunings at caning out; this is returned to the soil when the prunings are flailed between rows. There is evidence in many fruit crops that fruit Ca concentration is related to firmness or storage life. It is difficult to get Ca into the fruit as this nutrient is mobile only in the xylem and does not move from the leaves to the fruit. Foliar applications to leaves are ineffective at increasing fruit Ca (Vance *et al.*, 2017). If Ca is needed, based on leaf tissue levels, fertilizer can be applied as a granular product (e.g. lime or pelletized lime) or through fertigation during the early growing season.

Magnesium (Mg)

If magnesium is needed based on soil (Table 11.1) and tissue levels (Table 11.2), but the soil pH is in the desired range, then apply magnesium sulfate in the late winter to promote availability through rainfall washing in the fertilizer. If soil pH is at the lower range for blackberry, apply 2.2 tonnes/ha (1 ton/acre) of dolomite lime in the fall. Refer to the sections on soil pH above. Less than 0.5 kg/tonne (1 lb/ton) of Mg is removed in harvested fruit and 6 kg/ha (5 lb/acre) of Mg is in the prunings.

MICRONUTRIENTS

Micronutrients are required by plants in very small quantities. For example, less than 8 g/tonne (0.25 oz/ton) of each micronutrient is removed in harvested fruit and in the prunings in addition to any required for growth. However, micronutrients are often key catalysts in enzymatic reactions and are important to the normal functioning of plants. For this reason, it is important to make sure plants are not deficient in any of these nutrients. Note that many of these micronutrients can be added to pesticide/fungicide applications during the growing season.

Boron (B)

Small amounts of boron are critical for bud break and fruit set of blackberries. Boron deficiency results in small fruit, decreased yields, and, in severe situations, cane dieback. Tables 11.1 and 11.2 provide soil and tissue sufficiency levels for B. Note, however, that soil tests are less effective than tissue tests at predicting B needs for fruit crops.

Boron applications without soil or tissue tests are not recommended because B is toxic to plants at very low concentrations. In an Oregon trial, continued application of B reduced yields two years in five when tissue B was adequate.

If tissue B is low and soil B is less than 0.5 ppm, apply 2 kg/ha of B (1.8 lb/acre; Borax is a good source) in the late winter. If tissue B is low and soil B is 0.5–1.0 ppm, apply 1 kg/ha (0.9 lb/acre) of B. Do not apply B to the soil in bands along the berry rows. Either broadcast B or apply it as a foliar spray. It is best to broadcast solid B-containing fertilizers throughout the plantation to avoid high concentrations within the crop rows. Foliar B applications in fall, just prior to leaf fall of the primocanes or in spring, just prior to bloom, are effective. If tissue levels of B are deficient, spray sodium pentaborate (20% B) at a concentration of 1 kg of sodium pentaborate per 935 l of water per ha (2 lb/100 gallons/acre).

Other micronutrients

Other than B, micronutrient applications have not been shown to increase growth or yield of blackberries. However, soils that are suitable for blackberry production vary greatly, and supplemental micronutrient applications to soils or plants may be required in some regions. Base the need for foliar micronutrient application (suggested method of application) on leaf tissue levels. Be cautious with granular applications of micronutrients, as there is a fine line between sufficiency and toxicity.

Iron nutrition can be complicated, particularly on alkaline and/or poorly drained soils. Iron (Fe) concentrations in soil and tissue often correlate poorly with plant performance. Under certain conditions, both soil and tissue tests may show abundant iron concentrations while the plants show symptoms of iron chlorosis – yellow or dead tissues between dark green veins on the leaves.

Keeping soil pH values neutral to moderately acidic (5.6–7.0) and providing adequate water drainage usually prevents iron deficiency problems. On alkaline soils where iron and other micronutrient availability is limited, foliar fertilizer applications often are used for fruit production.

ORGANIC FERTILIZERS

Fish emulsion (applied as tank spray or through the drip system) or feather meal (granular) are good sources of N and other nutrients depending on the fertilizer (Table 11.3). Fish fertilizer should be diluted in water 1:10 (v/v fish:water) before applying to avoid burning the plants (straight fish has an EC of ~20 dS/m and blackberry plants are sensitive to an EC greater than 2 dS/m). We have found the N in these products to be available quickly. Apply fish at the same rate (of N) and timing as for the inorganic products. Apply feather meal (at same total rate of N) in a split application as described above; applying this product too late reduces N availability to the plants. Other organically approved products may be available. Products also differ in the proportion of N available during the season (Table 11.3). Organic sources of fertilizer can vary tremendously in the cost per kg of actual N applied (Strik and Bryla, 2015a).

Use of fresh manures is not recommended as a fertilizer source in berry crops. Using manure to supply plant nutrients requires handling more material compared to commercial fertilizers. For example, an application of 70 kg/ha (62 lb/acre) of N as urea (46% N) requires 152 kg/ha (135 lb/acre) of fertilizer material. The same amount of N from manure, assuming 1% N, requires 7000 kg/ha (6230 lb/acre). Additional manure (5000–7000 kg/ha (4450–6230 lb/acre)) is required the first year, as not all the N in manure is initially available (e.g. Sullivan, 2008), often less than 25% of the N applied is available the first year (Strik and Bryla, 2015a). The typical nutrient content of fresh manures is presented in Table 11.4.

Losses of N exceeding 50% can occur during manure storage or after application to the soil surface. Nitrogen loss is least when fresh manure is spread and worked into the soil immediately. Manure can serve as a source of weeds and pests. Unless livestock have been fed on weed-free feed, use aged manure that has been composted at temperatures high enough to kill seeds.

Table 11.3. Typical nutrient content of common organic fertilizers.

Organic fertilizer product[a]	N	P_2O_5	K_2O	Ca	Mg
			(%)		
Pelletized chicken litter	4	5	3	11	1
Soy bean meal	7	1	3	<0.5	<0.5
Feather meal	12	1	0.5	1	0
Fish emulsion hydroyzate (liquid)	4	1	5	0	0
Corn steep liquor blend (liquid)	25	3	2	<0.5	<0.5
Bat guano	10	3	1	na[b]	na
Blood meal	12–15	1	1	na	na
Bone meal	1–4	12–24	0	na	na
Cotton seed meal	6–7	2	1	na	na
Rock phosphate	0	25–30	0	0	0
Kelp meal	1	<0.5	2–5	na	na

Note that nutrient content can vary widely among products and even among batches and years. Adapted from Collins *et al.* (2013) and Fernandez-Salvador *et al.* (2015a,b).
[a]The following fertilizers were tested in a laboratory for nutrient content: Pelletized chicken litter = Nutri-Rich 4–3–3 (Stutzman Environmental Products, Canby, OR); Soy bean meal = Leafy Green 8–1–2 (California Organic Fertilizers, Hanford, CA); Feather meal = PROpellit Feathermeal 12–0–0 (Planet Natural, Bozeman, MT); Fish emulsion = TRUE 4–0–2 (True Organic Products, Spreckels, CA) a fish hydrolysate and fish emulsion with added molasses mixture; Corn steep liquor blend = Agrothrive LF 2.5N–1.1P–1.2K (AgroThrive, Inc., Morgan Hill, CA) a corn steep liquor and ground fish waste fertilizer blend.
[b]na = not available, as products not tested in a lab.

Table 11.4. Typical nutrient content of fresh animal manures. Note that nutrient and moisture content can vary widely, depending on handling, bedding, and age of the manure. (Adapted from Collins *et al.*, 2013.)

Type of manure	N	P_2O_5	K_2O	Moisture	Weight
	kg/m³[a]			(%)	kg/m³[a]
Chicken with litter	14	17	18	30	534
Dairy manure, separated solids	2	1	1	80	831
Horse	4	2	4	70	831
Sheep	8	4	15	75	831
Rabbit	7	4	6	75	831
Beef	5	4	7	75	831

[a]These values on an 'as-is' basis, meaning wet material with a moisture content typical for manure stored under cover. 1 kg/m³ = 1.69 lb/yd³.

Centipede-like organisms called symphylans have been reported to be a problem in some areas (e.g. western Oregon) when infested manures were applied to fields. Symphylans feed on germinating seeds and young roots. High-temperature composting should reduce problems associated with manure-introduced symphylans. However, aged and composted manure contains lower N concentrations than fresh manure (Collins *et al.*, 2013).

Organic growers may use compost (yard debris or chicken) as a surface-applied mulch or nutrient source. It is best to test any composted materials to insure that you know the pH and what nutrients are being applied to the field.

The conversion of unavailable N to the available form occurs throughout the growing season. If plant demand exceeds the rate of conversion, a deficiency occurs. Conversely, if conversion to available forms is high late in the season, unwanted late-season growth might be a result which can also increase risk of winter cold injury.

REFERENCES

Anderson, N., Hart, J., Sullivan, D., Christensen, N., Horneck, D. and Pirelli, G. (2013) *Applying Lime to Raise Soil pH for Crop Production (Western Oregon)*. Oregon State University Extension Publication EM 9057. Oregon State University, Corvallis, Oregon.

Archbold, D.D., Strang, J.G. and Hines, D.M. (1989) Yield component responses of 'Hull Thornless' blackberry to nitrogen and mulch. *HortScience* 24(4), 604–607.

Bolda, M., Gaskell, M., Mitcham, E. and Cahn, M. (2012) *Fresh Market Caneberry Production Manual*. Agriculture Natural Resources Publication 3525, University of California, Davis, California.

Bushway, L., Pritts, M. and Handley, D. (eds.) (2008) *Raspberry & Blackberry Production Guide for the Northeast, Midwest, and Eastern Canada.* Plant Life Science Publication, NRAES-35, Cooperation Extension, NRAES, Ithaca, New York.

Clark, J.R., Buckley III, J.B. and Hellman, E.W. (1988) Seasonal variation in elemental concentration of blackberry leaves. *HortScience* 23(6), 1080 (abstract).

Collins, D., Miles, C., Cogger, C. and Koenig, R. (2013) *Soil Fertility in Organic Systems.* Washington State University, Oregon State University and Idaho State University Publication PNW 646. Available at: https://catalog.extension.oregonstate.edu/pnw646 (accessed May 4, 2017).

Dale, A. (1989) Productivity in red raspberries. *Horticultural Reviews* 11, 185–228.

Dixon, E.K., Strik, B.C. and Bryla, D.R. (2016) Weed management, training, and irrigation practices for organic production of trailing blackberry. III. Accumulation and loss of aboveground biomass, carbon, and nutrients. *HortScience* 51(1), 51–66.

Fernandez, G. and Ballington, J. (1999) *Growing Blackberries in North Carolina.* North Carolina State University Extension Service Publication AG-401. North Carolina State University. Available at: https://content.ces.ncsu.edu/growing-blackberries-in-north-carolina (accessed September 20, 2017).

Fernandez-Salvador, J., Strik, B.C. and Bryla, D.R. (2015a) Liquid corn and fish fertilizers are good options for fertigation in blackberry cultivars grown in an organic production system. *HortScience* 50(2), 225–233.

Fernandez-Salvador, J., Strik, B.C. and Bryla, D.R. (2015b) Response of blackberry cultivars to fertilizer source during establishment in an organic fresh market production system. *HortTechnology* 25(3), 277–292.

Fernandez-Salvador, J., Strik, B.C., Zhao, Y. and Finn, C.E. (2015c) Trailing blackberry genotypes differ in yield and postharvest fruit quality during establishment in an organic production system. *HortScience* 50(2), 240–246.

Gutser, R., Ebertseder, T., Weber, A., Schraml, M. and Schmidhalter, U. (2005) Short-term and residual availability of nitrogen after long-term application of organic fertilizers on arable land. *Journal of Plant Nutrition and Soil Science* 168, 439–445.

Harkins, R.H., Strik, B.C. and Bryla, D.R. (2014) Weed management practices for organic production of trailing blackberry. II. Accumulation and loss of biomass and nutrients. *HortScience* 49(1), 35–43.

Hart, J., Strik, B.C. and Rempel, H. (2006) *Caneberries. Nutrient Management Guide.* Oregon State University Extension Service, EM8903-E. Oregon State University, Corvallis, Oregon.

Krewer, G., Smith, B. Brannen, P. and Horton, D. (1999) (2015 revised) *Commercial Bramble Culture.* Cooperative Extension Service University of Georgia Bulletin 964. University of Georgia, Athens, Georgia.

Kuepper, G.L., Born, H. and Bachmann, J. (2003) *Organic Culture of Bramble Fruits: Horticultural Production Guide.* Appropriate Technology Transfer for Rural Areas, IPO22. Available at: https://attra.ncat.org/attra-pub/download.php?id=15 (accessed May 4, 2017).

Malik, H., Archbold, D. and MacKown, C.T. (1991) Nitrogen partitioning by 'Chester Thornless' blackberry in pot culture. *HortScience* 26(12), 1492–1494.

Mohadjer, P., Strik, B.C., Zebarth, B.J. and Righetti, T.L. (2001) Nitrogen uptake, partitioning and remobilization in 'Kotata' blackberries in alternate year production. *The Journal of Horticultural Science and Biotechnology* 76(6), 700–708.

Naraguma, J. and Clark, J.R. (1998) Effect of nitrogen fertilization on 'Arapaho' thornless blackberry. *Communications in Soil Science and Plant Analysis* 29, 2775–2783.

Naraguma, J., Clark, J.R., Norman, R.J. and McNew, R.W. (1999) Nitrogen uptake and allocation by field-grown 'Arapaho' thornless blackberry. *Journal of Plant Nutrition* 22(4/5), 753–768.

Nelson, E. and Martin, L.W. (1986) The relationship of soil-applied N and K to yield and quality of 'Thornless Evergreen' blackberry. *HortScience* 21(5), 1153–1154.

Pereira, I., Picolotto, L., Gonçalves, M., Vignolo, G. and Antunes, L. (2015) Potassium fertilization affects floricane mineral nutrient content, growth, and yield of blackberry grown in Brazil. *HortScience* 50(8), 1234–1240.

Rempel, H., Strik, B.C. and Righetti, T. (2004) Uptake, partitioning and storage of fertilizer nitrogen in red raspberry as affected by rate and timing of application. *Journal of the American Society of Horticultural Science* 129(3), 439–448.

Rincon, A.R. and Salas, J.A.M. (1987) Influence of the levels of N, P, and K on the yield of blackberry. *Acta Horticulturae* 199, 183–185.

Strik, B.C. (2008) A review of nitrogen nutrition of *Rubus. Acta Horticulturae* 777, 403–410.

Strik, B.C. (2015a) Nutrient management of brambles. New England Small Fruits and Vegetable Conference, Savannah, Georgia. Available at: www.newenglandvfc.org/2015_conference/7_2_Strik.pdf (accessed May 11, 2017).

Strik, B.C. (2015b) Seasonal variation in mineral nutrient content of primocane-fruiting blackberry leaves. *HortScience* 50(4), 540–545.

Strik, B.C. and Bryla, D.R. (2015a) Nutrient management of blueberry – assessing plant nutrient needs and designing good fertilizer programs. In: *Proceedings of the Oregon State University Blueberry School*, March. Oregon State University, Corvallis, Oregon.

Strik, B.C. and Bryla, D.R. (2015b) Uptake and partitioning of nutrients in blackberry and raspberry and evaluating plant nutrient status for accurate assessment of fertilizer requirements. *HortTechnology* 25(4), 452–459.

Strik, B.C. and Vance, A.J. (2017) Seasonal variation in mineral nutrient concentration of primocane and floricane leaves in trailing, erect, and semi-erect blackberry cultivars. *HortScience* 52(6), 836–843.

Strik, B.C., Righetti, T. and Rempel, H. (2006) Black plastic mulch improved the uptake of ^{15}N from inorganic fertilizer and organic prunings in summer-bearing red raspberry. *HortScience* 41(1), 272–274.

Sullivan, D. (2008) *Estimating Plant-available Nitrogen from Manure.* Oregon State University Extension Publication, EM 8954. Oregon State University, Corvallis, Oregon. Available at: https://catalog.extension.oregonstate.edu/em8954 (accessed June 6, 2014).

Vance, A.J., Jones, P. and Strik, B.C. (2017) Foliar calcium applications do not improve quality or shelf-life of strawberry, raspberry, blackberry, or blueberry fruit. *HortScience* 52, 382–387.

12

Pruning and Training

Bernadine C. Strik,[1,]* Fumiomi Takeda,[2] and Gary Gao[3]

[1]*Oregon State University, Corvallis, Oregon, USA;* [2]*Appalachian Fruit Research Station, USDA-ARS, Kearneysville, West Virginia, USA;* [3]*South Centers and Department of Horticulture and Crop Science, Piketon, Ohio, USA*

INTRODUCTION

Pruning and training of blackberry varies considerably among the three types, trailing, erect, and semi-erect, because of differences in growth and development (Chapter 2). Trailing blackberry cultivars are predominantly grown in the western USA, parts of Europe, Australasia, and in Chile, and are grown for machine-harvested processed markets and for hand-picked fresh markets. Erect and semi-erect cultivars are grown in all blackberry production regions in the world for fresh and processed markets (Strik et al., 2007). In this chapter, we will address pruning and training practices for each type of blackberry, with important differences by production region noted.

TRAILING BLACKBERRIES

Training during establishment

Trailing blackberries are grown at an in-row spacing of 1–2 m (3–6 ft) with 3 m (10 ft) between rows. Plantings are typically established using tissue-cultured-propagated plugs (e.g. in 10-cm-square pots; 4 inch). Growers either train the new primocanes in the planting year, or re-cut them to crown height the following winter and train in the second year; the latter method is most common in Oregon, but the amount of primocane growth may vary in production regions. New primocanes emerge from the crown and these are trained to the trellis as they grow by bundling the canes using bailer's twine and tying the bundles to the trellis wires (see below). The first year of production is in year 2 or 3.

* Corresponding author: Bernadine.Strik@oregonstate.edu

Trellis

Trailing blackberry plantings require a sturdy, yet simple trellis, particularly considering many are harvested by machine. Treated, wooden or metal posts (organic fields) that are 10-cm (4 inch) or more in diameter are used at the end of rows. These should be 2.4 m (8 feet) long and be placed 0.6 m–0.9 m (2–3 feet) deep into the soil, and are stabilized by a screw anchor or other system, depending on row length (Fig. 12.1). Metal T-posts or smaller diameter wooden posts are used in the row at a spacing of about 6–10 m (20–30 feet). The number of wires on the trellis varies among growers. In all training systems, an upper 10- or 12-gauge high-tensile wire is attached to the posts at 1.7 m (5.5 feet) above ground. A small proportion of growers train all the primocanes to this one wire. However, most growers use a trellis that has two training wires, one at 1.2 m (4 feet) and the other at 1.7 m (5.5 feet) from the ground. Some growers may add another training wire below 1.2 m. Growers that irrigate using drip systems will usually install a lower wire at about 30 cm (1 foot) to support the drip line. Wire tighteners are used so that tension may be adjusted as needed.

Fig. 12.1. 'Karaka Black' trailing blackberry grown with multiple wire trellis in Australia (B. Strik).

Every-year and alternate-year production systems

Trailing blackberries can be grown in every-year (EY) or alternate-year (AY) production systems. In EY production, fruit is produced each year on a given area, so the primocanes, necessary for next year's crop must be carefully managed during the growing season. In the spring, while the floricanes (trained on the trellis) are flowering and fruiting, new primocanes that have emerged from the crown are trained under the canopy (Fig. 12.2). Primocanes may first be pruned through 'primocane suppression' (see below) to delay growth. In general, primocanes are first trained after they 'flop' or fall over and start trailing (Chapter 2). Different methods may be used to keep primocanes under the canopy including wire hoops and stakes. It's important the primocanes not be in the aisle where they can be damaged by the wheels of tractors or machine harvesters. After fruit harvest is finished (typically in early August in the western USA), the floricanes are allowed to fully senesce. The dead floricanes are then removed by pruning ('caning out'). Growers may use hand secateurs or a combination of hand pruning and a de-caner machine (chops up dead floricane tissue between wires such that it can easily be removed from the trellis). The

Fig. 12.2. 'Marion' trailing blackberry grown in an every-year (EY) production system with primocanes trained under the canopy of the floricanes (B. Strik).

floricane prunings are raked between the rows and are chopped with a flail mower such that the nutrients may be returned to the soil (Chapter 11). The primocanes may be trained onto the trellis wires in late summer (Fig. 12.3) or late winter. In Oregon, most growers train primocanes in February, leaving canes more protected from cold nearer the soil surface through most of the winter (Fig. 12.4). In some cultivars, August training (northern hemisphere) leads to increased yield (Bell et al., 1995a). However, training time may have little effect on yield in other cultivars or years, and August training can increase risk of cold injury in cold-sensitive cultivars (Dixon et al., 2015). When training, the primocanes produced by each plant are divided into two bundles. Half the canes are then looped in one direction from the upper to lower trellis wires, bringing them back towards the plant with one or two twists; the other half is then looped in the opposite direction. It is near impossible to train the long primocanes without damaging or kinking some canes during the process; however, taking care during training will improve yield. Canes that are too short or thin are removed during primocane training. In February, training must be done before the buds swell or too much damage occurs.

In Oregon, some growers are training primocanes up and over the old, dead fruiting canes (not pruned out). This system, 'new over old' has much lower training costs, less primocane breakage during training, and increased

Fig. 12.3. 'Marion' trailing blackberry grown in an every-year (EY) production system showing August-training of primocanes (B. Strik).

winter cold tolerance of primocanes (Strik, unpublished). The number of years this can be done without decreasing fruit quality (especially machine harvest contaminants) has not yet been studied.

In areas with a long growing season (e.g. some regions of New Zealand) it may be possible to remove primocanes (de-sucker) until the fruit harvest is almost underway and still get enough growth to train up a full canopy for the next year's harvest. This method keeps the primocanes out of the way for fruit harvest.

In AY production systems, plants fruit every other year. In the 'on-year,' floricanes produce a crop, but primocanes are not managed or trained. However, primocane pruning or primocane suppression (see below) may occur in machine-harvested systems. In autumn, once the plants are dormant, the dead floricanes and the primocanes are pruned off at the crown using a sickle bar or a cutting blade. Pruning before the plants are dormant will encourage re-growth from the crown, reducing the carbohydrate and nutrient reserves in the roots and crown, weakening the plants. The canes are chopped with a flail mower and are left between the rows in the field. The plants go through the winter with no canes present, ensuring no risk of winter cold damage in the western USA. In the following 'off-year,' primocanes are trained to the trellis as they grow. Growers will bundle the primocanes using bailer's

Fig. 12.4. Trailing blackberry grown in an every-year (EY) production system showing primocanes before training in February (B. Strik).

twine and will go through the field two or three times, depending on primocane vigor; the first pass is done just before the canes flop over (at
approximately 1 m tall) (Fig. 12.5). The bundled canes are tied to the trellis
wires. Once the bundle of primocanes is about 1 m (3 feet) taller than the top
trellis wire, the bundle is divided in half with each half trained in opposite
directions on the trellis in a similar pattern to that described for EY production.
Primocanes often develop short branches at this point and continued training
('re-tucking') will be needed into early autumn (Fig. 12.6). A grower would
have half of their planted area in the off-year and half in the on-year in a given
year.

The yield of an AY field is about 80% of an EY field over a two-year period,
depending on cultivar (Julian *et al.*, 2009; Strik and Finn, 2012). In an AY
system, primocanes that grow during the off-year, without the presence of
floricanes, are more cold-hardy than those that grow with floricanes, as occurs
in an EY system (Bell *et al.*, 1995b; Cortell and Strik, 1997). In addition, AY
systems are easier and less expensive to prune and train, and have less cane
disease. Some growers cut the first flush of primocanes back to the crown in
the off year when they are about 30 cm (~1 foot tall) as this increases yield and
cold hardiness in the subsequent flush of canes (Bell *et al.*, 1995a,b). In Oregon,

Fig. 12.5. 'Marion' trailing blackberry grown in an alternate-year (AY) production
system showing primocanes being trained as they grow in the 'off-year' (B. Strik).

25–40% of the trailing blackberry production is grown in AY systems. In most of the other production regions worldwide, trailing blackberry are grown in EY systems.

In the eastern United States, western trailing blackberries are not commercially grown. The canes of western trailing blackberries, in particular those trained vertically on T-posts or wooden posts with one to several wires, are damaged or even killed to the ground when the temperatures drop to −13.8°C (7°F). Also, the floricanes held in a vertical orientation in winter are vulnerable to wind desiccation and cane water content declines from 70% in late fall to less than 30% by late winter (Takeda *et al.*, 2008). The research by Takeda and Phillips (2011) showed that the lack of cold hardiness in western trailing blackberries can be mitigated by training the canes on the Rotating Cross-Arm (RCA) trellis system, lowering the canes close to the ground by rotating the cross-arms, and covering them with a floating row cover (see below). When 'Siskiyou' and 'Black Diamond' were grown in this production system, they survived winters in which the temperatures dropped to −20°C (−5°F) and produced 6 kg (13.2 lb) per plant the following summer whereas the plants that were left upright and uncovered during the winter produced less than 3 kg (6.6 lb).

Fig. 12.6. Trailing blackberry grown in an alternate-year (AY) production system showing primocanes trained to the trellis in winter (B. Strik).

Primocane suppression

This form of summer pruning is a common tool in EY systems and in the fruiting year (on-year) of AY systems to improve harvest efficiency (Stanley *et al.*, 1999; Strik and Finn, 2012). Primocane suppression involves removing the first flush of primocanes (Chapter 2) when they are about 15–30 cm tall (6–12 inches). In trailing blackberry, primocane suppression may be of benefit for next year's yield, but does not increase yield of the current-season floricane crop (Strik, unpublished). Most commonly, primocane suppression is done using contact herbicides labeled for this purpose in each region (Fig. 12.7). The contact herbicide also burns back the lower fruiting laterals on the floricanes and provides weed control. This lower fruit often would not be harvested (may touch soil and thus a food safety risk or would be below the catcher plates on the machine harvester – Chapter 14). Primocane suppression delays growth, making them shorter and less likely to interfere with the catcher plates on the machine harvester, thus increasing machine harvest efficiency. In high-value fresh-market production systems, the first flush of primocanes may be removed by hand (using secateurs) to improve hand harvest efficiency or next year's yield (Strik and Finn, 2012). Use of chemical suppressants for many years may

Fig. 12.7. 'Marion' trailing blackberry grown in alternate-year (AY) production. Primocane suppression has occurred with contact herbicide (B. Strik).

reduce planting vigor and is also not recommended in weaker or stressed plantings.

SEMI-ERECT BLACKBERRIES

Semi-erect blackberries (e.g. 'Chester Thornless.' 'Loch Ness') are commonly grown in the western USA (mainly Oregon and California) and in Europe. The production systems described here are commonly used in all of these production regions (Strik and Finn, 2012).

Semi-erect blackberries are typically grown at an in-row spacing of 1.2–1.8 m (4–6 ft) with 3–3.6 m (10–12 ft) between rows. Plantings are commonly established in the spring with full or mature production in year 3. Primocanes are usually tipped by hand during the growing season to encourage branching and increase yield the following year. Primocane tipping height varies from 0.45–1.8 m (1.5–5 ft), depending on grower or region. Soon after fruit harvest or in winter, the senesced floricanes are removed by pruning the dead canes just above the crown. In the winter, growers either shorten the primocane branches by hand using secateurs or they train all of the branch growth. Primocanes that are too short or thin are removed in winter when pruning. In regions with cold winters, growers should delay pruning until late winter when the extent of cane injury can be observed and removed.

Primocanes are either trained on a multiple wire trellis or to a divided canopy. Single-canopy, multiple-wire trellises are similar to what is described above for trailing blackberry, although growers will always use at least three training wires. Divided canopy trellises range from simple systems with wooden end posts and a wooden cross-arm at the top (at a height of about 1.7 m; 5.5 feet) with wires at each side of the 'T' (Figs. 12.8 and 12.9) to more complex 'V'-shaped systems with multiple wires on each side or the Rotating Cross-Arm (RCA) trellis (see below; Takeda et al., 2003a, b). The improved light exposure to primocanes and floricanes in a truly divided canopy system improves yield (Chapter 2). However, maintaining the divided canopies is labor intensive and is typically only done by growers with high-value fresh fruit markets (Chapter 14). In some regions, semi-erect cultivars are grown for processing or are 'cleaned-up' by machine harvest (Chapter 14); in this case growers will use a trellis that is suited for machine harvesting, typically a single-canopy system.

The fruiting season of semi-erect blackberries may be extended in tunnels and greenhouses using container production and pruning techniques (Bal and Meesters, 1995). For example, 'Loch Ness' can be pruned to produce two crops on the floricane by cutting back the fruiting laterals immediately after the last fruit is harvested; secondary laterals emerge from basal buds on branches or the main cane resulting in a second crop about two months later, although with a low yield (Oliveira et al., 2004; Pitsioudis et al., 2009).

Fig. 12.8.   'Čačanska Bestrna' semi-erect blackberry in Serbia (B. Strik).

Fig. 12.9.   'Chester Thornless' semi-erect blackberry in Oregon (B. Strik).

ERECT BLACKBERRIES

Erect blackberries (e.g. 'Navaho' and 'Ouachita') are commonly grown in the USA, Mexico, Australasia, and Europe. The production systems described here are commonly used in all of these production regions (Strik and Finn, 2012). Erect blackberries can be biennial or annual fruiting (Chapter 2). We have included 'Tupy,' the most common cultivar grown in Mexico, as an erect type, even though its parentage includes erect and trailing types (Strik and Finn, 2012). The specialized production system used for 'Tupy' in Mexico is described in Chapter 14.

Plants are established 0.8–1.2 m (2.5–4 feet) apart in rows 3 m apart. This is the only type of blackberry that produces primocanes from buds on the roots. Thus these cultivars may be grown in hedgerows or a solid row of canes (Fig. 12.10). The number of primocanes produced from roots decreases as the planting ages (Chapter 2). Most cultivars will produce primocanes that grow prostrate, along the ground, in the first year. Primocanes grow upright in year 2.

Fig. 12.10. Erect blackberry showing vigorous upright primocane growth during fruiting, California (B. Strik).

Biennial- or floricane-fruiting

Plantings are commonly established in the spring with full or mature production in year 3. Primocanes are tipped by hand during the growing season to encourage branching (Fig. 12.11) and increase yield the following year. Primocane tipping height varies from about 0.8–1.2 m (2.5–4 ft), depending on grower or region. Many growers will tip the vigorous branches to encourage more cane development (sub-branches) within the canopy. This may be done by machine in some regions, but is more typically done by hand. After fruit harvest, the senesced floricanes are removed by pruning just above the crown. This is typically done immediately after harvest to minimize cane disease. Any primocanes that emerge outside the in-row area (in the aisles) should be removed by tilling.

In the winter, growers will shorten the primocane branches by hand using secateurs or by machine (hedger or sickle bar) to maintain a canopy width of about 0.6 m (2 feet) (Fig. 12.12). In a young, establishing planting, most growers will train all of the branch growth to the trellis (Fig. 12.13). In the western USA, some growers train all the branch growth, regardless of planting age. Training more cane growth may increase yield, but may lead to a reduction in berry weight, depending on the cultivar. In vigorous cultivars, the number of

Fig. 12.11. 'Navaho' erect blackberry with tipped primocane and subsequent branching in foreground, North Carolina (B. Strik).

Fig. 12.12. Erect blackberry primocane branches being shortened in winter by machine, Georgia, USA (B. Strik).

Fig. 12.13. Young erect blackberry field trained to 'V'- trellis, winter Georgia, USA (B. Strik).

Fig. 12.14. 'Ouachita' erect blackberry in spring, Oregon (B. Strik).

primocanes per plant or per meter of row may be thinned. In regions with cold winters, growers should delay pruning until late winter when the extent of cane injury can be observed and removed.

Trellis types used vary tremendously by region and even with growers within the region. The most simple trellis used by some growers has metal or wooden end posts with metal T-posts in the row and one or two cross-arms (usually 0.3–0.45 m; 1–1.5 feet wide); canes are retained between 2–4 wires held on the cross arm(s) (Fig. 12.14). More complex trellises may be used with the plants either maintained as a single canopy (approximately 0.6 m wide; 2 feet) or as a divided canopy. The more elaborate divided canopy systems range from a 'V'-shaped system with multiple wires on each side to the Rotating Cross-Arm (RCA) trellis (see below) (Takeda et al., 2003a,b). The complex systems and the required hand labor to train them are commonly only used by growers with high-value fresh fruit markets (Chapter 14).

Annual- or primocane-fruiting blackberries

Primocane-fruiting, erect blackberries can be grown for a double-crop (floricane in early summer plus primocane in late summer through autumn) or a single-crop (primocane only). Whether plantings are managed for a double crop depends on the quality and fruiting-season or the potential market of the floricane crop relative to other floricane-fruiting cultivars that are available and adapted to the region. Management of the primocane crop, particularly related to modifying the fruiting season (Chapter 14), is limited when double-cropping (as the floricanes are present). In addition, the costs of primocane tipping are higher when double cropping (floricanes in the way) and yield of the primocane crop is lower than when managed to crop only on the primocanes (Strik et al., 2008).

Plants are generally established 0.8–1 m (2.5–3 feet) apart in rows 3 m apart or closer (in tunnels). These cultivars may be grown in hedgerows, but the number of primocanes produced from roots decreases as the planting ages (Chapter 2). Pruning systems vary depending on whether these are grown to produce fruit only on the primocanes or on both primocanes and floricanes.

Primocane crop only

Primocane yield of the most commonly grown commercial cultivars to date has been limited in many production regions of the USA by their late fruiting season because canes do not have much time to fruit prior to the first frost or heavy rains in autumn (Chapter 14). Yield can be increased in some of these regions by planting earlier-fruiting cultivars (e.g. 'Prime-Ark® Traveler') or by advancing the growth of primocanes and fruit harvest (Chapter 14). In the USA, coastal California is the largest production region for primocane-fruiting blackberry at present.

Primocanes are tipped during the growing season as yield is increased more than threefold on a cane that is tipped once to encourage branching compared to an un-tipped cane (Strik et al., 2008, 2012). Most growers soft-tip primocanes removing approximately 5 cm (2 inches) of the tip portion of the cane. However, removing more of the cane when tipping to 1 m tall, 'hard-tipping' (removing 30–45 cm; 1–1.5 feet), produced more branches and fruit than soft-tipped canes (Strik and Buller, 2012). Canes that are tipped to a shorter height will produce longer branches than those that are tipped at a taller height (Strik and Thompson, unpublished).

Canes need to be tipped soon after they reach the appropriate height to increase branching. Also, tipping canes after they have formed flower buds will reduce yield (Chapter 2). Growers need to go through the planting several times to catch all the flushes of cane growth. Primocane tipping heights range from 0.5 m to 1.5 m (1.5–5 feet). 'Double-tipping,' which has been shown to increase yield considerably over a 'single-tip' (Thompson et al., 2009), is commonly used in large production regions. Growers that tip only once should tip canes to about 1 m tall, whereas those that plan to double-tip should tip shorter to 0.45 m so that the branches produced are long enough to tip. Once branches reach about 0.5 m in length, they should be soft-tipped to 0.45 m long.

Thompson *et al.* (2009) found that this method of double tipping increased yield compared to a single tip at 0.45 m and led to a more compact plant growth and uniform presentation of fruit, increasing picking efficiency. Double-tipping did not reduce fruit size. In fact fruit size increased over 30% when double-tipped compared to a single tip. In addition, tipping method had no effect on primocane bloom or harvest date in Oregon (Thompson *et al.*, 2009). Late-emerging primocanes should not be tipped since they are unlikely to produce fruit before frost.

While hedging shows potential for reducing labor costs for a single-tip pruning method, growers must use caution when hedging to insure that there are not too many canes in the row that have already formed a flower bud and to hedge as early as possible (Strik and Buller, 2012). Growers that want to double-tip would have to tip the branches by hand.

In plants where primocanes were mowed back to ground level when height averaged about 0.5 m, fruit production was delayed by about 4 weeks (Thompson *et al.*, 2009). Of course, this production system is only useful in a warm region with a long-enough growing season to harvest the later primocane crop. In a cooler climate (shorter growing season), the primocane crop may be advanced using spun-bound polypropylene row covers placed over the row from late winter through early tipping or by growing plants in a tunnel with plastic on all season (Chapter 14).

Other than the summer tipping of primocanes, plantings are left unpruned until late winter when all the canes are mowed back down to just above ground or crown height. This method of pruning repeats the cycle in a primocane-only cropping system. Growers generally use a very simple trellis (Fig. 12.15) as described above for floricane-fruiting erect blackberry.

Double-cropping

When growers want to double-crop annual-fruiting blackberry, winter pruning is typically done using a hedger (sickle bars) to remove the portion of the primocanes that fruited last autumn. Any buds on the primocanes that did not break will potentially produce fruit on the floricane the following season (Chapter 2). Yield of the floricane crop is dependent on the cultivar grown, the vigor of the stand, how the canes were managed when they were primocanes, winter pruning method, and growing region (Chapter 14). After fruit harvest, the floricanes senesce and would either need to be removed by hand or be left in the row (dead canes) to reduce pruning costs. In Oregon, producing a floricane crop reduced the number of primocanes per meter of row and thus would be expected to reduce yield relative to a primocane-only crop (Strik *et al.*, 2008).

Double-cropping is not very common in the currently grown primocane-fruiting blackberry cultivars for several reasons: it is difficult and labor-intensive to properly summer prune the primocanes to get the best possible

Fig. 12.15. Trellis in primocane-fruiting blackberry (B. Strik).

primocane crop; the floricane crop reduces yield of the primocane crop which is typically a high-value fruiting season; and the floricane-fruiting season of these cultivars is similar to currently available alternative floricane-fruiting cultivars. Still, some growers are interested in double-cropping and might reduce pruning labor costs by rotating between double cropping and single cropping.

ROTATING CROSS-ARM TRELLIS AND TRAINING SYSTEM

In the United States, blackberry production in the region from the Northeast to the Great Plains has been limited due to recurring injuries to floricanes from low winter temperatures that commonly reach −20°C or lower. Since 2009, there has been a rapid increase in commercial blackberry production for fresh market in the eastern United States following the introduction of the Rotating Cross-Arm (RCA) trellis system (Fig. 12.16). Commercial, fresh market production of blackberries using the RCA trellis system currently totals about 200 ha (~500 acres) ranging in size from 0.4 ha (1 acre) to 9.3 ha (23 acres) (Fig. 12.17). These farms are located in Massachusetts and Pennsylvania in the Northeast; Florida and Georgia in the Southeast, and across the Ohio River Valley into Kansas and Iowa in the central plains area, and Texas and

Fig. 12.16. A new blackberry planting using the RCA™ trellis (G. Gao).

Fig. 12.17. A commercial blackberry farm in Georgia using the RCA trellis and cane training system. With this system, fruit can be positioned on one side of the row. The right-hand side shows the fruiting side, while the row on the left is showing the back side without any fruit. The rows are oriented northeast–southeast and were photographed at 9 am in mid-May in Georgia, with the sunrise occurring to the left. Note the back side of the row (left) is still in the shade. In the afternoon, the side of the row with the fruit will be in shade (F. Takeda).

Oklahoma in the Southwest. When the primocanes are properly trained on the RCA trellis and the articulating cross-arms are rotated at the proper time, more than 95% of fruit are positioned on one side of the row. The growers using the RCA trellis system report an increase of 30% in harvest efficiency and report better spray coverage by having almost all of the fruit located on one side of the canopy.

In the northern states, the RCA trellis system is used primarily to provide protection from low winter temperatures and reduce cane and bud injuries (Takeda et *al.*, 2008; Takeda and Phillips, 2011). Erect and semi-erect black-berry cultivars such as 'Natchez,' 'Ouachita,' and 'Chester Thornless' have been grown successfully on the RCA trellis system and exhibited little or no damage due to low winter temperatures. Winter survival of blackberry plants is improved because the over-wintering floricanes trained on the RCA trellis system can be positioned close to the ground by the rotation of the cross-arms and covered with a floating row cover. For example, in 2014 and 2015, black-berry growers in OH using the RCA trellis system suffered only a 5% crop loss, while growers using the T- or I-post trellis lost virtually 100% of their crop due to weakened polar vortex, allowing the cold air from the North Pole to pour down in some years across Canada and into the USA. In the southern states, the RCA trellis system can be used to position fruit away from direct sunlight in the afternoon to mitigate fruit damage due to high light intensity and tempera-tures (Takeda et *al.*, 2013). Direct exposure to sunlight especially in the after-noon increases the number and severity of white drupelets (Chapters 2, 5). For example, in the San Joaquin Valley of California, white drupelet formation in 'Apache' blackberry was eliminated when the fruit was positioned on the north side of rows that were oriented east–west.

The RCA trellis system (Harper et *al.*, 1999; Takeda and Peterson, 1996, 1999) is similar to the limited arm-rotation shift-trellis (Stiles, 1999), but the RCA trellis system uses a unique primocane-training procedure which permits the entire plant canopy to be rotated as much as 130° with little effort, but more importantly, without breaking the floricanes.

Establishing blackberry plants on the RCA trellis system

The rows should be established in east–west, north–south, or southeast–northeast directions. The RCA-trellised rows have plants established at a spacing of 1.5 (5 feet)–1.8 m (6 feet). The rows are spaced 3.4 m (11 feet) wide. The commercial RCA trellis available from Trellis Growing Systems (Fort Wayne, Indiana) is constructed of fiberglass-reinforced plastic components manufactured by a pultrusion process (square tubing and plates) (Fig. 12.18). A few growers have constructed their own RCA trellis using strut channels made of sheet steel and rectangular aluminum tubing with dimensions similar to the RCA trellis distributed by Trellis Growing Systems. The RCA trellis post is

Fig. 12.18. The Rotating Cross-Arm (RCA) trellis. The commercial version is constructed of fiberglass reinforced plastic components manufactured by the pultrusion process. The trellis consists of a post (~50 cm (20 in)) (a) which has two plates (b) attached at the top. A long (c) and a short (d) cross-arm is secured between the two plates with detent pins. There are two cane-training wires (e1 and e2) that are threaded through holes in the plates. Additional trellis wires (f) are threaded through both cross-arms and secured to end trellis assembly arms. The wires in the foreground are connected to a wooden tie-back post (g). The primocanes are placed on the training wire below the short cross-arm (e1). Wires terminate at the wooden tie-back post and on end trellis assembly arms on the first and last posts of each row with a 'Quik-End' termination tensioner (h) which has internal spring-loaded clamps. In early winter the canes are pushed over to the training wire under the long cross-arm (e2). The cross-arms are kept exactly perpendicular to the row by adjusting the tension to three wires going back to the tie-back post. Assembly and installation instructions are available from Trellis Growing Systems, Inc. (www.trellisgrowingsystem.com).

1 m (40 in) long and the cross-arms are 1.5 m (5 feet) and 0.6 m (2 feet) long; both are square tubings (5 cm; 2 in. square) (Fig. 12.18).

The posts (Fig. 12.19) are installed 8 m (26 feet) apart within the row, with two of its sides aligned parallel to the row direction and 0.5 m (20 in) in the ground. Two plates are attached with two bolts to the top of the post. The plates are positioned on the post in such a way that the long cross-arms, when they are in the upright position, point to the east on a south–north oriented row, to

Fig. 12.19. A close-up of an assembled RCA trellis components that consist of fiberglass material, metal pins and bolts, and high tensile wires. The post is made of square fiberglass tube. Two fiberglass plates are attached slightly offset to the top of the post with two bolts (a). A long cross-arm (b) and short cross-arm (c) are attached to the plates using detent pins. These arms rotate or pivot about these pins. The location of the hole (d) is precisely positioned on the plate so that when the long arm rests on top of the post (harvest position), the angle of the cross-arm is 20° from vertical. Another detent pin is inserted at the top of the plate to prevent the long cross-arm from rotating to the left. The short cross-arm is held at 20° outward with a detent pin. The pin is removed in winter and the cross-arms are allowed to rotate down. When the long cross-arm is rotated to the left and all the way to horizontal it will rest on the butt end of the short arm. Two training wires (f and g) are installed through the plates. The wire (f) under the short cross-arm is used for training primocanes. Floricanes are secured to the wire (g) (under the long cross-arm. Each wire on the cross-arm is tensioned to the end-row trellis post assembly with a 'Quik End' termination tensioner (h). A fiberglass sleeve (i) is placed on both short and long cross-arms and wires connects cross-arms to a tie-back post. (F. Takeda).

the north in east–west oriented rows, and to the northeast on southeast–northwest oriented rows. The cross-arm direction is important in positioning the fruit away from direct sun in the afternoon. Next, the short and long cross-arms are placed between the plates and secured to them with detent pins. Both cross-arms are thus rotatable. The detent pin that secures the long arm is placed through the holes on the plates located 7.5 cm (3 in) below the top of the post and 5 cm (2 in) from the post.

The long arm can be rotated up and 20° beyond vertical. Additional rotation of the cross-arm is prevented because the long arm at that point is designed to rest against the post. Another detent pin is inserted opposite the post to prevent the cross-arm from being pushed away from the post and engineered to withstand strong winds. Thus, during the growing season, the long arm remains fixed away from vertical and training wires on the long arm create an inclined surface for positioning the lateral branch canes. The rotation in the opposite direction can be more than 90° from vertical if the detent pin is removed. The long cross-arm rotation can be limited to 0° (horizontal) or 30° above horizontal with insertion of a detent pin in additional holes in the plate located to provide these angles. The short arm is held in the plates with detent pins through the holes located 15 cm (6 in) from the post and 2 cm (0.8 in) below the holes for the long arm. There is another set of holes above and slightly to the outside so that the short arm can be angled 30° outward during the growing season and released in the winter to hang down. As a result, the long-arm can be rotated to horizontal in winter.

The RCA trellis system requires installation of multiple wires on the cross-arms and through the plate at the top of the trellis post. There are two primocane-training wires that are threaded through holes in the plates. Additional trellis wires are threaded through both cross-arms and secured to end trellis assembly arms (Fig. 12.19). The wires in the foreground are connected to a wooden tie-back post. These wires terminate at the wooden tie-back post and are tensioned between the end trellis assembly arms on the first and last posts of each row and the wooden tie-back post with a 'Quik-End' tensioner which has internal spring-loaded clamps. Additional assembly and installation instructions are available from Trellis Growing Systems, Inc. (www.trellis-growingsystem.com).

Primocane training and RCA trellis operation

Primocane training and management of lateral branches are the most critical tasks in growing blackberries on the RCA trellis system. The primocanes emerge in the spring from the crown and roots and vigorously grow upward. The first three primocanes to reach beyond the training wires at 50 cm (20 in) height are trained (Fig. 12.20). Those primocanes that emerge later are removed. When the primocanes reach about 60 cm (2 ft) height, they are placed between the two wires running through the plates at the top of the post and secured to the wire located below the short cross-arm. The terminal portion of each primocane is carefully bent at the wire height, without moving the distal portion laterally, and tied to the training wire. The bending of the

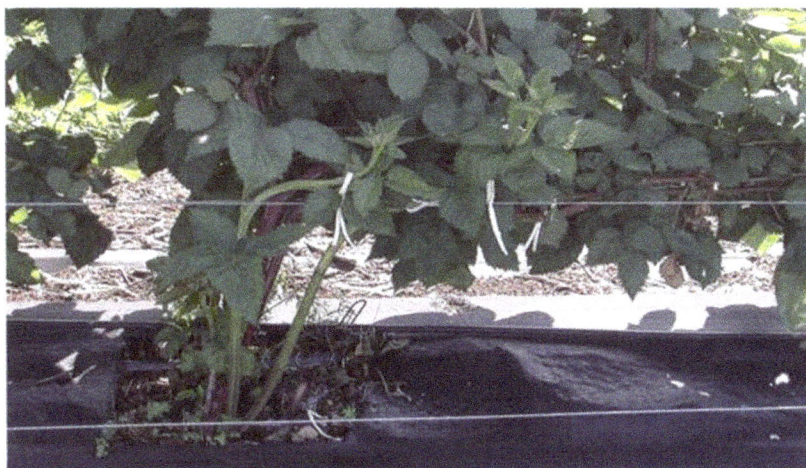

Fig. 12.20. Numerous primocanes emerge from the crown in spring. Typically the first three primocanes to reach the training wire are bent and trained to the wire on the side of the short cross-arm. The floricanes (dark colored) are in the background. Shown here are two primocanes positioned between the two training wires, bent and tied to the training wire. The third primocane (smaller diameter) has not grown tall enough to be tied. Timely tying of primocanes is required to maintain a horizontal extension growth. Note the terminal section of the secured primocanes has bent up (F. Takeda).

primocanes is always done to one direction. Also, an effort is made not to push the primocanes laterally during tying and bending, because at this point of their development it is more likely that the primocane will break at the base. As these three primocanes continue to grow in length, a tie is placed 10 cm (4 in) from the apex and secured to the training wire at 3- to 4-day intervals (Figs. 12.21 and 12.22). This insures that the three primocanes are trained horizontally. If tip-tying is not performed at these short time intervals, the terminal will bend upward (Fig. 12.20). Additional effort is required to bend them down and secure the tip to the training wire or one risks having primocanes that are horizontally not straight. The tips of the primocanes are soft-tipped when they have grown to the next plant in the row. An exception to this primocane-training protocol is in year 1, when the primocanes can be bent and trained to either wire because the wire under the long wire is not occupied by the floricanes.

Additional canopy manipulation is performed during the primocane-training period. Once the trained primocanes have grown horizontally for 20 cm (8 in) and cane breakage has not occurred at the bend or at the base, all the leaves and axillary buds or shoots that have emerged below the bend are removed. As the three trained primocanes grow 1.5 m (60 in) to the adjacent plant, axillary buds along the horizontally developed section of the primocane

Fig. 12.21. Bent primocanes (a) have grown about 1.7 m on the training wire. The portion must be secured on the training wire (b). Each primocane is tipped when the growing point reaches the adjacent plant. Secondary shoots (e.g. lateral canes) (c) develop from axillary buds on the primocanes. The laterals are kept in the 'V' or between the two cross-arms and tied to wires on the long arm. All lateral canes that emerge from nodes below the bend point are broken off. A floricane (d) and its lateral canes (e) that bear the current year's crop can be seen behind this year's primocanes (a) and secured to the training wire underneath the long cross-arm (F. Takeda).

begin to break. The lateral branch canes that develop from bent primocanes generally grow upright, regardless of their position on the bent primocanes and within the 'V' created by the long and short cross-arms (Fig. 12.22). A few that grow outside the wires on short and long cross-arms can be pulled back inside the V and laid on the training wire on the trellis arms. When lateral branch canes that rest on the wire located at the end of the short cross-arm are 1-m (40-in) long, they are pushed toward the long cross-arm and tied to a wire on the long cross-arm.

Many lateral branch canes grow 1.5 m (5 feet) to the top wire on the long cross-arm and some will grow further and eventually can reach the ground by late August (Fig. 12.23). Takeda *et al.*, (2003a,b) reported that 'Chester Thornless' blackberry plants with only two primocanes trained on the RCA trellis yielded as much as 82% of yield obtained with plants in which as many as six

Fig. 12.22. Primocane development on the RCA trellis in late summer after harvest is completed and the floricanes have been pruned. Three primocanes (a) have been bent (arrow) and secured to the training wire. The entire section of the primocane (b) beyond the bend point is oriented horizontally. Note many lateral canes (c) have developed from the axillary buds along the entire 1.5 m (5 ft) cane length. The leaves on bent primocanes have been removed to reveal the origin (F. Takeda).

primocanes were retained. Although the plants with two trained primocanes had fewer nodes and produced fewer racemes than those with six trained primocanes, larger racemes (e.g. more flowers per lateral) and larger fruit were recorded when plants had only two trained primocanes. Another contributing factor to yield compensation is that the number of long lateral branch canes is highest in the first primocane (ten branch canes/primocane), tall enough to be trained compared to about only three for the sixth primocane. These findings indicated the later-emerging primocanes do not contribute much to plant productivity. Despite the slight reduction in productivity, retention of only two or three primocanes on each plant can potentially reduce labor costs for summer cane training and tying by 50% (Harper *et al.*, 1999). Also, retention of just a few primocanes means that only those primocanes emerging from the crown by early May require tying and cane training can be completed by late June, prior to fruit harvesting. If more primocanes are trained, then training of additional primocanes will continue into July and August at the time when fruit is being harvested (Takeda *et al.*, 2003a,b). Blackberry cane management strategies that will not conflict with labor-intensive harvest operation are needed and assist in mitigating the scarcity and high cost of labor.

The lateral branch canes developing from the bent primocanes can reach a length of more than 3.5 m (11 feet) by late summer. The lateral branch canes that grow beyond the top wire should not be tipped in summer but delayed

Fig. 12.23.   The development of primocanes on the RCA trellis. By late summer, the lateral canes have grown beyond the top wire on the long cross-arm and have reached the ground. These primocanes should not be pruned at or near the top wire (arrows) after late fall. Shortening them earlier can lead to the development of tertiary lateral canes proximal to the cut (F. Takeda).

until late fall. Tipping of lateral branch canes in summer will stimulate the growth of tertiary branch canes from axillary buds just below the cut. The lateral branch canes should be shortened to a length of 1.5 m (5 feet) in late fall. Leaving the long lateral branch canes unpruned or shortening to 2.5 m (8.2 feet) leads to more than a 50% increase in nodes per cane compared to branch lateral canes shortened to 1.5 m (5 feet), but plant yield is not increased (Takeda and Rose, 2015). Their findings indicated that the buds located proximal to the bent main cane are more fruitful than those buds located toward the distal end. The diameter of the lateral branch cane and the size of the axillary bud gradually decrease with increasing distance from the base. Inflorescence size (number of flowers) and fruit size are highly correlated with cane diameter at which the flower shoot is located (Takeda, 1988, 2002).

In early winter after the lateral branch canes are pruned back to the top wire of the long cross-arm, the three canes that had been tied to the training wire under the short cross-arm are relocated as a bundle to the training wire under the long cross-arm. A twine is looped around the three canes and around the training wire under the long cross-arm and then pulled tightly to move the bundle closer to the wire under the long cross-arm. By transferring the bundle of canes to the wire underneath the long cross-arm, the wire under the short cross-arm becomes available for training of primocanes the following season.

Manipulation of rotating cross-arms on the RCA trellis system

In the northern areas (USDA Plant Hardiness Zones 7a and below) where low winter temperatures can injure blackberry plants, the long cross-arms are rotated towards the side with the short cross-arm in early winter (Fig. 12.24). In this operation, a curtain of lateral branch canes that have been tied to the wires on the long cross-arm in a nearly upright orientation in the summer are rotated to a horizontal orientation. The rotation lowers the curtain of lateral branch canes close to the ground during the winter months. The rotation of the canopy does not place any stress on the lateral branch canes. The rotation is possible because the horizontally oriented section of the main canes twists about the bend point, while the basal section of the main cane remains upright. If the main canes had been trained upright on the RCA trellis system, transferring them from vertical to horizontal orientation would require them to bend after they have become woody (Fig. 12.25). At that stage of cane maturation, the main canes are difficult to bend with a small arc. Any attempt to bend them from vertical to horizontal close to the ground will result in either the cane breaking or displacement of the post on the RCA trellis outward. Once the canopy has been lowered, a floating row cover ranging in weight from $100\,g/m^2$

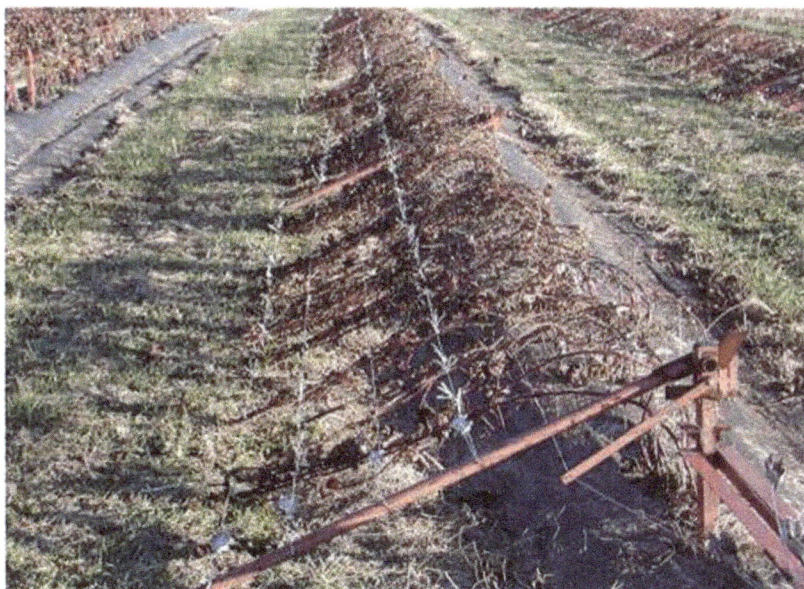

Fig. 12.24. A row of 'Siskiyou' blackberry trained on the RCA trellis photographed in November soon after the lateral canes have been pruned at the top wire of the long cross-arms. The long cross-arms have been rotated down to ground level to position the lateral canes near the ground for the winter (F. Takeda).

Fig. 12.25. When primocanes are not bent to the training wire but allowed to grow upright in the summer, proper trellis arm manipulation cannot be performed in the winter. The stiffness of upright floricane prevents the rotation of the cross-arms to less than 45° above horizontal (red arrow). Note some of the lateral branch canes (green arrows) have developed on the main canes. Lateral branch canes should be pruned at their juncture with the main cane. When a row cover is placed over them, the fabric will likely tear and insulation will be decreased. The resistance of stiff main canes to be bent also creates a large arc (red arrow) and could even push the post outward (yellow arrow) (F. Takeda).

($3\,oz/yard^2$) to $168\,g/m^2$ (5 oz/yard^2) can be placed over each row and edges secured with sand bags or rocks.

To keep plants dormant, winter covers may need to be vented, whenever daytime temperatures are predicated to be higher than 15°C (59°F) for more than two consecutive days. In USDA Plant Hardiness Zones 7a and higher where winter temperatures below −17.8°C (−5°F) are unlikely, the row cover application is not necessary and the rotation of the cross-arm can be delayed until bud break in the spring.

Fruiting year

The floating row cover is left in place over the blackberry plants until the probability of temperatures falling below −18°C is low or until just prior to bud break. At bud break, the long cross-arms can be rotated up to vertical so that any needed pesticides may be applied, but should be returned to a horizontal position or less than 30° above horizontal from 4 to 6 weeks. With the lateral branch canes oriented horizontally, the shoots that emerge from axillary buds will grow upward (Fig. 12.26). When the lateral branch canes are oriented less than 30° above horizontal from bud break to shortly before the onset of anthesis (petal fall), nearly all of the fruiting laterals (flower shoots) that emerge will grow upward above the trellis wires (Fig. 12.27) (Takeda and Peterson, 1999). The cross-arms should remain in a horizontal position or less than horizontal.

The long arm can be rotated up to its harvest position or 20° beyond vertical as early as when a few flower buds reach the 'popcorn' stage or later, when most flowers are at post-anthesis stage (Fig. 12.28). When the canopy rotation is performed at about the time of anthesis, the axis of the fruiting laterals will remain straight and point downward. However, the rotation must be performed before any of the primocanes reach the height of the training wire and grow through the canopy of lateral branch canes. If not, primocanes that are growing through the floricane canopy will be damaged during the rotation of the long arm to bring the lateral canes with fruiting laterals from a horizontal to a vertical position. However, if the rotation is performed during the time fruiting laterals are still elongating, the inflorescence axis tends to curve up as a geotropic response. Fruiting laterals that are still elongating can actually arc 180° and grow upward above the 20° inclined lateral canes. The rotation of the long

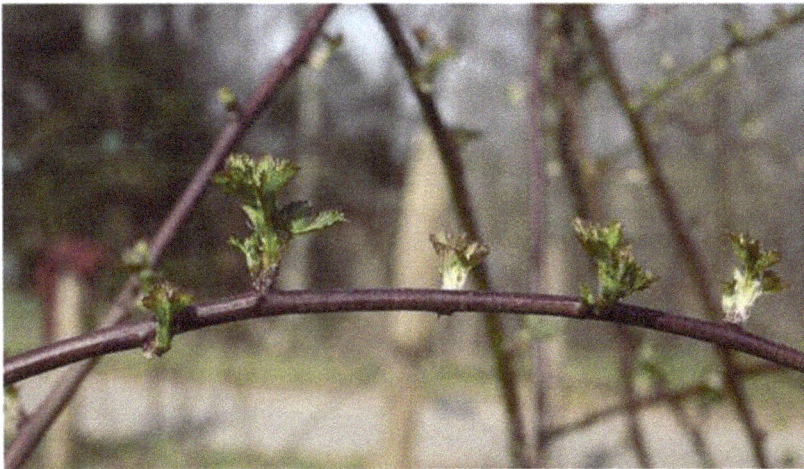

Fig. 12.26. Flower shoot development on horizontally-oriented lateral canes. Note that the axillary buds are positioned all around the circumference of the lateral cane. Initially, the flower shoots emerge at the angle of the bud, but they all grow upward, some arcing as much as 180°. Flower shoots emerging from uprights and canes that are oriented more than 45° from horizontal are less likely to grow upright as shown in the background, and radiate outward, thus positioning them on one or other side of a curtain of canes is not possible (F. Takeda).

Fig. 12.27.   Blackberry plants on the rotatable cross-arm trellis at bloom. The long cross-arms are oriented horizontally. Note the top of the gray-colored post on left side. The lateral canes that were secured to the wire on the long cross-arms have produced flower shoots and all have grown upward (F. Takeda).

Fig. 12.28.   Soon after all the flower shoots have a few opened flowers the cross-arms can be rotated upward and beyond vertical. By this time, the rachis (inflorescence axis) is woody and will not curve upward. The upward rotation of the cross-arms positions the fruit on one side of the row (F. Takeda).

Fig. 12.29. Ripening fruit on blackberry plants trained to RCA™ trellis during the third year. (Courtesy of Trellis Growing Systems, LLC, Fort Wayne, Indiana).

cross-arm beyond vertical positions all the bent floricane, lateral cane branches and developing fruiting laterals on one side of the row and away from the short cross-arm (Fig. 12.29). The training wire under the short cross-arm then becomes available for training and tying of new primocanes.

REFERENCES

Bal, E. and Meesters, P. (1995) Year-round production of blackberries. In: *Proceedings of the North American Bramble Growers Association*. North American Bramble Growers' Association, Pittsboro, North Carolina, pp. 49–61.

Bell, N., Strik, B. and Martin, L. (1995a) Effect of date of primocane suppression on 'Marion' trailing blackberry. I. Yield components. *Journal of the American Society for Horticultural Science* 120(1), 21–24.

Bell, N., Strik, B. and Martin, L. (1995b) Effect of date of primocane suppression on 'Marion' trailing blackberry. II. Cold hardiness. *Journal of the American Society for Horticultural Science* 120(1), 25–27.

Cortell, J. and Strik, B. (1997) Effect of floricane number in 'Marion' trailing blackberry. I. Primocane growth and cold hardiness. *Journal of the American Society for Horticultural Science* 122(5), 604–610.

Dixon, E.K., Strik, B.C. Valenzuela-Estrada, L.R. and Bryla, D.R. (2015) Weed management, training, and irrigation practices for organic production of trailing blackberry: I. Mature plant growth and fruit production. *HortScience* 50(8), 1165–1177.

Harper, J.K., Takeda, F. and Peterson, D.L. (1999) Economic evaluation of improved mechanical harvesting systems for eastern thornless blackberries. *Applied Engineering in Agriculture* 15(6), 597–603.

Julian, J., Seavert, C., Strik, B. and Kaufman, D. (2009) *Berry Economics: Establishing and Producing 'Marion' Blackberries in the Willamette Valley, Oregon*. EM 8773. Oregon State University, Corvallis, Oregon.

Oliveira, P.B., Lopes-da-Fonseca, L. and Jennings, D.L. (2004) Summer pruning effect on reproductive yield components of 'Triple Crown' blackberry. *Acta Horticulturae* 649, 277–281.

Pitsioudis, F., Odeurs, W. and Meesters, P. (2009) Early and late production of raspberries, blackberries, and red currants. *Acta Horticulturae* 838, 33–37.

Stanley, C.J., Harris-Virgin, P.M., Morgan, C.G.T. and Snowball, A.M. (1999) Boysenberry primocane management for improved productivity. *Acta Horticulturae* 505, 79–86.

Stiles, H.D. (1999) *Limited arm-rotation shift-trellis (LARS) and primocane management apparatus (PMA) for raspberries and blackberries (Rubus cvs. or crops)*. Bulletin Series 99-1. Virginia Agricultural Experiment Station, Blacksburg, Virginia.

Strik, B.C. and Buller, G. (2012) The impact of severity and time of tipping and hedging on performance of primocane-fruiting blackberry in a tunnel. *HortTechnology* 22(3), 325–329.

Strik, B.C. and Finn, C.E. (2012) Blackberry production systems – a worldwide perspective. *Acta Horticulturae* 946, 341–347.

Strik, B.C., Clark, J.R., Finn, C.E. and Bañados, P. (2007) Worldwide production of blackberries, 1995 to 2005 and predictions for growth. *HortTechnology* 17(2), 205–213.

Strik, B.C., Clark, J.R., Finn, C.E. and Buller, G. (2008) Management of primocane-fruiting blackberry to maximize yield and extend the fruiting season. *Acta Horticulturae* 777, 423–428.

Strik, B.C., Clark, J.R., Finn, C.E. and Buller, G. (2012) Management of primocane-fruiting blackberry – impacts on yield, fruiting season, and cane architecture. *HortScience* 47(5), 593–598.

Takeda, F. (1988) Some factors associated with fruit maturity range in cultivars of the semi-erect, tetraploid thornless blackberries. *HortScience* 22(3), 405–408.

Takeda, F. (2002) Winter pruning affects yield components of 'Black Satin' eastern thornless blackberry. *HortScience* 37(1), 101–103.

Takeda, F. and Peterson, D.L. (1996) Mechanical harvester, trellis designs, and cane training for eastern thornless blackberry production. In: *Proceedings of the North American Bramble Growers Association Conference*. North American Bramble Growers' Association, Pittsboro, North Carolina, pp. 38–46.

Takeda, F. and Peterson, D.L. (1999) Considerations for machine-harvesting eastern thornless blackberries for fresh market: trellis designs, cane training system, and mechanical harvester development. *HortTechnology* 9(1), 16–22.

Takeda, F. and Phillips, J. (2011) Horizontal cane orientation and row cover application improve winter survival and yield of trailing 'Siskiyou' blackberry. *HortTechnology* 21(2), 170–175.

Takeda, F. and Rose, A. (2015) Lateral cane lengths affect yield components in 'Triple Crown' blackberry on rotating cross-arm trellis and cane training system. *International Journal of Fruit Science* 15(3), 281–289.

Takeda, F., Hummell, A.K. and Peterson, D.L. (2003a) Effects of cane number on yield components in 'Chester Thornless' blackberry on the rotatable cross-arm trellis. *HortScience* 38(3), 377–380.

Takeda, F., Hummell, A.K. and Peterson, D.L. (2003b) Primocane growth in 'Chester Thornless' blackberry trained to the rotatable cross-arm trellis. *HortScience* 38(3), 373–376.

Takeda, F., Demchak, K., Warmund, M.R., Handley, D.T., Grube, R. and Feldhake, C. (2008) Rowcovers improve winter survival and production of western trailing 'siskiyou' blackberry in the Eastern United States. *HortTechnology* 18(4), 575–582.

Takeda, F., Glenn, D.M. and Tworkoski, T. (2013) Rotating cross-arm trellis technology for blackberry production. *Journal of Berry Research* 3(1), 25–40.

Thompson, E., Strik, B.C., Finn, C.E., Zhao, Y. and Clark, J.R. (2009) High tunnel vs. open field: management of primocane-fruiting blackberry using pruning and tipping to increase yield and extend the fruiting season. *HortScience* 44(6), 1581–1587.

13

DISEASES, VIRUSES, INSECTS, AND WEEDS OF BLACKBERRIES AND THEIR HYBRIDS

Annemiek C. Schilder,[1],* Harvey K. Hall,[2] Ioannis E. Tzanetakis,[3] and Richard C. Funt[4]

[1]Michigan State University, East Lansing, Michigan, USA; [2]Shekinah Berries, Ltd, Tauranga, New Zealand; [3]University of Arkansas, Fayetteville, Arkansas, USA; [4]The Ohio State University, Columbus, Ohio, USA

INTRODUCTION

Blackberries and their hybrids are susceptible to a number of diseases caused by fungi, bacteria, and viruses. The control or elimination of these is a key factor in the development and management of a commercial block of blackberries and to insure longevity of the planting, quality and quantity of the fruit, long-term plant survival, and the overall success of the venture. Usually plants need to produce for more than the initial year, and if the initial development cost can be spread over a longer plantation life with ongoing high yields of high quality fruit, then the profitability of the enterprise will be greater, or from another perspective the cost of production will be lower, the profit margin will be higher, and the grower will have a stronger ability to withstand competition in the marketplace. Another important management factor is competition between plants, which compete for space, light, and the nutrition provided to grow the crop. These are usually described as weeds, even if they are plants from a crop previously grown in the field, such as potatoes, maize or another cereal crop, or straw that has been used for mulching the blackberries. The main weeds are pest species and they can include incursions of wild blackberries as well as grasses, seedlings of shelter trees, members of the Solanaceae, or other adventive species.

* Corresponding author: Schilder@msu.edu

FUNGAL AND BACTERIAL PATHOGENS *(Annemiek Schilder and Harvey Hall)*

Fungal and bacterial diseases tend to be more prevalent in the humid growing areas of the world. Blackberries in dry regions may be more prone to soilborne problems. Virus diseases will be considered in a separate section of this chapter. The relative prevalence of diseases varies by country and region, with warmer regions having different disease complexes than colder regions. In general, three conditions must be met for a plant disease to occur:

1. a susceptible cultivar must be present;
2. a virulent pathogen must be present; and
3. environmental conditions must be suitable for infection.

This is commonly referred to as the disease triangle. In the case of blackberries, the inherent cultivar susceptibility and weather conditions often determine whether a disease becomes a problem and to what degree. Most blackberry pathogens are either present in the planting or may spread from wild brambles in the vicinity. In some cases, non-certified blackberry planting material may be the source of new disease outbreaks. Certain cultural practices like close plant spacing and minimal pruning may increase disease pressure. Furthermore, marginal winter hardiness in cold climates can predispose plants to cane diseases, whereas severe cane and foliar diseases in turn may compromise winter hardiness. To effectively manage blackberry diseases, an accurate identification of the pathogen and a proper understanding of the disease cycle are crucial (Ellis *et al.*, 1991). An integrated approach to disease management, combining cultural practices and biological control with the use of properly timed fungicides, is more effective and sustainable than sole reliance on chemical control. To enhance the understanding and management of diseases of blackberries and hybrids, the major diseases are described below, followed by disease management recommendations.

Fire blight or anthracnose *Elsinoë veneta* (Burkholder) Jenk.

Anthracnose is a common bacterial disease of blackberries, characterized by numerous circular to oval lesions on the green canes and fruit stems (Ward *et al.*, 2012). The lesions are whitish-gray with purplish margins and may be scabby or sunken (Plant Village, 2017). Deep and numerous lesions may lead to girdling, cracking, and eventual cane death. Leaf lesions are small, often arranged along major veins, and may develop a 'shot hole' appearance as the centers drop out. Severe infections can result in defoliation, low vigor, and low yields of poor-quality berries. The fungus overwinters in infected canes in the planting or nearby wild blackberries. In the spring, spores are dispersed by rain and wind to young, actively growing plant tissues which are most susceptible to infection. Cool, wet conditions are conducive to disease development and infection risk is highest between bud break and fruit set. The first lesions appear about a week after infection.

Cane blight *Leptosphaeria coniothyrium* (Fuckel) Sacc.

This is a serious disease of blackberries in the southeastern USA caused by a fungus that causes stem canker in roses and other ornamentals. Following a primocane infection or after wounding, dark red to purple cane blight lesions form with irregular purple borders later in the growing season, similar to *Botrysphaeria* cane canker symptoms. As the lesions enlarge, canes develop tip dieback or become girdled, resulting in wilting and death of lateral shoots. Small black fruiting bodies (pycnidia) may be visible in older lesions, producing gray, smudgy spore masses under wet conditions. Cracking and breakage of infected canes is common. Floricane lesions become brittle in the spring and release spore masses often producing a silvery to gray surface on the cane. Second-year canes show reduced vigor, low yields, and sudden death of side branches between bloom and harvest. The pathogen overwinters in infected canes and, in the following spring, produces spores that are rain-splashed or windborne to nearby canes. Wounds due to pruning, wind damage, or insect feeding allow entry of the fungus. Infected canes left in the field may produce spores for several years.

Purple blotch *Septocyta ruborum* (Lib.) Petr., (1968)

This disease is characterized by irregular, dark green cane blotches of about 1–5 cm (0.4–2 in) in length. Over the winter, the lesions turn purplish brown with a red margin and may become large enough to girdle and kill the cane when infecting European and other tetraploid blackberries. In spring, small black fruiting structures (pycnidia) in mature floricane lesions produce spores that are dispersed by rain and can infect only primocanes. It is widespread internationally, in the USA, Europe, and Australasia. In 'Youngberry' the disease is particularly virulent, causing severe cane dieback, especially in wetter winters, in field areas that are very wet and where weeds are prevalent.

Leaf spot *Sphaerulina westendorpii* (Westendorp) Verkley, Quaedvlieg & Crous (*Septoria rubi*)

Leaf spot is very common and can be quite severe on blackberries. Leaf lesions are round with a light, tan-gray center and purplish border and are typically scattered across the leaf surface, in contrast to anthracnose lesions, which are more concentrated along leaf veins. Severe infections may cause premature defoliation, thereby weakening plants and predisposing them to winter injury. Spots on canes and petioles may be somewhat more oval than leaf spots. Tiny black fruiting bodies (pycnidia) form in the leaf spots and release spores, which are spread by rain splash and can cause repeated infections through the growing season. The fungus overwinters in infected cane and leaf debris and in the spring releases copious amounts of rain-splash-dispersed spores. Since the spores require an extended leaf wetting for infection, rainy periods are highly conducive to disease development.

Orange rust *Gymnoconia peckiana, Arthuriomyces peckianus* (Howe) Cummins & Y. Hirats, and *Gymnoconia nitens* (Schwein) F. Kern & Thurst.

This disease, one of the most serious diseases of blackberries, is primarily a problem in the northeastern and midwestern USA and Canada, but it has also been recorded in Europe, Asia, and Australia. Red raspberries are immune. The fungi become systemic in blackberries with newly formed leaves in spring becoming stunted, deformed, and yellowish, enabling infected plants to be easily spotted. Growth is reduced and infected plants appear to have a spindly appearance and reduced foliage. Plant growth is poor and infected plants produce few or no flowers. In late spring and early summer, orange blisters form on lower leaf surfaces. These first appear waxy but soon become powdery and bright orange, releasing wind-blown spores, which cause new infections on mature leaves of nearby plants. Rust pustules that develop on these leaves will turn dark brown in the fall, producing another type of spore, which can infect new buds and shoots at the crown. From there the fungus grows into the crown and roots, where it overwinters and starts the disease cycle anew in the spring. Orange rust is favored by relatively low temperatures (6–22°C; 43–72°F) and long periods of leaf wetness.

Growers in the Midwest have adopted fungicide control programs, starting before orange rust (aeciospores – bright orange spores) spores develop in the spring (mid-May in the northern hemisphere). The fungicide sprays should begin just before the spores are released and continued until the leaves die and dry up. Spore germination is very slow when temperatures are above 25°C (77°F). In late summer and autumn as teliospores infect the new shoots, sprays must be oriented to the bottom of the plant. This is the time when spores from plants outside the planting can affect the canes. This spray program prevents spread of the disease to healthy plants. Complete fungicide coverage of the plants is necessary. All weeds growing in the row and on the edges must be controlled to achieve full coverage. Infected plants need to be destroyed since the disease is systemic. Further, by destroying affected plants with herbicides in the area surrounding the planting, the disease pressure can be decreased. The affected area can be replanted since orange rust is not soilborne (Ellis, 2014).

Blackberry rust *Kuehneola uredinis* (Link) Arthur. (leaf and cane rust)

This is a minor disease of blackberries characterized by the development of powdery yellow urediniospores, which burst out of the cane bark in springtime. Unlike orange rust, this rust is not systemic. The first symptoms visible in the spring are yellow powdery pustules on floricanes that split the bark. Later on, smaller yellow pustules develop in leaves. The spores produced in these pustules are airborne and can cause repeated infections on primocanes and leaves. Moderate temperatures and rainy periods favor disease development.

Blackberry cane and leaf rust is common in tetraploid blackberries from Europe and the USA and in blackberry hybrids, but it also occurs occasionally in red and black raspberries. The hybrid cultivars 'Loganberry,' 'Boysenberry' and 'Youngberry' are immune, but cultivars like 'Olallie' are very susceptible, especially when grown in a warm, dry climate like California. 'Marion' also is susceptible, but the old Oregon cultivar 'Aurora' is resistant. The New Zealand cultivar 'Karaka Black' is susceptible, but in crosses with the thornless (spineless) Boysenberry 'Marahau,' the progeny segregate for susceptibility to the disease. Severe infection results in defoliation and loss of the commercial crop.

Blackberry rust *Phragmidium violaceum* (C.F. Schultz) G. Wint.

This rust disease is common in Europe and the Middle East. It has also been found in Australia, New Zealand, and Chile on wild European blackberries. It is a damaging foliar disease of 'Evergreen' blackberry in the Pacific Northwest USA. Blackberry rust occurs on leaves, flowers, and fruit of blackberries. When the rust is virulent, it has a significant effect on the fruit production potential of cultivated blackberries, especially those derived from 'Merton Thornless' and other European blackberries. In Australia, blackberry rust is being investigated as a potential control measure for wild European blackberries, but a nonvirulent strain was introduced illegally and it has become another disease of cultivated blackberries with little gain in terms of weed control.

Initial symptoms are the formation of yellow-red blotches, which change into purple or red spots, often having yellow to brown centers. Uredinia form on the underside of the spots, and when the disease is severe the leaves wither, curl and drop prematurely. On the lower leaf surface, creamy white to yellow-orange spore pustules are visible. Spores are windborne and need more than six hours of leaf wetting at relatively low temperatures ($10–15°C$; $50–59°F$)

for infection. New, actively growing tissues are most susceptible to infection and canes, flower buds, and fruit are all susceptible. As infected leaves age, they may wither away and fall off the plant, leading to premature defoliation. In late summer and early fall, some pustules turn black and contain the overwintering spores.

Blackberry late leaf *Rust Phragmidium americanum* (Farl.) Arthur

This disease primarily affects red, black, and purple raspberries, but it also is known to infect *Rubus ursinus*, the Pacific trailing blackberry in the USA and Canada. It is found on the West Coast and in the northern parts of central and eastern North America. Leaf infection can cause premature defoliation and on fruit uredinia develop on individual drupelets, which makes the berries unsuitable for fresh market sales.

Blackberry rosette (Double blossom) *Cercosporella rubi* (Wint.) Plakidas

This disease causes significant problems in susceptible genotypes in southern USA, especially under hot, humid conditions (Sanders and Kirkpatrick, 2008). The disease has limited spread, outside the southeast of the USA, but it was imported to New Zealand with plant material from Arkansas, and it has become epidemic in 'Boysenberry,' which is particularly susceptible.

Rosette is characterized by malformed petals, which make the blossoms appear doubled. The infection may also cause sepals to become enlarged and leaf-like and unopened flowers to enlarge and become reddish. The fungus may be visible as a whitish mat of mycelium and spores on flower parts. An unusual proliferation of shoots may result in rosettes or 'witch's brooms.' On infected parts of the plant, berries are absent or small and of poor quality. Many blackberries are very susceptible, including thorny upright blackberries in the east and southeast USA; the disease was initially a significant impediment to expansion of Arkansas cultivars in the southeast USA. This was resolved with the discovery that resistance from 'Merton Thornless' blackberry had been incorporated into the new thornless cultivar 'Navaho'.

Powdery mildew *Podosphaera aphanis* var. *aphanis* (Wallr.) U. Braun et S. Takamatu

Powdery mildew, formerly known as *Podosphaera macularis*, is widespread in blackberries and raspberries in the northern hemisphere but is not present in the southern hemisphere. Powdery mildew forms a mycelial net over leaves and particularly the growing tips of young blackberry canes, causing severe growth retardation and malformation of leaves and canes. It also infects the fruit of susceptible genotypes and may render the fruit unmarketable due to the white hyphal mat on the fruit surface, reducing yield and causing significant crop losses. Young, actively growing tissues are most susceptible to infection. Severely infected plants may be stunted and less productive. Disease development is favored by warm weather and moderate to high humidity. The fungus overwinters as mycelium in buds on shoot tips and starts the disease cycle anew in the spring. Powdery mildew is more prevalent under tunnels and glass. As the area of production is expanding significantly under protected cultivation, this disease needs to be considered carefully as part of pest and disease management protocols.

Downy mildew *Peronospora sparsa* Berk.

This fungal disease is widespread on blackberries and roses in Europe, Asia, North and South America, and Australasia and has been shown to be cross-transmissible from the genus *Rubus* to *Rosa*. Downy mildew is most severe in blackberries as a nursery disease where the gray sporangia can cover the underside of leaves. It is particularly virulent on recently weaned tissue cultured plants. Care should be taken to control this disease in the nursery or plant losses will be high and subsequently a significant percentage of the survivors will be systemically infected, making the disease a problem in commercial plantings. In mature plants downy mildew lesions are typically angular and bounded by the leaf veins and veinlets. However, in breeding of hybrid blackberries in New Zealand, the bounding of lesions by leaf veins was absent in some genotypes and they formed ink-blot-type infections that spread quickly in susceptible genotypes. Downy mildew also causes 'dryberry' symptoms in fruit, causing them to wither and dry up before ripening. Infection is especially severe in locations with high rainfall during growth and harvest. In these conditions, without fungicide protection, crop losses may be 100%.

Some thornless tetraploid blackberry varieties show significant resistance to downy mildew, with little infection on mature canes or leaves, but thorny Arkansas blackberries are susceptible (Table 13.1).

Gray mold *Botrytis cinerea* Pers.: Fr.

In blackberry and other *Rubus* cultivars, gray mold is a serious fruit-rot disease, when there are wet conditions during flowering and at ripening. It also causes blossom blight of blackberries and other brambles when plants flower and fruits ripen in humid or wet conditions, especially when temperatures cause

Table 13.1. Disease resistance or susceptibility reported for blackberries and blackberry hybrids.

Cultivar	Type	Known disease resistance or susceptibility
Choctaw	Thorny	Immune to orange rust
Kiowa	Thorny	Immune to orange rust
Shawnee	Thorny	Immune to orange rust
Arapaho	Thornless	Resistant to orange rust
Black Satin	Thornless	Resistant to anthracnose
Chester Thornless	Thornless	Resistant to cane blight but susceptible to fire blight
Dirksen	Thornless	Resistant to anthracnose, tolerant to leaf spot
Navaho	Thornless	Resistant to Rosette but susceptible to orange rust
Boysenberry	Hybrid	Resistant to Kuehneola rust, susceptible to Verticillium wilt, rosette and downy mildew
Youngberry	Hybrid	Resistant to Kuehneola rust, susceptible to Verticillium wilt, purple blotch, downy mildew and rosette
Olallie	Hybrid	Susceptible to Kuehneola rust but resistant to Verticillium wilt
Karaka Black	Hybrid	Susceptible to Kuehneola rust
Marion	Hybrid	Resistant to Verticillium wilt but susceptible to Kuehneola rust
Loganberry	Hybrid	Resistant to Verticillium wilt and downy mildew, susceptible to leaf spot
Waldo	Thornless-AT	Very susceptible to downy mildew and Botrytis
Austin Thornless	Wild Species	Very susceptible to downy mildew
Austin Mayes	Wild Species	Very susceptible to downy mildew
Aurora	Hybrid	Resistant to Verticillium wilt and Kuehneola rust

dew formation overnight. Fruit infections usually do not show up until harvest and manifest as soft, light brown, rapidly enlarging areas on the berries. Infected berries become covered with a gray, fluffy growth of the fungus and can infect neighboring healthy berries through contact. Large numbers of airborne spores are produced on rotting fruit. As berries ripen they become more susceptible to infection and picked berries are extremely susceptible, especially if bruised. Handling of infected fruit during picking may also spread the fungus to healthy berries. The gray mold fungus overwinters as mycelium and black mycelial masses called sclerotia on infected plant tissues. In early spring, large numbers of airborne spores are produced. When moisture is present, the spores germinate and infect susceptible tissues within a few hours. The fungus usually enters the fruit through flower parts, where it remains latent until the fruit ripens. However, a lot of infections also take place on ripe and overripe fruit.

Temperatures of 20–27°C (68–81°F) and wetness from rain, dew, or irrigation are ideal for disease development. The disease can develop at lower temperatures if foliage remains wet for long periods and mycelia may still grow in frozen fruit at temperatures lower than 0°C (32°F). Blackberries grown in tunnels with good air movement suffer considerably less from Botrytis fruit rot and have higher fruit quality than fruit produced in the open field. Gray mold is a disease that is promoted by lack of air movement and significant moisture either from rainfall or dew.

Usually canes and foliage of mature plants show some degree of resistance to gray mold infection, but small seedlings and densely planted tissue cultured plants may be very susceptible, especially when air movement is low. Site selection and plant spacing at planting are important factors for increasing air movement (see Chapters 5 and 14). When fruiting laterals are hanging in dense clumps and when foliage is dense, the leaves, canes, and fruit also are prone to gray mold infection. The cultivar 'Waldo' is particularly vulnerable to this disease and along with other 'Austin Thornless' derivatives, it may show gray mold infection on canes, green leaves, flowers, and green fruit.

Cane canker *Botryosphaeria dothidea* (Moug.: Fr.) Ces. & De Not

This disease causes reddish-brown to dark reddish-brown lesions around wound sites or around buds on mature canes, sometimes girdling them entirely. When an infection site encompasses a node, lateral buds do not grow shoots and a girdled cane is killed beyond that point. Well-developed cankers split open longitudinally, exposing the pith. Severe infections have necessitated the removal of entire blocks of susceptible cultivars. In an Arkansas trial, 'Triple Crown' and 'Arapaho' were found to be resistant (Stewart *et al.*, 2003). The primary means of control is through establishing plantings with disease-free stock.

Verticillium wilt *Verticillium albo-atrum* Reinke. & Berth. and/or *Verticillium dahliae* Kleb.

This is a soilborne fungal disease that causes wilt in a very wide range of hosts worldwide including many species from the Solanaceae and Rosaceae, such as strawberries, eggplant, tomatoes, potatoes, stone fruits, and peppers. It is one of the most serious diseases of blackberries, causing wilting, stunting, and eventually death of fruiting canes or the entire plant. It is usually a cool weather disease and is most severe in poorly drained soils and following cold, wet springs. The appearance of symptoms on new canes frequently coincides with water stress during hot, dry, midsummer weather. Symptoms appear in early summer. The lower leaves may at first have a dull green cast to them. Starting at the base of the cane and progressing upward, leaves wilt, turn yellow, and drop. Eventually, the cane may be completely defoliated except for a few leaves at the top. Blackberry canes may exhibit a blue or purple streak from the soil line extending up the cane to varying heights. In the spring following infection, many of the diseased canes will have died. The fungus overwinters in plant debris and in the soil as dormant mycelium or small black survival structures called microsclerotia. The fungus can survive in the soil for many years after infecting a susceptible host. When conditions are favorable, microsclerotia germinate and infect roots aided by breaks or wounds in the roots.

Verticillium wilt has been responsible for the demise of a significant part of 'Boysenberry' and 'Youngberry' production in California where the disease was a significant problem (Wilhelm and Thomas, 1950; Wilhelm et al., 1965). The resistant cultivar 'Olallie' was able to replace these cultivars. Other cultivars such as 'Black Logan,' 'Cascade,' 'Chehalem,' 'Loganberry,' 'Mammoth,' 'Marion,' and 'Pacific' have been reported to be resistant. Modern hybrid blackberry cultivars have all been derived from these and resistance is expected to be widespread.

Armillaria root rot *Armillaria mellea* (Vahl: Fr.) P. Kumm. and other *Armillaria* species

Armillaria root rot is an economically important disease of woody plants worldwide and it is effective in producing white root rot in blackberries where they have been grown on cleared woodlands containing the disease organism, which may remain in the soil for an extended length of time. Infected plants display a slow or a rapid decline and may have small leaves and fruit clusters, dying canes, and an overall unthrifty appearance. The plants may initially appear to suffer from a nutrient deficiency or pH problem. Leaf color of *Armillaria* infected plants can range from dull green to yellow or red, but frequently it can kill entire plants. Dead and dying plants may be scattered through the field or occur in patches. Areas where infected plants occur are generally not lower or higher than other areas in the field. To diagnose *Armillaria*, scrape the outer bark off the stems at the base of the plant. The most reliable symptom is the presence of a creamy white layer of fungal mycelium, called a mycelial fan, below the bark in the collar region. A white layer of mycelium may also be present in rotted roots. Rotting wood and roots show a brown internal discoloration and have a strong mushroom odor. Sometimes 'rhizomorphs,' black shoestring-like strands, may be attached to the roots and trunk. Rarely, the fungus produces clumps of mushrooms at the base of infected plants in late summer or early fall. These mushrooms are usually 2.5–12.5 cm (1–5 in) in diameter and have a conspicuous ring around the stem below the cap. They are actually considered a delicacy in Eastern Europe but have to be cooked because they are slightly poisonous when eaten raw. Also, some people are intolerant of them. The most important species is *Armillaria mellea*, also called the honey mushroom. *Armillaria* survives in the soil on remnants of infected plant roots, such as those from oak trees. Contact with infected root pieces or rhizomorphs can lead to colonization of blackberry roots. The fungus can then spread further by root-to-root contact between neighboring plants. The fungus can also survive on wood chips used for mulching.

Phytophthora

Blackberries are resistant to many of the *Phytophthora* species that cause root rot in red raspberries (Maloney et al., 1993). Blackberries are able to grow in heavier soils and wetter conditions than tolerated by raspberries; they rarely succumb to root rots from these organisms. However, occasionally thornless blackberries with 'Merton Thornless' may become infected with an aerial form of the disease, which has symptoms similar to fire blight.

Crown gall *Agrobacterium tumifaciens* (E.F. Smith & Towns.) Conn and *Agrobacterium rubi* (Hildebr.) Starr & Weiss.

Crown gall is a widespread soilborne bacterial disease of blackberries and other brambles, which induces tumor-like galls on the roots and crowns or canes of infected plants. Bacteria can only enter the plant through natural wounds like freeze injury, insect feeding, natural growth cracks, or mechanical damage. Once the bacteria enter the plant, they insert a portion of their DNA (called T-DNA), which induces overproduction of certain plant hormones and cell proliferation resulting in tumor growth. Young galls developing in the spring are creamy white in color with a rough surface and are sometimes confused with callus formation or insect damage. As galls age, they become woody, brown to black, and somewhat crumbly. Galls can disrupt water and nutrient flow, causing plants to become weak and stunted with low yields and poor berry quality. Infected plants may also show symptoms of nutrient deficiency and become prone to winter injury.

The spread of crown gall in blackberry is associated closely with propagation, especially propagation from infected fields and spread from wild blackberries. If an outbreak of the disease in a nursery situation is not controlled, the disease can spread widely. In New Zealand where many of the first plantings of 'Boysenberry' were set with dibs (rooted cane tips) from producing fields, this disease became rampant, causing significant gall production on roots, crowns, and canes and reducing production of infected plants. Reduction of crown gall in plants from commercial nurseries has occurred through dipping in a suspension of a deactivated Agrobacterium produced by mutagenesis.

Stamen blight *Hapalosphaeria deformans* (Syd.) Syd.

This fungal disease is prevalent in the western USA, affecting the wild trailing blackberry *R. ursinus*, 'Boysenberry,' 'Youngberry,' 'Marion,' and also the tetraploid blackberry 'Evergreen' and it has been described in raspberry in Europe. Stamen blight gains infection in buds on primocanes during the summer and is expressed in flowers, causing stamens to become swollen, go brown, and become flattened against the petals. Sporulation occurs before the flowers open and they have a white, powdery appearance when open. Sepals may show leaf-like structures at their ends and the flowers may have double the petal numbers. Some species of blackberry may also show a 'witch's broom' proliferation of inflorescences (Pscheidt and Ocamb, 2016).

Orange felt (orange cane blotch) *Cephaleuros parasiticus* Karst.

This disease is prevalent on blackberries in very hot, humid environments, especially in the coastal plain areas of the southeast USA. The parasitic alga appears to not limit production when colony formation is limited, but when environmental conditions are conducive it grows vigorously, girdling canes, causing decline and death (Brannen, 2012).

Rubus stunt

This disease is considered to be a phytoplasma. Infection with this disease in blackberries results in thin, spindly canes with excessive lateral branching, to form a 'witch's broom' appearance, proliferation of flowers, and phyllody. It is transmitted by leaf hoppers, spittle bugs, and by graft inoculation.

VIRUSES *(Ioannis Tzanetakis)*

There have been significant changes in blackberry virology since the turn of the 20th century, with the number of viruses known to infect the crop more than doubling. The reason for such a dramatic change is multilateral: new technologies that allow for the identification and characterization of the most recalcitrant of viruses; the movement and establishment of vectors to new areas and the production of blackberries to areas with limited information on virus populations and dynamics. This has led to a significant change in control strategies for virus diseases. In the past, fields, at least in North America, lasted for 20 or more years. Today in southern latitudes this tends to be between 5 and 7 years.

When there are virus complexes present, it is often unfeasible to control all vectors; producers should focus on the ones that are easiest to control so as to delay disease spread and prolong plant productivity. In the rare event of a single virus causing symptoms, traditional control strategies can be implemented and potentially eliminate disease spread.

When it comes to virus diseases, blackberry production has changed over time, and there is no near-term solution. There are no active breeding programs focusing on identifying resistance to virus complexes, an immense task itself given the number and variability of the viruses affecting the crop. For this reason, well-planned operational procedures should be in hand before any new plantings are established. Propagation material should be obtained from the best source possible and be tested for all targeted pathogens. The virus history of the area should be in hand as to identify potential risks and launch the appropriate control measures before diseases become epidemics. Although these approaches may not eliminate disease, they will allow for sustainable and profitable production, the most important factor behind any horticultural operation.

Sixteen viruses and virus-like agents were known to infect *Rubus* before the turn of the 21st century. The data on the agents was primarily based on their biological properties as they were studied, because they were readily transmissible to herbaceous indicators or caused symptoms after grafting onto susceptible cultivars. This practice resulted in the under-representation of viruses that are not mechanically transmissible or have fragile virions, making transmission recalcitrant, leading to several misconceptions regarding the causal agents of blackberry virus diseases observed around the globe. Certain diseases were assumed to be caused by a single virus, whereas today (2016) we know that the majority of blackberry virus diseases are caused by virus complexes, the individual components of which are usually asymptomatic in single infections (Martin et al., 2013).

The lack of molecular data on blackberry viruses has led to misconceptions when it comes to virus identity. Based on biological properties a virus studied in North America could be considered the same species as one studied in Europe and vice versa. As knowledge is gained on the population structure of viruses infecting a crop, it has become evident that what have been considered 'strains' of a virus may belong to different taxa, a fact that could have significant impact for production, movement of propagation material, and the ecosystem as a whole, given that different viruses differ in their host range and could affect crops other than blackberries. Further, the strawberry necrotic shock virus (SNSV), once thought to be a strain of tobacco streak virus (TSV), is known to be the more widely spread virus of the two in berry crops in North America (Tzanetakis et al., unpublished).

In antithesis, there are cases where strains of the same virus were labeled as different species based on indicator plant symptoms. The virus species concept is based on a genome sequence continuum; isolates and strains may have different effects on a plant genotype, but they behave similarly under most circumstances. Raspberry leaf spot virus and raspberry leaf mottle virus are identified today (2016) as strains of the same virus, but have long been listed in the literature as different entities (McGavin and MacFarlane, 2010).

The following section of this chapter will discuss the impact of major viruses and virus diseases, as well as ways to eliminate some of the misconceptions that still affect the movement of propagation material across borders.

Vectors

The better characterized blackberry viruses at the biological level are those transmitted by pollen, nematodes, and aphids. These viruses have been studied for more than 50 years and their geographic range and impact is well documented, with few exceptions as noted above. In more recent times, there have been a plethora of new viruses species identified, which are transmitted by novel groups of vectors in blackberries, including whiteflies, hoppers, thrips, and eriophyoid mites. These vectors have been proven after transmission experiments using specific vector species or after deduction based on their taxonomic grouping with viruses that are known to be transmitted by a specific taxon of vectors. A list of the viruses shown to infect blackberries in a natural setting (excluding experimental transmissions), their mode of transmission, and geographic range can be found in Table 13.2.

Viral diseases

Blackberry yellow vein disease
Tomato ringspot virus (TRSV) was discovered in blackberries in the 1960s; it causes dramatic symptoms including vein banding, mosaic, mottling, puckering, but most importantly plant decline. It was not until the turn of the 21st century that the symptoms of TRSV were questioned. A dramatic decline in blackberry production in North and South Carolina, USA, even in areas where TRSV was not widespread, initiated studies to dissect the possible agents associated with the disease. Nematode transmission with the virus onto virus-tested plants failed to cause visual symptoms indicating that TRSV is not the causal agent of the symptomatology, at least not as a single virus (Gergerich, unpublished). The research to identify agents associated with blackberry decline, and what was later named yellow vein disease (Figs 13.1 and 13.2), identified a suite of new virus species. More importantly, this research provided the concept that modern blackberry cultivars may be tolerant to single virus infections and symptoms are frequently caused by virus complexes, the identity of which is not as important as is timing of infection and the number of species infecting the crop, with some plants found to sustain infection of six or more viruses.

Table 13.2. List of viruses shown to infect blackberries in a natural setting (excluding experimental transmissions), their mode of transmission, and geographic range.

Virus name	Transmission	Regional occurrence					
		NA	SA	Europe	Africa	Asia	Australia/NZ
Apple mosaic	Pollen/seed ¶	Yes	Yes	Yes	Yes	Yes	Yes
Arabis mosaic	Nematodes/semi-persistent ☐◇	Yes*	Yes*	Yes	Yes*	Yes*	Yes*
Beet pseudo-yellows	Whiteflies/semi-persistent	Yes	Yes*	Yes*	Yes*	Yes*	Yes*
Blackberry chlorotic ringspot	Pollen/seed	Yes	N/A	Yes	N/A	N/A	N/A
Blackberry vein banding	Mealybugs/semi-persistent?	Yes	N/A	N/A	N/A	N/A	N/A
Blackberry virus E	Unknown (eriophyiod mites?)	Yes	N/A	N/A	N/A	N/A	N/A
Blackberry virus S	Unknown (leafhoppers?)	Yes	N/A	N/A	N/A	N/A	N/A
Blackberry virus Y	Unknown (eriophyiod mites?)	Yes	N/A	N/A	N/A	N/A	N/A
Blackberry yellow vein	Whiteflies/semi-persistent	Yes	N/A	N/A	N/A	N/A	N/A
Black raspberry necrosis	Aphids/non-persistent	Yes	N/A	Yes	N/A	N/A	N/A
Cherry leaf roll	Pollen/seed (nematodes?) ¶	Yes	Yes*	Yes	N/A	Yes*	Yes
Cherry rasp leaf	Pollen/seed (nematodes?)	Yes	N/A	N/A	N/A	N/A	N/A
Grapevine Syrah Virus 1	Unknown (leafhoppers?)	Yes	Yes*	N/A	N/A	N/A	N/A
Impatiens necrotic spot	Thrips/persistent	Yes	Yes*	Yes*	Yes*	Yes*	Yes*

Raspberry bushy dwarf	Pollen/seed	Yes	Yes	Yes	Yes	Yes	Yes
Raspberry latent	Aphids/persistent	Yes	N/A	N/A	N/A	N/A	N/A
Raspberry leaf curl	Unknown	Yes	N/A	N/A	N/A	N/A	N/A
Raspberry leaf mottle	Aphids/semi-persistent	Yes	N/A	Yes	N/A	N/A	N/A
Raspberry leaf blotch	Unknown (eriophyiod mites?)	N/A	N/A	Yes	N/A	N/A	N/A
Raspberry vein chlorosis	Aphids?	Yes?	N/A	Yes	N/A	N/A	Yes?
Raspberry ringspot	Nematodes/semi-persistent	No	No	Yes	N/A	Yes	No
Rubus canadensis 1	?	Yes	N/A	N/A	N/A	N/A	N/A
Rubus yellow net	Aphids/semi-persistent	Yes	N/A	Yes	N/A	N/A	N/A
Sowbane mosaic	Pollen/seed	Yes	Yes*	Yes	Yes*	Yes*	Yes*
Strawberry latent ringspot	Nematodes/semi-persistent	Yes*	No	Yes	Yes*	Yes*	Yes*
Strawberry necrotic shock	Pollen/seed	Yes	N/A	Yes?	N/A	Yes?	Yes?
Tomato black ring	Nematodes/semi-persistent ◊	No	N/A	Yes	N/A	Yes*	N/A
Tomato ringspot	Nematodes/semi-persistent ◊	Yes	Yes*	Yes*	Yes*	Yes*	Yes*
Tobacco ringspot	Nematodes/semi-persistent	Yes	Yes*	Yes*	Yes*	Yes*	Yes*
Wineberry latent/Blackberry calico	?	Yes	N/A	Yes	N/A	N/A	N/A
Blackberry leaf mottle	Eriophyiod mites?	Yes	N/A	N/A	N/A	N/A	N/A

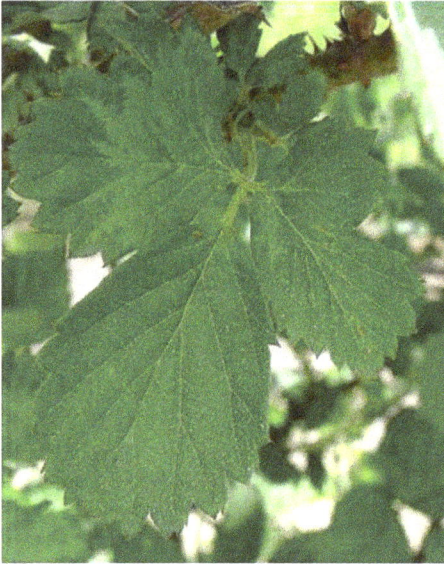

Fig. 13.1. Symptoms of blackberry yellow vein disease (I. Tzanetakis).

There are several viruses associated with the disease (Table 13.2, under-lined) and their presence is irregular, primarily based on vector incidence and topography. Blackberry leaf mottle associated virus and blackberry yellow vein associated virus (BYVaV) are the most widespread among all yellow vein disease associated viruses with more than 60% of the tested material being infected (Hassan et *al.*, 2017; Poudel et *al.*, 2013). Tobacco ringspot virus, blackberry chlorotic ringspot virus, blackberry leaf mottle associated virus, and blackberry virus Y were also found in a significant number of diseased plants but their presence was more localized (Martin et *al.*, 2013).

Given the significant number of viruses associated with the disease, and no viable eradication method being available, control of all viruses is non-economic. Control measures thus are limited to the planting of clean, high health stock and action to remove the vectors from the commercial fields. Since there is information on the presence of individual viruses in areas with signifi-cant disease impact, it is recommended that producers focus on control of spread of endemic viruses. If not all vectors can be controlled, it is advisable that control is aimed at those easiest to contain. As a working hypothesis, assume that the most prominent BYVD viruses in a production area are BYVaV, a virus transmitted by the greenhouse and banded-winged whiteflies, and blackberry vein banding associated virus, transmitted by the grapevine mealy-bug. The host range of the whiteflies and the fact that they are strong flyers make them a difficult target for control. On the contrary, the grapevine mealy-bug does not spread fast in commercial settings and it has a relatively narrow window where it is active in the field. In such a case, targeted sprays against the

Fig. 13.2. Symptoms of blackberry yellow vein (I. Tzanetakis).

mealybug can successfully eliminate the vector from the field. Although BYVaV may be present, it is asymptomatic in single infections and its effect on yield and field longevity is minimal, allowing for the slow spread of the disease as not all necessary components for symptom onset are present (Martin and Tzanetakis, 2015).

Aphid-borne viruses

Aphid-borne viruses (Table 13.2) present a major constraint for raspberry production in the temperate areas around the world, whereas their incidence in blackberries is sporadic. All *Rubus* aphid-borne viruses are transmitted by aphids in the genus *Amphorophora* (raspberry aphid) (Lightle et *al.*, 2014). Three of them, black raspberry necrosis virus (BRNV), raspberry leaf mottle virus (RLMV) and *Rubus* yellow net virus (RYNV) are part of the raspberry mosaic complex. None appear to cause symptoms in single infections and their effect in plant vigor and longevity is manifested when found in mixed infections with other members of the group or other viruses. Any combination of two mosaic viruses can induce symptoms, whereas a combination of RLMV and raspberry latent virus with raspberry bushy dwarf virus could cause severe drupelet abortion (Quito-Avila et *al.*, 2014). RYNV is of particular concern as it has the ability to integrate to the host genome (Kalischuk et *al.*, 2013). If this happens, none of the available methods for virus elimination (thermo-, cryo-, or chemotherapy, see Chapter 8) could eliminate the virus. Additionally, all the progeny of an infected plant will carry the virus. Even though the virus rarely causes symptoms in sole infections, it makes the plants susceptible for disease epidemics as all progeny and plant parts (including pollen and seed) could develop severe symptoms once plants are infected with additional viruses. For this reason, plants infected with RYNV should not be planted in commercial fields, nor should they be used for breeding purposes.

Nematode-borne viruses
There are eight viruses transmitted by nematodes or presumed to be transmitted by nematodes (Table 13.2). They are all pollen- and seed-transmissible and tend to have a wide host range that includes both monocots and dicots. Seven of these are transmitted by the needle and dagger nematodes (*Longidorus* and *Xiphinema*, respectively) with the vector of cherry leaf roll virus yet to be identified. Viruses persist in their respective vectors until they molt or feed on the roots of non-hosts where virus particles are depleted after continuous feeding. Their distribution is rather local and spread across fields, countries, or continents is primarily because of movement of infested soil. The major incidence areas of nematode-transmitted viruses can be seen in Table 13.2.

Traditionally, nematode-transmitted viruses were thought to cause significant damage, including severe decline and even plant death, but it is not clear whether they were the sole disease causing agent or were acting as part of a virus complex infecting plants (see TRSV in the section 'Blackberry yellow vein disease' above). The effect of nematode-transmitted viruses on modern cultivars is under-studied, primarily because nematodes have not been of significant importance for the industry due to the introduction of methyl bromide and field fumigation before planting. As the more potent nematode control chemicals are phased out, nematodes and the viruses they transmit may reemerge as a major issue for blackberry producers. The best and most effective control strategy is avoidance. Fields should be tested for the presence of vectors before planting. If vectors are present and fields are infested with weeds that are virus hosts, it is recommended that the site not be used for production or at least weeds are sampled prior to site preparation to determine if any of the nematode-transmitted viruses are present. If vectors are present without alternative virus hosts, production should not be affected as nematode vector species feeding does not cause economic damage.

Nematodes are microscopic creatures which cause damage to plant roots. They cannot be seen with the naked eye, and expert diagnosis is necessary to confirm their presence. The most common, *Pratylenchus penetrans*, has a wide host range including many pasture species. Nematodes are associated with decline of raspberries through feeding on the roots, with the spread of certain viruses (nepoviruses), and also with damage allowing crown gall to invade plants. Typically, damage is seen as a non-specific decline in vigor, and failure to respond to fertilizer or irrigation. Poor areas gradually extend downhill as the nematodes migrate in soil water. Small numbers can co-exist with raspberries without causing symptoms; in North America, a count of 500 nematodes per 500 cm² of soil is regarded as the economic damage threshold. Control relies on clean farming practices, as once they are established they are hard to eradicate. Planting nursery stock from high health sources, and planting into well-prepared soil that has been prepared by green manuring and fallowing are the best means of control.

Pollen- and seed-borne viruses (PSVs)

All nematode-transmitted viruses infecting blackberries are also pollen- and seed-transmitted and will not be discussed in this section. Raspberry bushy dwarf virus and the three ilarviruses infecting the crop will be discussed in more detail. Control of PSVs is focused primarily on avoidance, genetic resistance, and exclusion of pollen-feeding arthropods like thrips.

Raspberry bushy dwarf virus (RBDV)

RBDV is a cosmopolitan virus that in nature infects species in the genus *Rubus* and grapevine, although it has been experimentally transmitted to several other rosaceous hosts. The name was assigned after the symptomology of the plants used in its characterization (Jones et al., 2000). It was later discovered that those plants were also infected with black raspberry necrosis virus and symptoms were due to the synergistic effects between the two viruses. RBDV may cause leaf chlorosis on some blackberry cultivars (Fig. 13.3), including 'Marion,' but its major effect is drupelet abortion. Unlike raspberries where the fruit becomes crumbly, the major effect in blackberries is aesthetic, making fruit unmarketable. The fruit becomes misshapen and productivity is significantly reduced. Severe crumbly fruit of raspberry is caused by the synergistic effect of RBDV and two newly identified aphid-borne viruses, but this is yet to be investigated in blackberry (Quito-Avila et al., 2014). Severe crumbliness also can be caused by RBDV alone in raspberries when there is a high titer of the virus within the plant. Control of RBDV in raspberries can be effective in some regions through the incorporation of host resistance. Crosses have been done between hybrid blackberries and resistant red raspberries, but the resistance of the hybrids remains to be elucidated (Hall, personal observation).

Ilarviruses

The three ilarviruses infecting blackberries are (BCRV), an integral part of the yellow vein disease complex in the southeastern USA (SNSV) and (TSV). The host range of BCRV includes *Rubus*, rose, and apple (Poudel et al., 2014); SNSV infects strawberry and *Rubus*, whereas TSV infects more than 100 species of herbaceous and woody plants. After the molecular characterization of SNSV, a virus previously thought to be a strain of TSV, the presence of the latter has not

Fig. 13.3. Raspberry bushy dwarf virus (I. Tzanetakis).

been confirmed in blackberries and for this reason the two will be used interchangeably. BCRV was found to cause chlorotic ringspots and line patterns in 'Bedford Giant,' 'Himalaya Giant,' and 'Oregon Thornless,' whereas it is asymptomatic in 'Natchez.' SNSV/TSV cause necrosis when grafted onto blackberry indicators, but the effect on modern cultivars remains unknown. As found with other viruses, which encode a suppressor of gene silencing, it is assumed that the main effect of SNSV/TSV is in mixed infections with other viruses, as has been proven for BCRV.

WEED MANAGEMENT *(Richard C. Funt)*

Controlling weeds in the blackberry planting is an important cultural practice for commercial growers. Weeds compete for nutrients, sunlight, and water and harbor insects and diseases that are harmful to blackberry plants. Weeds can be annual, biennial (other brambles), and/or perennial plants. It is important to keep weeds from increasing in numbers by controlling them before they come into the seeding stage of growth. The removal of certain weeds before planting and to maintain a weed-free environment over the life of the planting is also important. In this section, information will be provided as to the reduction of harmful weeds and the types of weed management systems.

The clean culture system refers to hand weeding or a combination of cultivation using mechanical hoes, discs, or harrows, and hand weeding. In Maryland, between 1890 and 1930, black raspberry growers used hand weeding and cultivation using horse-drawn cultivators (Ross and Auchter, 1930). The fields were weeded three to four times per year. To minimize damage to the raspberry roots, cultivation was shallow and/or kept some distance from the crown. Animal manures were used, increasing the quantity of weeds, as seeds in the straw or hay bedding used for the animals got into the field. Cultivation, during and soon after harvest, loosens the soil, which was packed down by the people (pickers) who harvested the fruit. Further, cultivation during harvest can result in dropped fruit. Economically, this system over time destroys soil structure (see Chapter 9) and requires many hours of labor.

During the 1970s, growers in the Midwest and eastern USA converted to using herbicides before planting to control established weeds and after planting in the row. Grass (sod) is used as a permanent ground cover in the row middle or drive row. Grass-covered soil (row middle) is a biological weed control method; the sod is planted for rapid establishment and is maintained by mowing over the life of the planting. The herbicide/grass cultural method reduces hand weeding labor by 80%, reduces the destruction of soil structure, improves organic matter, and reduces erosion. Certain herbicides are used immediately after planting so as to not harm young roots. Different herbicides are recommended for year 3 and older plantings (Erf and Funt, 1984). New cultivars of tall fescues include dwarf types, which are drought-tolerant, require less mowing (reduced trash for mice and voles) and are usually 'endophytic,' unless the label specifies they are endophyte-free. Endophytes are also found in certain perennial rye grasses, tall chewing or hard fescues, and improve sod survival. Non-endophytic types do not do well under difficult conditions when the seeds become infected with a fungus (*Neotyphodium coenophialum*), which creates a toxin for insects and nematodes. Also, the endophyte infection causes early closing of the stomata to conserve moisture during droughts. Early closing of the stomata also conserves nutrients. These chemicals can reduce chinch bugs, thrips, bill bugs, sod webworms, and aphids that feed on grass stems. The endophyte also aids the fescues in storing more organic carbon (global warming) and nitrogen in the soil. Because endophytes also produce substances toxic to livestock, cattle or other animals should not be allowed to eat these types of fescues or grasses.

Another system used is an inter-row cover crop that is either killed by winter conditions, cultivation or mowing, or the use of herbicides. A seasonal cover crop, such as oats, is planted between the rows, usually in the summer, and is incorporated into the soil the following spring. This improves organic matter, improves soil structure, increases soil organic matter, and reduces levels of nematodes and other pests. A herbicide is used within the row of blackberries. Seasonal cover crops can also reduce excess nitrogen. In very fertile soils in the northeastern and midwestern USA, blackberries often grow vigorously and late into autumn, which can result in early freeze injury. Further, when the cover crop is incorporated into the soil in the spring, the nitrogen will be released back into the soil, reducing the need for nitrogen fertilizer.

Blackberries grown in a high tunnel can be grown in any of the systems mentioned above, although in four-season tunnels, sod cannot be grown between the rows without irrigation. In greenhouses (glasshouses), soil or potting media can include other non-soil-derived materials, such as sand, coir, peat moss, or vermiculite mixtures, plus nutrients for optimal plant growth, either for in-ground or pot culture. These mixtures may not contain weed seeds; however, weeds can come into these structures and hand weeding or the use of weed mats may be beneficial.

Blackberry growers may want to consider specialty buildings or shelters, such as a greenhouse or high tunnels, for production of fresh market fruit. Greenhouses are considered to be permanent structures and use heat and electricity. They may be used for blackberry production, giving the ability to extend the harvest season for earlier and later production. In many operations, these structures are used for annual bedding plant production or vegetable starter plants. Blackberry producers should consider a design for the placement of structures near water supplies, electrical sources, and waste management. New systems can recycle water or water containing fertilizers, reducing water and fertilizer usage.

The use of herbicides and cultivation by plowing or rototilling can greatly reduce the weed pressure before, during, and after planting. When planting into fields that have been in a monoculture (pasture, hay, or woody plants) for many years, it is wise to use a combination of herbicides and cultivation 1–2 years (years –1 to 0) prior to planting to reduce perennial weeds and their roots (Gao, 2000). Cultivating in the designated row just prior to setting plants can reduce weeds and cultivation will improve herbicide performance, allowing the transplants to adjust to their new environment. Using a safe herbicide in the row just after setting the plants can provide much less pressure of weeds in the second or third year. Hand weeding during the first year can be beneficial, particularly when weeds are young and the pulling of small weeds near the plant will reduce plant root injury as compared to tall, established weeds. A system of weed management benefiting many growers is the use of a herbicide in the row, and either seasonal cover crops or permanent sod in the drive row (alley). As described above, using certain grasses to establish a permanent sod can reduce insects in harsh environments. Further, if a thick sod is immediately formed after planting, fewer weeds will emerge, and thus, sod becomes a biological weed control method. Research in Oregon showed that the presence of weeds in the row reduced blackberry yield by more than half compared to bare soil (hoed) or weed mat used in the row (Dixon et *al.*, 2015). Weed mats have become popular in the USA in both conventional and organic production systems. Weed mats placed immediately after planting can control weeds for more than five years, reduce hand labor for clean culture and/or the use of herbicides in the row, and are economical during the early establishment years. In New Zealand the use of a mulcher to shred 'Boysenberry' prunings and a sweeper to shift the mulch underneath the plant row, is an effective practice to benefit the planting, providing organic matter and restricting the growth of weeds.

In the USA research and outreach programs have determined the safe use of herbicides and offer information on the timing and concentration of herbicides that are cleared for use. Websites in the east, midwest, and Pacific Northwest USA at the Land Grant Universities can provide useful information.

Control of blackberries as weeds

Both in the areas of natural dispersal and in countries where blackberries have become adventive as weeds, European blackberries become established as large clumps and dense thickets of tall thorny (spiny) bushes. In the United Kingdom (UK) and Europe several species of blackberries are widespread in the wild, including *Rubus (R.) ulmifolius* Schott var. *ulmifolius* in Western and Southern Europe, *R. armeniacus* (*R. procerus*), *R. anglocandicans* A. Newton, *R. cissburiensis* W.C. Barton & Ridd., *R. echinatus* Lindl., *R. erythrops* Edees & A. Newton, *R. laciniatus* Willd., *R. leightonii* Lees ex Leight., *R. leucostachys* Schleich. ex Sm., *R. phaeocarpus* W.C.R. Watson, *R. polyanthemus* Lindeb., *R. riddelsdellii* Rilstone, *R. rubritinctus* W.C.R. Watson, and *Rubus vestitus* Weihe in northern Europe, especially in Germany and in the UK. All of these have spread to some locations outside of Europe, including regions in Australia, New Zealand, North and South America. The reproduction of these species is spread through seed, dispersed by birds, and once established through sucking and/or top rooting. The majority of these species are apomictic, spreading as if by cloning through reproducing the female parent, regardless of the male parent involved. This excludes *R. ulmifolius* and other diploid species where seeds are produced by normal sexual reproduction. All these weedy species are thorny, discouraging browsing animals. In addition to becoming an environmental pest species, taking over farmland and any area of unoccupied land, they also can become an important reservoir of pests and diseases that can spread into commercial plantings.

Even when wild species are brought into an arboretum, they can spread from there and naturalize into the surrounding environment. A collection of blackberries was planted in New Zealand at the then HortResearch facility at Riwaka, along with a wide range of cultivated accessions from importation and from the blackberry and hybrid berry breeding program being conducted there. The wild and apomictic species spread into the local hedgerows and environment, while no spread was observed from the cultivated material, thornless or thorny, blackberry or hybrid berry. This is consistent with observations of large areas of cultivated blackberries planted on the east and west coasts of the USA, in Mexico, Chile, Spain, the UK, New Zealand, and Australia. While there may be wild types in the same area there was no evidence of spread to the cultivated types of 'Boysenberry,' 'Marion,' 'Young,' and other trailing blackberries, Arkansas blackberries, 'Tupy' ('Tupi'), 'Brazos,' and 'Loch Ness' or of the cultivated types to the wild. Long-established plantings of 'Smoothstem' and 'Thornfree' blackberries in New Zealand are similarly free of spread into the surrounding environment.

When the cultivated varieties are left abandoned in a field, plants may remain for some time in the absence of intensive grazing, but rarely is anything seen when the area has been effectively grazed, especially thornless types. Spread of these plants is limited to tip rooting or suckering. In New Zealand, plants of *R. argutus* left abandoned in old goldfield workings for around 100 years have remained, but spread very little and only by vegetative means. Even after this time, they could easily be removed with the use of grazing or herbicide. Clearly the path for the future, in terms of weed control, is to actively remove the apomictic adventive types that are already established. This action should not stop the introduction of valuable types of blackberries for cultivation due to unfounded beliefs on weed potential.

INSECTS *(Richard C. Funt and Harvey K. Hall)*

Insects (arthropods) may be beneficial, most are neither beneficial nor harmful, and some are harmless. Certain insects can be vectors of diseases, particularly viruses. In general, growers should protect the beneficial insects, as pollinators (bees) and predators of other pests, and protect plants from the harmful insects that attack the roots, canes, and fruits (Finn, 2008). Mites are not insects, but are found on the underside of the leaves and can cause economic damage, particularly under hot, dry conditions and under cover. Entomologists around the world have studied insects and their control and provide information on degree days for emergence and thresholds (amount of insects to cause economic loss to fruit or plant damage) as to the best control strategies for predators and/or insecticide use.

Insects that affect the root, crown, or cane

Raspberry crown borer (*Pennisetia marginata* Harris)

The raspberry crown or raspberry root borer is a serious pest of all *Rubus* fruit in North America, but does not exist in Europe. It looks like a yellow jacket (Family Vespidae), and is a clear-winged moth with black and yellow bands on the abdomen. It takes two years to complete its life cycle (Fig. 13.4). Adults appear in mid- to late summer (late July and August in the midwestern USA). Females can be seen on the foliage, where they lay reddish brown eggs (Fig. 13.5) on the underside of the leaf; the eggs take 30–60 days to hatch (Ellis *et al.*, 2004). The adults die in about one week. After hatching, the larvae move to the base of the plant; the next spring the larvae enter the crown and roots and girdle the new canes before they go to the root. In the second winter, they

Fig. 13.4. Adult raspberry crown borer. (Courtesy: Ohio State University.)

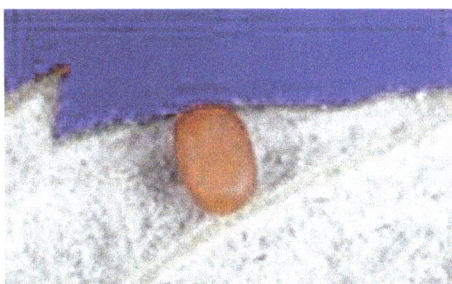

Fig. 13.5. Raspberry crown borer egg. (Courtesy: Ohio State University.)

are in the root and by the following summer, the crown can be damaged. The larvae transform into pupae and the adults emerge in mid- to late summer. The first symptoms that appear are wilting, dying of the cane foliage, and half-grown fruit (Bushway *et al.*, 2008). Growers should destroy infected canes. All wild bramble plants, in and near the planting, should also be destroyed. If insecticides are available, a heavy application of insecticide (drenching) in the early spring should kill most larvae and a spray to the soil of an insecticide in early fall (mid-October to mid-November in the USA) can kill adults.

The raspberry cane borer (*Oberea bimaculata* Olivier)

The raspberry cane borer is found in the eastern parts of North America and Europe, is a slender black beetle, 1.2 cm (½ in) in length with prominent antennae and usually two black dots on a yellow prothorax, has a two-year life cycle; adults appear in early to late summer, feeding on the tender green epidermis of cane tips and leave brown patches or scars (Figs 13.6 and 13.7) (Jennings, 1988). The female creates two puncture rings around the cane, about 1.2 cm (½ in) apart and about 15 cm (6 in) from the cane tip, and deposits an egg between the rings (Bushway *et al.*, 2008) These grubs (larvae) then feed inside the cane and bore down to the base of the cane by fall and into the crown by the next summer. They feed at the crown during the entire second season and into the second winter.

Fig.13.6. Raspberry cane borer damage. (Courtesy: Ohio State University.)

Fig. 13.7. Raspberry cane borer damage. (Courtesy: Ohio State University.)

Control can be accomplished by removing and destroying the infected portion of the stem a few inches below the wilted part immediately after the first site of damage. Further, destroy damaged canes during dormant pruning. Insecticides should be applied for the adult control in a late prebloom application.

Strawberry clipper or strawberry bud weevil (*Anthonomus signatus* Say)
This clip or weevil is a small 0.25 cm (0.10 in) long adult, reddish brown in color with rows of pits or punctures along its back and two white spots with dark centers. The adult overwinters in fence rows, mulch, or wooded areas and emerges as temperatures rise above 16°C (60°F). They move into raspberry

Table 13.3. Bramble plant symptoms and arthropod identification in the northern hemisphere.*

Plant symptoms	When and how to look	Arthropod description	Arthropod common name
Galls on primocanes with longitudinal splits in bark	April – split open galled cane to see larva	Worm 1/2–3/4", thin legless, flattened head	Red-necked cane borer (RNCB)
Shot holes in primocane leaflets and black frass	Early May to early June look on primocanes, leaves for RNCB adults	Beetle, 1/4", thin black body, orange-red neck	
Terminal leaves curled, malformed (worse if pines nearby = overwinter site)	Bud swell, inspect plants for curled leaves in perimeter	Adults, 1/10", waxy yellow to brown, clear wings	Blueberry psyllid
Stem of flower bud girdled, bud dies and falls off	When flower buds first swell, look for symptoms	Weevil, 1/10", snout, dark-reddish brown body	Strawberry bud weevil or Strawberry clipper
Poor drupe development and deformed fruit	April to harvest, weekly check 100 fruit for damaged buds or fruit	Worm, 1/8", yellow, legless; Bug (adult or nymph), 1/2–3/4", brown to green, suck	Blueberry midge and Stink bugs
Defoliating leaves and ripening fruit	June and July check for damaged leaves and beetles July feed only on ripe fruit	Beetle, 1/2", brown wing, green neck Beetle, 1", green wing	Japanese beetle Green June beetle
Canes are wilted, low vigor or dead from soil with terminal a shepherd's crook	June or July, look for larva by splitting open root crown of low-vigor canes	Worm, 1", white legs, thick body and amberhead	Raspberry crown borer

*From: Johnson and Lewis, 2015.

fields and feed on immature pollen by puncturing the blossom bud with their snouts (Bushway *et al.*, 2008). Some infected flower buds are girdled, dry up, dangle from the stem (Fig. 13.8) and eventually fall to the ground. In some areas, economic damage may occur and pesticide sprays may need to be applied at the prebloom stage (same time as for raspberry fruit worm) (Ellis *et al.*, 2004).

Fig. 13.8. Strawberry clipper damage on raspberry. (Courtesy: Ohio State University.)

The red-necked cane borer (*Agrilus rificollis* Fabricus)

The red-necked or flat-headed cane borer appears in eastern North America, is seldom a serious pest, and is a slender metallic black beetle about 0.6 cm (0.25 in) long with a reddish or coppery thorax (neck) and short antennae; the larvae are white, legless 1.9 cm (0.75 in) long and flat-headed and reach full size by fall (Figs 13.9a and 13.9b) (Ellis *et al.*, 2004). Galls are formed in symmetrical swellings about 1.2–2.2 cm (0.5–1 in) in diameter and about 30–1 m (1–3 ft) above the soil line and canes often break at the swelling (Fig. 13.10).

If fewer than 5% of the canes are affected, prune out these infected canes and burn, bury, or destroy them (Ellis *et al.*, 2004). Apply an insecticide in late prebloom (same as for the raspberry cane borer) if more than 5% of the canes are affected and another spray at or before petal fall until no more adults are found.

The raspberry cane maggot (*Pegomya rubivora* Coq.)

The raspberry cane maggot is smaller than a housefly. The larvae (maggot) tunnel down the cane and girdle the cane from the inside which appears like the raspberry cane borer, but with this insect there are no girdling marks. Damage may occur each year and the cane should be cut off below the infection when the symptoms appear in the summer; also cut and remove during dormant pruning. No insecticide is recommended. It is found in the USA, Canada (British Columbia), and the UK (particularly England) and occasionally found on red raspberries (Jennings, 1988).

Tree crickets (*Oecanthus* sp. Order Orthoptera)

Tree crickets are small green-white insects with a slender body and dark antennae, which can be longer than its body. Both nymphs and adults can be seen on canes in summer. In late summer, females lay eggs with 30–80 eggs in the canes about 0.5 m (1.5 ft) from the tip; this weakens canes and canes can be broken (Bushway *et al.*, 2008). Diseases can then enter the injured part of the plant. Prune out and destroy damaged canes (particularly spent canes) after harvest. If damage is severe, an insecticide may be applied in late summer

Fig. 13.9. a: Red-necked cane borer adult. (Courtesy: Ohio State University.) b: Red necked cane borer adult on leaf (H. Hall).

Fig. 13.10. Red-necked cane borer damage on cane. (Courtesy: Ohio State University.)

(August and September in the midwestern USA) or as nymphs appear in the spring (Ellis *et al.*, 2004).

The raspberry cane midge (*Resseliella theobaldi* Barnes)

The raspberry cane midge is a serious pest in the UK and Europe and plays a role in midge blight. The first generation of adults emerges from the soil in spring (May and June in Scotland) and lays its eggs in splits caused by internal growth stress on primocanes of certain cultivars; larvae are found two weeks later. The larvae are translucent at first and turn pink to orange as they mature. The pupae spend 2–3 weeks in cocoons; the second generation appears in mid-summer; a third generation in late summer (August–September) is rare. No recommendations have been given on control (Jennings, 1988).

Insects that affect the leaves and fruit

Blackberry aphid (*Amphorophora rubi* Kaltenbach)

The blackberry aphid is common in Europe. It is indistinguishable from the large raspberry aphid (*Aphis idaei*), but has a different chromosome number.

Japanese beetles (*Popillia japonica* Newman), rose chafer (*Macrodactylus subspinosus* Fabricius), and green June beetle (*Cotinis nitida* Linnaeus)

These insects (Scarab beetles) eat the leaves, flower buds, and/or berry fruit. All three of these insects have one generation per year (Fig. 13.11). Larvae or grubs develop in pastures, lawns, or other types of turf. The Japanese beetles are about 1.25 cm (0.5 in), copper-colored with metallic green markings and white hairs on the bottom, emerge from pupal chambers in the soil in midsummer (June and July in the USA). The rose chafer is light brown, 1.25 cm (0.5 in) long with long legs, emerges in early spring and feeds most commonly on white flowers and foliage, sometimes destroying flower buds and reducing fruit yields. Sandy soils can promote high populations. They feed on grasses and weeds. Green June beetles are 2.5 cm (1 in) long, metallic green on top and brown on the sides; larvae are soft white grubs with six legs and a curled brown head and are most noticeable in the planting or near a site where manure or compost has been spread. These areas attract the egg-laying females and serve as ideal areas for larvae (Ellis et al., 2004). Chemical sprays may be needed at the time that these beetles first appear or at late prebloom before the blossoms open (Bushway et al., 2008) or during harvest where it can be forecast that the insects will destroy approximately 20% of the leaves. Insecticides may need to be applied more than once in a few days. Preharvest spray restrictions must be obeyed.

New Zealand grass grub (*Costelytra zealandica* White)

This Scarab beetle also feeds on the foliage of raspberries. The most serious damage is done by the larvae on the roots, where a severe infestation will eat all the finer roots and the root epidermis of the plant up to the ground level, killing the plants. Control methods for grass grub include the use of insecticides and the use of bio-control agents, including *Serratia entomophila*, which prevents the insect larvae from surviving.

Fig. 13.11. Japanese beetle on leaf. (Courtesy: Ohio State University.)

Picnic beetle or sap beetles (*Glischrochilus quadrisignatus* Say and *G. fasciatus* Olivier)

These beetles are about 0.5–0.6 cm (0.2–0.25 in) long, black with four orange-yellow spots on the back or wing covers. There is one generation per year (Fig. 13.12). The larva is white with a brown head and about 0.6–0.9 cm (0.25–3/8 in). They will overwinter in many different plant covers and as temperatures reach 16–18°C (60–65°F) in the spring, they feed on fungi, pollen, or sap of plants. Adults feed on ripe and overripe raspberries and other fruit and/or any other fermenting material (Bushway *et al.*, 2008). One method of control is to place a bucket of overripe fruit and allow this to attract the beetles and trap them. Sanitation is the best method of control, with the berries picked frequently or at close harvest intervals so that overripe fruit does not occur. Keep ripe berries off the ground and bury culled fruit near packing plants (Ellis *et al.*, 2004).

The raspberry sawfly *(Monophadnoides geniculatus* Hartig)

The larvae are spiny pale-green worms, about 5–6 mm in length, that chew on the edges of leaves as young larvae; older larvae chew on everything except larger veins, causing a skeletonized appearance. The damage can result in considerable loss of yield. The adult (about 6 mm or 0.25 in) is a black four-winged fly with a yellow band on the abdomen and red markings (Ellis *et al.*, 2004). The female lays eggs singly on the top and bottom of the leaf and has one generation per year. It is not a common pest in North America and is not found in Europe (Jennings, 1988). However, a second species (*Priophorus morio*), known as the small raspberry sawfly, has an appearance similar to *M. geniculatus*, but has two or more generations per year and has been reported to be a problem in greenhouse grown raspberries in northeastern USA and has become a problem in Europe (Bushway *et al.* 2008). Chemical control is suggested at the early prebloom and late prebloom stages; this application will also control the fruitworm.

Fig. 13.12. Adult picnic beetle. (Courtesy: Ohio State University.)

The raspberry fruitworm (*Byturus unicolor*), the eastern USA raspberry fruitworm (*Byturus rubi* Barber), the raspberry beetle (*Byturus tomentosus* Degeer) in Europe, or the western raspberry fruitworm (*Byturus bakeri* Barber) in North America

These insects are small, light-brown beetles about 0.3 cm (1/8 in) in length. As an adult, it emerges (Fig. 13.13) in early spring (late April in the USA and May in Britain), feeding first on the growing point of the primocane as the leaves begin to open and then on the flower buds and young fruit, keeping the druplet from developing and condemning any sample for fresh or processing markets (Jennings, 1988). After the larvae hatch, they enter the blossom or young fruit. The larvae, which feed for about 30 days, are fully grown by early summer (July in USA) (Ellis *et al.*, 2004). Early-ripening cultivars may be more susceptible to the eastern raspberry fruitworm than late-ripening ones. Because the larvae fall to the ground in early summer, autumn-fruiting raspberries often escape injury (Bushway *et al.*, 2008). Cultivation of the ground in late summer can control the larvae; however, cultivation that injures roots can cause more disease. Chemical control is mostly used at early prebloom as blossom buds appear and late prebloom before blossoms open. These sprays should control sawflies as well. Biological control may be available in some locations.

Spotted wing drosophila (SWD) (*Drosophila suzukii* (Matsumura) and Insecta: Diptera Drosophilidae)

The drosophilidae were first detected in the USA in August 2008. It is native to southeast Asia and widely distributed in China, India, Korea, Myanmar, Russia, and Thailand. In Europe, it is widely distributed in France, Italy, and Spain. Drosophila adults are small (3–4 mm) yellowish-brown flies with red eyes (Fig. 13.14). The adults have a pale brown or yellowish brown thorax with black bands on the abdomen. The antennae are short and stubby with branched arista. Males have a distinguishing dark spot along the front edge of each wing. Spotless males are possible, but rare. The females have a serrated ovipositor

Fig. 13.13. Raspberry fruit worm adult. (Courtesy: Ohio State University.)

with which they penetrate the fruit skin. The serrations are much darker than the rest of the ovipositor. The eggs are translucent milky white, and glossy. The eggs develop and hatch within the fruit in which they are laid. The larvae are milky white and cylindrical with black mouth parts (Fig. 13.15). To test for larvae, place ripe fruit in salt water; the larvae will leave the fruit. Internal

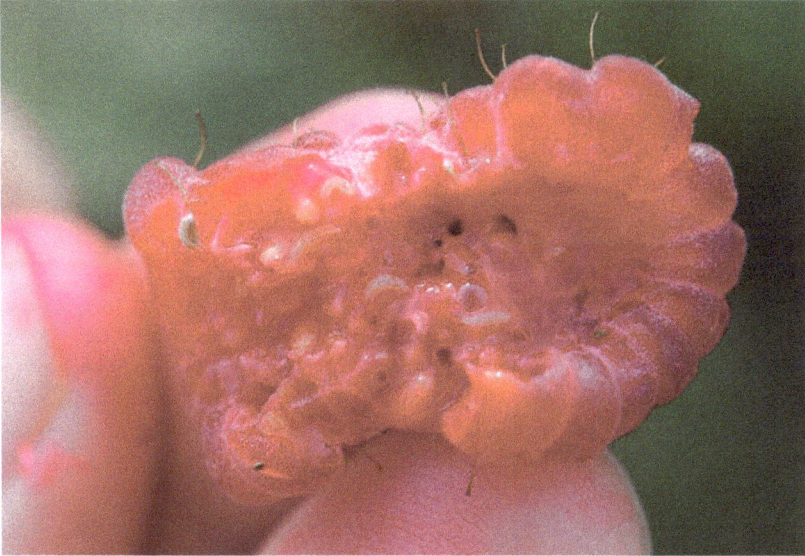

Fig. 13.14. Spotted wing drosophila in raspberry fruit (C. Welty).

Fig. 13.15. Spotted wing drosophila in blackberry salt water test (J. Jasinski).

organs are visible after they eat some fruit. Larval development occurs in the fruit and there are three instars before pupation. They are associated with ripe or rotted fruit (Fig. 13.16). Growers need to monitor fields in the final weeks before harvest with traps. It is necessary to pick all ripe fruit in the field; do not leave ripe or rotten fruit on the ground or in storage or packing areas. Ripe and ripening fruit and insects can be placed in a plastic bag in the warm sun as a control method. Insecticides can also be used carefully; precautions are to be observed so fruit don't have illegal residues.

Tarnished plant bug (*Lygus lineolaris* Palisot de Beauvois)

Tarnished plant bug adults are about 0.6 cm (1/4 in) in length, oval, somewhat flat, greenish- (coppery) brown with reddish-brown markings on the wings and a distinguishing small, yellow-tipped triangle on the back. Nymphs resemble aphids but are more active (Bushway *et al.*, 2008). They are pale green and about 25 mm when they first hatch in early spring. Adults and nymphs have piercing/sucking mouth parts. These insects are present on many plants, such as apple and peach trees and strawberries, until the first frost in autumn. Control weeds and do not mow forage crops, such as alfalfa, when brambles are flowering, since mowing encourages tarnished plant bugs to move to the berry crop and nearby weeds, where they feed on the flowers and developing fruit. To monitor the tarnished plant bug population, sample 50 plants in the morning

Fig. 13.16.   Spotted wing drosophila damaged blackberry fruit (J. Jasinski).

by tapping one flower or fruit cluster over a tray. Control is advised if there are 0.5 insect per cluster. When that threshold is reached, apply insecticides just before blossoms open and again before the fruit begin to color.

The two spotted spider mite (*Tetranychus urticae* Koch)
This mite is a serious bramble pest in North America and Australia, but is somewhat less of a pest in Europe. It is most commonly found in red raspberries, where it increases in numbers in hot weather (Jennings, 1988). The mites are from pale green to crimson red, have eight legs, and are about 0.05 cm (1/50 in) long. Adult females are oval shaped and marked with two large spots, one on each side; they can produce as many as ten generations per year (Bushway *et al.*, 2008). Under heavy infestation, leaves are marked by white stippling or bronzing after the mites have fed. Severe damage can reduce yield and fruit quality, reducing consumer appeal due to their appearance, which looks like brown dust on the fruit (Ellis *et al.*, 2004). The number of mites may be reduced under heavy rain or soaking sprays under high pressure. Natural enemies (predators), including predatory mites, lady beetle, and lacewings, can be purchased from suppliers in the USA. Miticide sprays are applied as the population increases. Reducing certain types of insecticide spray can reduce the loss of some predators and increase the control of mites. Using a 10× magnifier, monitor the underside of leaves; if there are 10–15 mites per leaf, chemical sprays may be justified. Red spider mites (*Panonychus caglei* Mellott and *P. ulmi* Koch) are found in North America and Europe, respectively, and are frequently found as greenhouse pests but are not generally found in the field (Jennings, 1988). Mite damage in the greenhouse may look like mosaic virus or other disease.

Blackberry psyllid (*Trioza tripunctata*; order Homoptera family Psyllidae)
The blackberry psyllid curls and stunts the growth of newly developing shoots and leaves of cultivated and wild blackberries. The infected leaves are greener than unaffected leaves and damage is due entirely to the adult females. Since the distortion shows up after the feeding, insects may not be present and this may be misdiagnosed as another malady, such as virus. There is one generation per year. The adult is 0.15 cm (1/16 inch) long, yellowish brown with brown shadowing along the vein in its wings. It is often found more in southern Ohio than other Ohio areas. The blackberry psyllid overwinters in pines, spruces, hemlock, and cedars. Do not plant blackberries near these types of trees. If blackberry psyllids appear, they can be controlled by insecticides in the spring after migration but before egg-laying.

Potato leafhopper (*Empoasca fabae* Harris)
The potato leafhopper occurs in North America. It is bright green and about 0.3 cm (1/8 in) long. Young nymphs are smaller and light green; adults are identified when they move side to side. Eggs are hatched within the leaves and

stems. Nymphs on the underside of leaves and adults are very mobile and attack over 140 species of plants (Bushway *et al.*, 2008). Potato leafhopper injury can be mistaken for herbicide injury, nutrient deficiency, or symptoms of viral infection (Fig. 13.17). Margins of affected leaves develop a light yellow color; new growth can be curled downward and stunted. Generally, damage can be found after the mowing of adjacent alfalfa fields in summer when the insect moves into blackberry plantings.

Yellowjackets (*Paravespula*, *Vespa*, and *Vespula* species)

Yellowjackets feed on ripe and/or injured fruit, particularly when weather conditions are dry. Because of their activity, picking can be difficult, particularly for pick-your-own customers. They generally build their nests underground or in old logs. Their numbers peak in late summer (Fig. 13.18). They can be discouraged by sanitation, picking all ripe berries, removing overripe fruit, and not allowing pickers to bring sugary drinks, lunches, or other attractants into the field. Traps can be placed around the planting before harvest.

Yellowjackets cannot be effectively controlled by insecticides. There are several species of this group of wasps in midwestern USA. Yellowjackets are yellow and black wasps about 1.2 cm (½ in). Whitejackets and bald-faced hornets are close relatives, are black and white, and aggressive, nasty stingers (Ellis *et al.*, 2004).

Fig. 13.17. Potato leaf hopper and damage (H. Hall).

Fig. 13.18. Yellowjacket (wasp). (Courtesy: Ohio State University.)

Plague thrips (*Thrips imaginis*)

Plague thrips cause damage to flowers by sucking all flower parts. They are slender insects, up to 1.2 mm long, and are best detected by shaking flowers upside-down over a sheet of paper or white handkerchief; alternatively, they can be made to run around inside flowers by gently breathing warm air into the flower. They look like bristles cleaned from a shaving razor, except that they move. They may cause total loss of young berry crops if not controlled. They seldom damage 'Silvan' blackberry flowers, but damage 'Silvan' fruit by sucking the drupelets adjacent to the calyx, producing a speckled silver appearance. They migrate on the wind over large distances. Control starts with weed-free plantings. Extreme care must be used when spraying so as not to destroy the bees during pollination.

Slugs and snails

Slugs and snails can cause damage to *Rubus* crops, more commonly to raspberries than blackberries. Control begins with rigorous weed control. If they persist, bait may be spread on the ground under the crop. Consult your farm chemical supplier for products registered for use in raspberry and blackberry crops.

Birds

Birds can cause total devastation of *Rubus* crops. Blackberries are more commonly affected than raspberries. Bird predation is specific to each farm and its surrounding ecosystem. In areas adjacent to cities, the imported nuisance birds or native birds can be a serious threat. If the problem is severe, the only proven remedy is to net entire crops. It is important to net before birds start feeding; once they have developed a taste for a crop they will try hard to find access through holes in nets.

SPRAY EQUIPMENT SELECTION, SPRAY COVERAGE, AND WORKER SAFETY

Blackberries are subjected to many insects and diseases. Control of weeds and nematodes before planting can reduce disease pressure and improve yield and quality of the fruit. Starting with disease-free and essentially virus-free, high-quality plants can be one of the most important decisions a blackberry grower can make. Setting the plants into a well-drained soil that is weed-free and maintaining good weed control throughout the life of the planting is vital for early returns on the investment. Annual insect, disease, and weed control will be necessary as well as drip irrigation before, during, and after harvest in areas deficient in rainfall. The harvesting of ripe fruit at the proper time and maintaining short harvest intervals can be one of the best methods to reduce insects and diseases within the planting. Maintaining good sanitation practices, as mentioned in this chapter, and utilizing refrigeration will be beneficial for providing quality fruit to the customer.

The selection of equipment for the application of pesticides is a long-term management decision. The purpose of spray equipment is to deliver nutrients (foliar) and pesticides to the desired target, and ultimately produce a quality and safe product for the consumer (Handley and Pritts, 1989). There have been many new developments in sprayer technology from many countries around the world during the latter half of the 20th century (Fig. 13.19). Therefore, the manager is faced with a variety of sprayer technology decisions, such as tank size, a pump that has the ability to produce a volume of water, a fan that can disperse the air in the canopy and replace that volume with droplets of spray with the correct size of nozzles to cover the area in a timely manner. Thus, this operation must be capable of completely covering the vegetation in the row width without causing drift onto other crops or fields. A grower may have several types of spray equipment, from a small 12 l (3 gal) hand-held knapsack or backpack type to a 200 l (50 gal) 3-point hitch power take off (PTO) herbicide sprayer to a large 1200 l (300 gal) trailing (1 axle) motor driven air blast sprayer (with or without a tower). Small backpack sprayers may be used for spot treatments of weeds in established plantings or on newly set plants.

The size and type of pump (piston, centrifugal, diaphragm, or diaphragm/piston) must be capable of delivering sufficient sprays per minute per a set travel speed at a specific pressure to the nozzle and then to the canopy and area covered. Recommendations for an amount of insecticide or fungicide are made per hectare (acre) of plants. Recommendations for herbicides are made as per treated hectare (acre) since only the row area is treated and is only a portion of an area of plants. In many cases, one-third of a hectare (0.8 acre) is covered with an herbicide, thus an amount per hectare can cover three hectares (7.5 acres) of plants. Accurate weighing and measuring are important;

Fig. 13.19. Three-point hitch sprayer on a narrow tractor designed for spraying raspberries in narrow rows, which can apply sprays to two sides in one pass. (Courtesy: Ohio State University.)

chemicals can be wasted by making concentrations unnecessarily strong or by making them too weak and ineffective. Furthermore, if chemicals are too strong they may cause damage to the foliage.

Tractor size must be matched to the size and type of sprayer. As an example, the tractor power take off (PTO) sprayer may require 25 hp to be effective. The engine size or PTO horsepower (hp) rating of the tractor needs to be 25 hp for the sprayer, plus 15 hp or more (total 40 hp) so that it can move itself either on level ground or on slopes. The tractor may have to supply electrical, hydraulic, or pneumatic external services to the sprayer as well as having a protective cab, which will need air filtration to protect the driver. Sprayers are capable of covering one, two, three, or four sides of the row(s) (Fig. 13.20). It is recommended that a grower use a separate sprayer of any size for herbicides and one for insecticides, fungicides, and foliar nutrients. Applying pesticides indoors, such as high tunnels and greenhouses, will utilize different equipment. Regardless of being indoors or out in the fields, worker protection is vital for human safety.

Covering the plant and soil or leaf canopy with the proper amount of spray is the first step in effective control of pests and weeds. Always follow label directions for restrictions and concerns for the environment and wildlife. Pesticides can degrade rapidly once in the spray tank and should be applied immediately after mixing. Further, the pH of the water has an effect on the effectiveness of the pesticide. A high pH water can be treated to reduce the pH safely and not harm the spray or the targeted plant. Application timing is equally important. As mentioned previously, the application of a spray during the dormant (no leaf) stage and the prebloom or late prebloom stage of growth can be very critical to producing a quality fruit for the consumer as well as having effective and efficient control of the pest. Proper pruning, tying, and row widths that allow good air movement can also allow for good coverage by pesticides (see insect and disease control in this chapter). Pre-emergence herbicides are best applied before weeds emerge in the row; this may be in late fall or early spring, depending on the type of herbicide.

Fig. 13.20. Sprayer applying herbicides to young raspberry plants to two sides of the row plus drive row in New Zealand. (Courtesy: Enfield Farms, Washington State University.

Safe use of sprayer equipment applies to the worker who needs to be protected from moving parts, such as the PTO shaft, fan, or pump, to the tractor operator who needs to be protected from spray drift, and to people and plants which can be affected by spray drift. Keep all PTO shafts in good order and covered with guards, have the tractor driver wear protective clothing or be in a cab that is properly filtered, and insure that the spray does not become windblown or produce large droplets, thereby reducing drift onto off-target people and plants (Beasley *et al.*, 1997). When using chemicals, it is important to follow the manufacturer's instructions precisely. Only use pesticides on the fruits that are listed on the manufacturer's label (ALWAYS READ THE LABEL). In the USA, private or commercial applicators can be fined or imprisoned by state or local regulators if pesticides are improperly handled or applied.

REFERENCES

Beasley, E.O., Criswell, J.T. and Crockett, M. (1997) *Applying Pesticides Correctly: A Guide for Private and Commercial Applicators*. Oklahoma State University/Oklahoma Cooperative Extension Service, Division of Agricultural Sciences and Natural Resources. Stillwater, Oklahoma.

Brannen, P.M. (2012) Orange felt (orange cane blotch) of blackberry. University of Georgia Cooperative Extension Circular 892. University of Georgia, Athens, Georgia.

Bushway, L., Pritts, M. and Handley, D. (2008) *Raspberry and Blackberry Production Guide*. NRAES-35. Natural Resource, Agriculture, and Engineering Service (NRAES), Ithaca, New York.

Dixon, E.K. and Strik, B.C. (2016) Weed control increases growth, cumulative yield, and economic returns of machine-harvested organic trailing blackberry. *Acta Horticulturae* 1133, 323–328.

Dixon, E.K., Strik, B.C., Valenzuela-Estrada, L.R. and Bryla, D.R. (2015) Weed management, training, and irrigation practices for organic production of trailing blackberry. I. Mature plant growth and fruit production. *HortScience* 50(8), 1165–1177.

Ellis, M.A. (2014) Cabrio, Pristine and Rally for orange rust control. *Midwest Small Fruit and Grape Spray Guide*. Bulletin 506B. Ohio State University, Columbus, Ohio, p. 47.

Ellis, M.A., Converse, R.H., Williams, R.N. and Williamson, B. (1991) *Compendium of Raspberry and Blackberry Diseases and Insects*. American Phytopathological Society Press, St. Paul, Minnesota.

Ellis, M.A., Welty, C., Funt, R.C., Doohan, D., Williams, R.N., Brown, M. and Bordelon, B. (2004) *Midwest Small Fruit Pest Management Handbook*. Extension Bulletin 861. Ohio State University, Columbus, Ohio.

Erf, J.A. and Funt, R.C. (1984) Effect of herbicides on newly planted 'Brandywine' purple raspberry . *Fruit Crops, 1984: A Summary of Fruit Research*. Research Circular 283. Ohio State University Agricultural Research and Development Center, Columbus, Ohio, pp. 63–65.

Finn, C.A. (2008) Rosaceae. pp. 748–751. Available at: www.ars.usda.gov/ARSUserFiles/1718/PDF/2008/Encylopedia%20of%20Fruit%20and%20Nuts%20Rubus%20spp-blackberry%202008.pdf (accessed May 25, 2017).

Gao, G. (2000) *Brambles – Production, Management, and Marketing*. Ohio State University Extension Bulletin 782-99. Ohio State University, Columbus, Ohio.

Handley, D.T. and Pritts, M.P. (eds.) (1989) *Bramble Production Guide*. Northeast Regional Agricultural Engineering Service, Ithaca, New York.

Hassan, M., Di Bello, P.L., Keller, K.E., Martin, R.R., Sabanadzovic, S. and Tzanetakis, I.E. (2017) A new, widespread emaravirus discovered in blackberry. *Virus Research* 235, 1–5.

Jennings, D.L. (1988) *Raspberries and Blackberries: Their Breeding, Diseases, and Growth*. Academic Press, London.

Johnson, D. and Lewis, B.A. (2015) Bramble plant symptoms and arthropod identification. Blackberry Pest Symptoms, Scouting, ID (handout). Department of Entomology, University of Arkansas, Fayetteville, Arkansas.

Jones, A.T., McGavin, W.J. and Mayon, M.A. (2000) Comparisons of some properties of two laboratory variants of *Raspberry Bushy Dwarf* Virus (RBDV) with those of three previously characterized RBDV isolates. *European Journal of Plant Pathology* 106, 623–632.

Kalischuk, M.L., Fusaro, A.F., Waterhouse, P.M., Pappu, H.R. and Kawchuk, L.M. (2013) Complete genomic sequence of a *Rubus* yellow net virus isolate and detection of genome-wide pararetrovirus-derived small RNAs. *Virus Research* 178, 306–313.

Koike, S.T., Bolda, M.P., Gubler, W.D. and Bettiga, L.J. (2015) Orange rust. In: *Caneberries: Pest Management Guidelines for Agriculture*. UC ANR Publication 3437, p. 39. Available at: http://ipm.ucanr.edu/PDF/PMG/pmgcaneberries.pdf (accessed May 25, 2017).

Lightle, D., Quito-Avila, D.F., Martin, R.R. and Lee J.C. (2014) Seasonal phenology of *Amphorophora agathonica* and spread of viruses in red raspberry in Washington. *Environmental Entomology* 43, 467–473.

Maloney, K.E., Wilcox, W.F. and Sanford, J.C. (1993) Raised beds and metalaxyl for controlling phytophthora root rot of raspberry. *HortScience* 28(11), 1106–1108.

Martin, R.R. and Tzanetakis I.E. (2015) Control of virus diseases of berry crops. *Advances in Virus Research* 91, 271–309.

Martin, R.R., MacFarlane, S., Sabanadzovic, S., Quito-Avila, D.F., Poudel, B. and Tzanetakis, I.E. (2013) Viruses and virus diseases of *Rubus*. *Plant Disease* 97, 168–182.

McGavin, W.J. and MacFarlane, S.A. (2010) Sequence similarities between *Raspberry leaf mottle virus*, *Raspberry leaf spot virus* and the Closterovirus *Raspberry mottle virus*. *Annals of Applied Biology* 156, 439–448.

Plant Village (2017) Blackberry. Available at: www.plantvillage.org/en/topics/blackberry (accessed January 24, 2017).

Poudel, B., Wintermantel, W.M., Cortez, A.A., Ho, T., Khadgi, A. and Tzanetakis, I.E. (2013) Epidemiology of *Blackberry yellow vein associated virus*. *Plant Disease* 97, 1352–1357.

Poudel, B., Ho, T., Laney, A., Khadgi, A. and Tzanetakis, I.E. (2014) Epidemiology of *Blackberry chlorotic ringspot virus*. *Plant Disease* 98, 547–550.

Pscheidt, J.W. and Ocamb, C.M. (2016) Blackberry (*Rubus* spp.)-Stamen Blight. PNW Plant Disease Management Handbook. Available at: https://pnwhandbooks.org/plantdisease/host-disease/blackberry-rubus-spp-stamen-blight (accessed May 25, 2017).

Quito-Avila, D.F., Lightle, D. and Martin, R.R. (2014) Effect of *Raspberry bushy dwarf virus*, *Raspberry leaf mottle virus*, and *Raspberry latent virus* on plant growth and fruit crumbliness in 'Meeker' red raspberry. *Plant Disease* 98, 176–183.

Ross, H. and Auchter, E.C. (1930) A production and economic survey of the black raspberry in Washington County, Maryland. Bulletin No. 322. University of Maryland Agricultural Experiment Station, College Park, Maryland, pp. 1–28.

Sanders, S. and Kirkpatrick, T. (2008) Management of important blackberry diseases in Arkansas. Cooperative Extension Service Bulletin FSA7563. University of Arkansas, Fayetteville, Arkansas.

Stewart, P.J., Clark, J.R. and Fenn, P. (2003) *Botryosphaeria dothidea* in Eastern U.S. blackberry cultivars. AAES research series 520 Horticultural Studies, pp. 32–34.

Ward, N.A., Hartman, J.R, and Hershman, D.E. (2012) *Anthracnose of Brambles.* Cooperative Extension Service Plant Pathology Fact Sheet PPFS-FR-S-17. University of Kentucky College of Agriculture, Lexington, Kentucky. Available at: http://plantpathology.ca.uky.edu/files/ppfs-fr-s-17.pdf (accessed May 25, 2017).

Wilhelm, S.P. and Thomas, H.E. (1950) Verticillium wilt of bramble fruits with special reference to *Rubus ursinus* derivatives. *Phytopathology* 40, 1103–1110.

Wilhelm, S.P., Bringhurst, R.S. and Voth, V. (1965) Origins of *Rubus* resistant to Verticillium wilt. *Phytopathology* 55, 731–733.

14

CROP PRODUCTION

Bernadine C. Strik[1,]* and Michele Stanton[2]

[1]Oregon State University, Corvallis, Oregon, USA; [2]University of Kentucky College of Agriculture, Food and Environment, Covington, Kentucky, USA

INTRODUCTION

Blackberries are grown in diverse production systems worldwide (Strik *et al.*, 2007; Strik and Finn, 2012). Continued growth of blackberry production will be greatly affected by cultivar development (see Chapter 6), but also by innovation in production systems that improve quality, fruiting season or markets, and development or expansion of production areas through soilless cultivation (substrate) in protected environments such as tunnels.

In this chapter, we address production systems used by growers of trailing, erect, and semi-erect blackberry (see Chapter 2) in the USA and other important production regions in the world.

Blackberry plantings generally have a life span of 15–20 years; however, some 'Marion' trailing blackberry commercial plantings in Oregon are over 50 years old. Plantings are generally established in the spring using plants propagated by tissue culture, root cuttings, or tip layers, depending on type of blackberry grown (see Chapter 2). Blackberries are tolerant of a wide range of soil pH (4.5–7.5) and soil types, although growth is improved under conditions of good drainage (see Chapter 10) and a soil pH of 5.6–6.8 (see Chapter 11).

TYPES OF BLACKBERRIES

Trailing blackberries

Trailing blackberries are grown at an in-row spacing of 1–2 m (3–6 ft) with 3 m (10 ft) between rows (Figs 14.1 and 14.2). Plantings are commonly established in the spring with full or mature production in year 2 or 3. A trellis is

---

* Corresponding author: Bernadine.Strik@oregonstate.edu

Fig. 14.1. 'Black Diamond' trailing blackberry field at bloom, Oregon (B. Strik).

Fig. 14.2. 'Marion' trailing blackberry field at bloom, Oregon (B. Strik).

needed to support canes and the crop (see Chapter 12). Yield ranges from 8 to 18 tonnes/ha (3.5–12 tons/acre) depending on cultivar, growing region, production system, and whether hand or machine harvest methods are used. Approximately 90% of world production of this type of blackberry is processed and most is produced in Oregon using machine harvest (see below). The most widely grown cultivars for processing are 'Marion,' 'Black Diamond,' and 'Columbia Star' (Oregon), 'Thornless Evergreen' (Oregon; Serbia), and the hybrid 'Boysen' (Oregon and California; New Zealand; Chile). The most commonly grown trailing cultivars for the fresh market are 'Obsidian' (Oregon and California; northwestern Europe) and 'Karaka Black' (Australia; New Zealand) (see Chapter 5).

Trailing blackberries can be grown in every-year (EY) or alternate-year (AY) production systems (see Chapter 12). The yield of an AY field is about 80% of an EY field over a two-year period, depending on cultivar, but AY fields have reduced labor requirements and lowered disease and insect management costs. In addition, AY fields offer greater cold hardiness in cold-sensitive cultivars. In Oregon, 25–40% of the trailing blackberry production is grown in AY systems. In most of the other production regions worldwide, trailing blackberries are grown in EY systems. Other than pruning and training methods (see Chapter 12), AY and EY fields have similar production systems.

Semi-erect blackberries

Semi-erect blackberries (see Chapter 2) are typically grown at an in-row spacing of 1.2–1.8 m (4–6 ft) with 3–3.6 m (10–12 ft) between rows (Fig. 14.3). Plantings are commonly established in the spring with full or mature production in year 3. Primocanes are usually tipped by hand during the growing season to encourage branching and increase yield the following year (see Chapter 12). Yield ranges from about 20–30 tonnes/ha (9–13 tons/acre). The most common cultivars grown are 'Chester Thornless' and 'Triple Crown' (Oregon; Europe; Canada), 'Čačanska Bestrna' and 'Thornfree' (Serbia), 'Hull Thornless' (USA), and 'Loch Ness' (USA; Australasia; Europe) (see Chapter 5). Primocanes are usually tipped during the growing season to encourage branching and canes are trained to various multiple-wire trellis systems (see Chapter 12). While this type of blackberry is most commonly hand-harvested for fresh market, hand- and machine-harvest for processed markets is used in some regions (see section 'Mechanical harvest').

Erect blackberries

Erect blackberry cultivars may be biennial- or annual-fruiting (see Chapter 2). The plants of both types are established 0.8–1.2 m (2.5–4 ft) apart in rows 3 m

Fig. 14.3. Semi-erect blackberry field, California (B. Strik).

apart (Fig. 14.4). Erect blackberries produce primocanes from buds on the roots as well as the crown allowing them to be grown in hedgerows (see Chapter 12) (Fig. 14.5). Primocanes are tipped to encourage branches and these are often further pruned to produce sub-branches (see Chapter 12). Fields have a mature crop in year 2 or 3, depending on region. The most common floricane-fruiting cultivars grown are 'Tupy' (Mexico), 'Ouachita' (Fig. 14.6) and 'Navaho' (USA; Australasia; Europe), and 'Natchez' and 'Arapaho' (USA). Yields range from 8 to 14.5 tonnes/ha (3.5–6.5 tons/acre) for all regions, except for Mexico where 'Tupy' can produce multiple crops per year.

Primocane-fruiting erect blackberry is a relatively new crop, with the first commercial cultivars released in 2004 (see Chapter 5). These blackberries may be grown for a double-crop (floricane in early summer, plus primocane in late summer through autumn) or a single-crop (primocane only). Whether plantings are managed for a double crop depends on the quality and fruiting season or the potential market of the floricane crop relative to other floricane-fruiting cultivars that are available. Management of the primocane crop, particularly related to modifying the fruiting season (see below), is limited when double-cropping (as the floricanes are present) and management costs are higher (see Chapter 12).

Fig. 14.4. Newly established 'Osage' erect blackberry field, North Carolina (B. Strik).

Fig. 14.5. Erect blackberry field, Oregon (B. Strik).

The most commonly grown primocane-fruiting cultivar to date ('Prime-Ark® 45') has had reported primocane yields of 20–22.5 tonnes/ha (9–10 tons/acre) in the central coastal area of California when double-tipped and grown in vented tunnels (E. Thompson, personal communication). However, this cultivar has had a relatively low yield in many other production regions of the USA where yield is limited by the late fruiting season – canes do not have much time to fruit prior to the first frost or heavy rains in autumn. Reported yield in open, field-grown plantings has thus been low in Oregon (4.5–6.7 tonnes/ha; 2–3 tons/acre) and Arkansas (4.5–9 tonnes/ha; 2–4 tons/acre). Yield can be increased in some of these regions by planting earlier-fruiting cultivars (see Chapter 5), by advancing the growth of primocanes using spun-bound polypropylene row covers placed over the row from late winter through early tipping (see Chapter 2), various pruning techniques (see Chapter 12), or by using tunnels to extend the season in colder climates (see section 'High tunnels'). While yield for tunnel-grown (unheated, open ends) primocane-fruiting blackberry was still low in Michigan (MI) (1–3.4 tonnes/ha; 0.5–1.5 tons/acre), yields were much higher in Oregon for this type of production system (4.5–19 tonnes/ha; 2–8.5 tons/acre), especially when primocanes were double tipped (see Chapter 12). In addition, growing these in an unheated tunnel extended the harvest season about 3 weeks in Oregon's temperate climate (Thompson *et al.*, 2009). In these cooler regions, it is clear

Fig. 14.6. 'Ouachita' erect blackberry at harvest time, North Carolina (B. Strik).

that yield is limited by the weather – plants have many buds and flowers present on the first frost date.

When primocane-fruiting cultivars are 'double-cropped,' yield of the floricane crop is dependent on the cultivar grown, the vigor of the stand (number of floricanes/length of row), how the canes were managed when they were primocanes, winter pruning method (see Chapter 12), and growing region (Fig. 14.7). Yield of floricanes was 4.5–6.7 tonnes/ha (2–3 tons/acre) in Oregon, but has been reported as 6.7–24.7 tonnes/ha (3–11 tons/acre) in smaller, research plots and 6.7–9 tonnes/ha (3–4 tons/acre) in commercial fields in the coastal region of California (E. Thompson, personal communication). The fruiting season of 'Prime-Ark® Traveler' and 'Prime-Ark® 45' is similar to 'Natchez.'

All types of blackberry are grown with a permanent grass cover crop between the rows (which is mowed as needed) or a bare soil, tilled area, depending on region (Figs 14.1–14.8).

IRRIGATION

In most regions, drip irrigation systems are used in fresh market plantings with irrigation scheduled based on estimated crop needs (estimated crop

Fig. 14.7. 'Prime-Ark® 45' field, North Carolina (B. Strik).

evapotranspiration) or using other soil moisture sensing tools (see Chapter 10). In some regions, under- or over-canopy mist or micro-jet sprinklers are used to reduce heat stress damage to fruit. Cultivars that are grown for processing are irrigated using moveable pipe with overhead sprinklers, big gun irrigation systems, or drip. In Oregon, there are some farms that are not irrigated; trailing blackberry have been found to grow and yield well when withholding irrigation after fruit harvest, likely due to their deep root system (see Chapter 2).

FERTILIZATION

Growers typically base fertilization decisions on results of tissue analysis of primocane leaves taken in late July to late August (northern hemisphere), soil tests every few years, and observations of annual growth, yield, color of leaves, and fruit quality (see Chapter 11). Some growers are collecting data on nutrient concentration in leaf tissue throughout the season with a goal of better managing plant growth, yield, and quality through improved fertilization although tissue nutrient levels are extremely variable throughout most of the season (see Chapter 11).

SPECIALIZED METHODS/SEASON EXTENSION

Specialized production of floricane-fruiting erect blackberry

In Mexico, specialized production systems have been developed to extend the season for 'Tupy,' the most widely grown floricane-fruiting cultivar (Fig. 14.8). In winter, plantings are mowed to crown height and are burned to help control weeds and pests. Irrigation and fertilization programs encourage vigorous primocane growth. Flower buds are stimulated using cultural methods and applications of phosphoric acid and ethrel once the primocane has grown sufficiently and hardened. About 5–7 months after primocane emergence, growth is slowed using foliar applications of copper sulfate, urea, and mineral oil. Plants are hedged and then defoliated using a combination of urea, ammonium sulfate, copper sulfate, and mineral oil and then gibberellic acid and a cytokinin are used to promote bud break. Fruit harvest begins 90–100 days after defoliation. After the first crop is finished, growers prune to remove the portion of the cane that fruited, and repeat the bud-break stimulation treatments to obtain a second crop; this may be repeated for a third crop; however, yield is reduced for each successive crop (Lopez-Medina, personal communication). Using these methods, the Mexican fruiting season extends from mid-October to early May for the export market and May through June for local markets (Strik *et al*, 2007).

Fig. 14.8. 'Tupy' field in Mexico (B. Strik).

Protected culture

Blackberry production occurs under protected structures or tunnels in many regions worldwide. The types of protected structures used include greenhouses (limited and mainly in Western Europe), high tunnels (in most regions; Fig. 14.9), and shade structures (in southern hemisphere; Fig. 14.10). All structures are intended to protect plants and fruit from adverse environmental conditions or extend the growing season. In some regions, windbreaks are needed to protect blackberry plantings when grown in open field production or with tunnels (Fig. 14.11).

Overhead shading

Use of shade structures and net is relatively common in berry crops, depending on growing region. In Australia, an enclosed structure may be used to protect against bird depredation, but also to shade the crop improving fruit quality (Fig. 14.10), particularly by reducing sunburn or white drupelets (see Chapter 2). Shade structures may also offer advantages in lowering canopy temperature, improving plant growth and possibly fruit quality, especially in hot regions and in heat sensitive cultivars (Stanton *et al.*, 2007).

Fig. 14.9. 'Triple Crown' semi-erect blackberry grown in tunnel to protect fruit from sun damage, Oregon (B. Strik).

High tunnels

In addition to extending the season (longer favorable temperatures) of primocane-fruiting blackberry, high tunnels are also used to shelter plants from the environment, particularly rain and ultraviolet light, and to create diffusing light to improve crop growth, yield, and quality. Tunnels may also be used to advance, delay, or extend the harvest season and to exclude or reduce some pests. For example, blackberry plants are grown in tunnels in California to greatly reduce the incidence of downy mildew. Tunnels can be used to extend the fruiting season, simply by protecting the crop against adverse weather, so more of the crop can be harvested, or to advance the season by placing plastic over the tunnel at the end of the dormant period to advance growth. Adding heat to the tunnel would advance the season further, but is cost prohibitive in most regions or situations.

In the northwestern USA and in British Columbia, Canada, tunnels are commonly used to protect late-fruiting, floricane-fruiting cultivars from rainfall (e.g. 'Chester Thornless'). In Oregon and Australia, tunnels are also used to protect blackberry fruit from sun damage (see Chapter 2) (Fig. 14.9). In Mexico, the only way to grow and harvest blackberries in the rainy season is to use tunnels. Primocane-fruiting blackberry grown in a tunnel had a similar

Fig. 14.10. 'Loch Ness' semi-erect blackberry, Victoria, Australia (B. Strik).

Fig. 14.11. 'Boysen' hybrid trailing blackberry grown with a windbreak, New Zealand (B. Strik).

fruit quality (soluble solids, pH, color) to those grown in the open field, but fruit were heavier (Thompson et *al.*, 2009).

High tunnels that are covered in winter, allow floricane-fruiting blackberry cultivars, that would normally not be sufficiently cold hardy, to survive winter (Demchak, 2009). Floricane-fruiting blackberries showed improved winter survival in a closed tunnel than in open field production.

Research on growing floricane-fruiting blackberries in tunnels for off-season production started in the early 1980s in Belgium (Bal and Meesters, 1995). Yield was increased an average of 88% compared to open field as a result of less winter injury, protection from rain and wind, reduced fungal pathogens, and protection of fruit allowing for more marketable fruit harvest. They also conducted trials on manipulating the fruiting season of semi-erect blackberries by using potted, cold-stored plants that were placed in the greenhouse to alter the fruiting season. While blackberries are grown in pots ('substrate') in a few production regions, substrate production of blackberry lags significantly behind that of raspberry. This is most likely because primocane-fruiting raspberry are relatively easy to grow in substrate to get a significant crop on the primocane. Substrate production of blackberry may become more common or economical as new primocane-fruiting blackberry are developed and released or as production methods such as 'long-cane' blackberry (storage of floricanes) are developed.

SPECIAL PRACTICES FOR ORGANIC GROWERS

Oregon is a leading producer of blackberry in the USA accounting for 45% of the 6063 ha (14,982 acres) grown (USDA, 2010). There has been a significant amount of research done on development of production systems in certified organic blackberry (Dixon and Strik, 2016; Dixon et *al.*, 2015, 2016a,b; Fernandez-Salvador et *al.*, 2015a,b,c; Harkins et *al.*, 2013, 2014; Strik, 2016). Certified organic acreage of blackberry in Oregon grew 96% from 2008 to 2014 (USDA, 2014). Development of organic blackberry has been slower in most other blackberry production regions because of climate and thus pest/disease limitations. The climate in Oregon offers advantages for organic production, including relatively mild winters and dry, warm summers that reduce weed and disease management costs and provide optimal conditions for fruit ripening (Strik, 2016).

The most important limiting factors to certified organic production of blackberry are weed, disease (particularly in the southeastern USA), and insect (particularly spotted wing drosophila (SWD) and raspberry crown borer). In organic production systems, grass is commonly grown between the rows, mowed as needed (Fig. 14.12). Controlling weeds in the row area is very important for good blackberry growth and production; weed competition in the row reduced yield by half in trailing blackberry (Dixon et *al.*, 2015; Harkins et *al.*,

Fig. 14.12. Organic 'Chester Thornless' semi-erect blackberry field, Oregon (B. Strik).

2013). In trailing and semi-erect blackberry use of a black plastic woven polyethylene ground cover (weed mat) in the row is possible because these blackberries do not produce primocanes from the roots (Fig. 14.13). Weed mat is an economical method of weed management in organic blackberry (Dixon and Strik, 2016). Control of SWD in organic production is limited to use of approved pesticides, which are presently much more limited than for conventional production and management practices to reduce refuges – growers are encouraged to mow any border areas of fields that might harbor SWD adults or have alternate fruiting hosts for the fruit larvae, trap for the adult flies to better time the first pesticide application, use good sprayers for good coverage and control, and practice good field sanitation (picking fruit clean). Carefully following the pesticide label recommendations for effective control and to stay within the rules for maximum product applied per season is very important. There are no effective organically approved controls for raspberry crown borer or many of the important diseases found in many blackberry production regions. For these reasons, climate and field management to minimize disease presence are critical to success.

Cultivars differ in their adaptation to organic production systems, likely due to cultivar susceptibility to disease and insect pests (Dixon *et al.*, 2015; Fernandez-Salvador *et al.*, 2015c). In trials, yield of processed and fresh

Fig. 14.13. Organic 'Triple Crown' semi-erect blackberry field, Oregon (B. Strik).

blackberry cultivars grown in the best organic production systems was similar to what would be typical for conventional production (Dixon et al., 2015; Fernandez-Salvador et al., 2015b).

FRUIT HARVEST

Mechanical harvest

Mechanical harvest is most commonly done in Oregon where about 90% of blackberry production is specifically grown for processed markets, including individually quick frozen (IQF), bulk frozen, freeze dried, puree, juice, or concentrate. Most of the processed fruit in this region consists of high-quality, aromatic, small-seeded, trailing cultivars. These are harvested exclusively by over-the-row, self-propelled machine harvesters (Figs. 14.14 and 14.15). The most common harvesters used in blackberry have two, free-wheeling, rotary heads (one per side) containing flexible fiberglass rods. Beater speed (rpm) of these rods (also called 'fingers') can be controlled by the machine's driver. The amplitude of the horizontal movement of the fingers may be adjusted by changing the weights placed on top of the heads. The forward speed of the machine typically ranges from 1.6 to 3.2 kph (1–2 mph). The severity of the

Fig. 14.14. Machine harvest of 'Marion' trailing blackberry, Oregon (B. Strik).

Fig. 14.15. Machine harvest of 'Black Diamond' trailing blackberry, Oregon (B. Strik).

harvest or the intensity of the pick can be adjusted by modifying ground speed, beater speed, and amplitude of the fingers. Growers will pick more lightly at the beginning of the harvest season by having a faster ground speed, light weights, and a slower beater speed. As the harvest season goes on, ground speed will slow and weights and beater speed will be increased. The fingers on the heads vibrate the laterals, causing ripe fruit to fall onto overlapping catcher plates that fit tightly around the plant's base. Fruit roll down the slanted catcher plates and are then carried to a conveyor of cups that carry fruit to the top of the machine. Fruit fall onto a short moving sorting belt after a fan removes leaves, unripe fruit, and other debris. Sorters remove any obvious poor quality fruit before the moving belt drops fruit into shallow plastic lugs. Most machine harvesters are operated with two or three additional workers (other than the driver) who sort fruit and move empty and full lugs onto a holding area on the back of the harvester. Once the holding area is full, the platform on the machine is lowered and the full pallets of fruit are loaded onto a flat-bed truck using a fork lift and are transported to the processing plant.

Plantings are machine harvested about every 5–7 days, depending on cultivar and weather conditions. Harvesting begins when the first fruit are at dull black stage (see Chapter 15) and will release easily with gentle vibration of the fingers on the harvester. Blackberries are often machine harvested at night to improve fruit release. The goal of the grower is to harvest fruit that is of sufficient quality for the highest value processed market, IQF.

Machine-harvested fruit is considered to have high soluble solids and have more uniformity than hand-harvested fruit, as all the fruit are generally ripe. However, quality is affected by cultivar, weather, and pests and other contaminants. The cultivar requirements for machine harvest include fruit that are firm enough to maintain good integrity after machine harvesting (for IQF quality) and have good flavor, small seed size, and color and texture after freezing. The fruit must also release from the plant easily once ripe, but not so easily that immature fruit fall. Fruiting laterals must not be brittle and break during harvesting. Cultivars differ in their suitability for machine harvest (see Chapter 5).

Weather may reduce fruit quality and ease of fruit harvest when machine harvesting. For example, heat and intense light (ultraviolet) may cause fruit to soften or change color, including turning some drupelets white (see Chapter 2); this fruit is unmarketable for high-quality processed markets. Under conditions of high heat, ripe fruit do not release well from the plant. Once fruit become overripe and lose moisture they often remain on the plant. Machine harvesting at night improves fruit release. Plants cannot be machine harvested when fruit are wet from dew, overhead irrigation, or rainfall. Fortunately, it rarely rains during the fruiting season of processed trailing blackberry in Oregon. Growers who use overhead irrigation will schedule such that fruit are not wet for the next harvest.

The best quality is obtained from fields that are 'clean' of insects and fruit fungal problems. Fungal pathogens (e.g. botrytis) and insect pests that infect or damage fruit (e.g. larvae of spotted wing drosophila or leaf rollers) directly reduce fruit quality. However, any insects that are up in the plant canopy during machine harvest may fall into the fruit during machine harvest. Insect contaminants lead to a downgrade of the fruit quality or rejection of the fruit at the processing plant. Growers thus practice good integrated pest management (IPM) programs, using traps and scouts to assess pest management needs.

Plant parts may also be contaminants in the harvested product. Leaves and other plant parts are typically blown out of the machine by the fan. However, thorns can be a serious contaminant in thorny cultivars. Individual thorns may rub off canes or the non-senescent thorny petioles (see Chapter 2) may fall into the fruit during harvest. These thorns can quickly turn the color of fruit from leaked juice and be very difficult to detect. This type of contaminant may be a tremendous liability for processing companies. Growers may use machine harvesters equipped with brushes to remove most of the remaining leaves and petioles on primocanes in late winter and thus minimize the risk of thorn contamination (Strik and Buller, 2002). However, growing thornless cultivars eliminates this risk.

The efficiency of machine harvest (the proportion of total yield harvested by the machine) is affected by cultivar, the machine harvester operator, and the planting design or management. The trellis should be well constructed and weeds controlled to improve performance of the machine. Primocane suppression is a common tool to improve machine harvest efficiency as having the catcher plates fit tight around the plant increases recovery of fruit (see Chapter 12). Despite all of these methods, yields of machine-harvested fields are about 15–20% less than hand-harvested fields, depending on cultivar. In addition to fruit losses on the ground, immature fruit is sometimes inadvertently harvested, laterals may be broken, and canes may be damaged.

Fields need to be designed for machine harvest in advance of planting. Headlands at the ends of rows need to be wide enough for the machine to turn. Plantings should not be on a slope that hampers safe operation of the harvester or being able to maintain a consistent speed. Rows should be straight to improve harvest efficiency and should be labeled so the harvester goes in the same direction down the row each time it is harvested; changing the direction of harvest would lead to lateral breakage. Fields should be sub-soiled to about 1 m (3 ft) deep to combat soil compaction.

Hand harvest

In Serbia, most of the blackberries grown for processing are harvested by hand as are all the blackberries currently grown for fresh market. Fruit are harvested carefully by hand using a breaking motion, at the shiny black (more firm) to dull black stage (see Chapter 15) every 3–5 days, depending on cultivar and

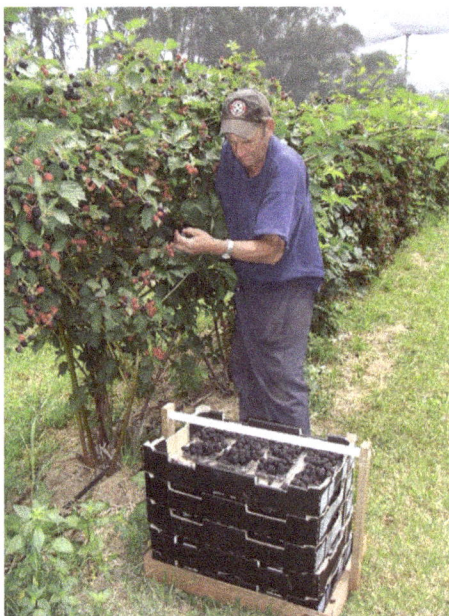

Fig. 14.16. Hand harvest of 'Loch Ness' semi-erect blackberry, Victoria, Australia (B. Strik).

climate. Hand pickers will field pack (into the final containers or clam shells) (Fig. 14.16) or pick into shallow buckets, which will be hand packed in a packing shed.

U-pick (pick-your-own)

In some areas of North America, an alternative harvest method involves the use of customer/unpaid labor in what is commonly called a U-pick operation. In this production style, berries are grown similarly to other production methods, with the main difference coming at time of harvest. During the harvest season, a producer opens up the planting to members of the public who wish to come pick and purchase berries for their own consumption. This may completely replace other harvest methods, or be used as a supplement to them.

There are several advantages to a U-pick operation. First, a producer can harvest the crop with fewer workers, a significant benefit in areas where seasonal labor is not readily available. Another advantage is that retail customers will purchase berries with imperfections that might not be acceptable to wholesalers. Because customers pick fruit they immediately purchase, the producer's need for harvesting and processing equipment, refrigeration, etc., can be greatly reduced. The producer can tailor the number of days the farm is open for picking, thus bringing as many customers as the harvest size demands. Additionally, U-pick customers will pay retail prices for freshly picked fruit.

Customers are drawn to U-pick operations because they value fresh produce, and because they also value the recreational and educational aspects of harvesting their own fruit. Customers often include families with young children, and they are familiar with the concept of U-pick farms.

Possible disadvantages to the U-pick system include berries not picked because they are overlooked by inexperienced pickers, and damage to canes or irrigation components because of pedestrian traffic.

The management of U-pick operations varies slightly from more traditional farms. U-pick operations are most successful when located close enough to cities and major transportation routes to insure sufficient customer base. The farm must have on-site parking to allow for buses or large numbers of cars. Lack of customer access will limit the viability of U-pick operations.

Although U-pick operations can be successful when blackberries are the only product offered for sale, profits are increased when the farm also offers other produce for sale, or uses blackberries to bridge harvest dates of other crops – raspberries, blueberries, or early apples, for example. Conversely, there should not be multiple blackberry U-pick operations within too close a radius.

Management of U-pick operations varies in other ways. Rows may be up to 5 m (16 ft) wide to allow for easier pedestrian movement through the planting. Pesticide sprays for spotted wing drosophila or other pests must be timed to allow for safe public access such as every-Saturday pickings.

Labor needs are also different. Needs of a small-scale U-pick farm might be handled by a few seasonal employees or family members. Large-scale operations may require many workers on harvest days: to direct parking and vehicular traffic, to provide customers with buckets or containers as they arrive; to direct customers to rows where fruit is ripest, or to rope off rows of later-ripening cultivars. Employees should be available to answer questions, to prevent pickers or their children from straying into other parts of the orchard, into farm ponds, outbuildings, etc., and to handle sales.

Facilities will need to include check-out lanes and weighing equipment if berries are not sold by volume. Local laws may require farms to provide bathrooms and potable water. Liability insurance is recommended.

Advertisement for U-pick operations may be necessary at first, but as customers learn of the farm, word-of-mouth 'advertising' may be sufficient. Customers can become very loyal and many return year after year. North American U-pick farms often have websites or post their picking dates and other information on Facebook or other social media outlets (L. Ayres and D. Rouster, personal communication, 2016).

REFERENCES

Bal, E. and Meesters, P. (1995) Year-round production of blackberries. In: *Proceedings of the North American Bramble Growers Association.* North American Bramble Growers' Association, Pittsboro, North Carolina, pp. 49–61.

Demchak, K. (2009) Small fruit production in high tunnels. *HortTechnology* 19(1), 44–49.

Dixon, E.K. and Strik, B.C. (2016) Weed control increases growth, cumulative yield, and economic returns of machine-harvested organic trailing blackberry. *Acta Horticulturae* 1133, 323–328.

Dixon, E.K., Strik, B.C., Valenzuela-Estrada, L.R. and Bryla, D.R. (2015) Weed management, training, and irrigation practices for organic production of trailing blackberry. I. Mature plant growth and fruit production. *HortScience* 50(8), 1165–1177.

Dixon, E.K., Strik, B.C. and Bryla, D.R. (2016a) Weed management, training, and irrigation practices for organic production of trailing blackberry, II. Soil and aboveground plant nutrient concentrations. *HortScience* 51(1), 36–50.

Dixon, E.K., Strik, B.C. and Bryla, D.R. (2016b) Weed management, training, and irrigation practices for organic production of trailing blackberry. III. Accumulation and removal of aboveground biomass, carbon, and nutrients. *HortScience* 51(1), 51–66.

Fernandez-Salvador, J., Strik, B.C. and Bryla, D.R. (2015a) Liquid corn and fish fertilizers are good options for fertigation in blackberry cultivars grown in an organic production system. *HortScience* 50(2), 225–233.

Fernandez-Salvador, J.A., Strik, B.C. and Bryla, D. (2015b) Response of blackberry cultivars to fertilizer source in an organic fresh market production system. *HortTechnology* 25(4), 277–292.

Fernandez-Salvador, J., Strik, B.C., Zhao, Y. and Finn, C.E. (2015c) Trailing blackberry genotypes differ in yield and post-harvest fruit quality during establishment in an organic production system. *HortScience* 50(2), 240–246.

Harkins, R.H., Strik, B.C. and Bryla, D.R. (2013) Weed management practices for organic production of trailing blackberry. I. Plant growth and early fruit production. *HortScience* 48(9), 1139–1144.

Harkins, R.H., Strik, B.C. and Bryla, D.R. (2014) Weed management practices for organic production of trailing blackberry. II. Accumulation and loss of biomass and nutrients. *HortScience* 49(1), 35–43.

Stanton, M.A., Scheerens, J.C., Funt, R.C. and Clark, J.R. (2007) Floral competence of primocane-fruiting blackberries Prime-Jan and Prime-Jim grown at three temperature regimens. *HortScience* 42(3), 508–513.

Strik, B.C. (2016) A review of optimal systems for organic production of blueberry and blackberry for fresh and processed markets in the northwestern United States. *Scientia Horticulturae* 208, 92–103. Available at: www.sciencedirect.com/science/article/pii/S0304423815303186 (accessed May 12, 2017).

Strik, B. and Buller, G. (2002) Reducing thorn contamination in machine-harvested 'Marion' blackberry. *Acta Horticulturae* 585, 677–681.

Strik, B.C. and Finn, C.E. (2012) Blackberry production systems – a worldwide perspective. *Acta Horticulturae* 946, 341–347.

Strik, B.C., Clark, J.R., Finn, C.E. and Bañados, P. (2007) Worldwide production of blackberries, 1995 to 2005 and predictions for growth. *HortTechnology* 17(2), 205–213.

Thompson, E., Strik, B.C., Finn, C.E., Zhao, Y. and Clark, J.R. (2009) High tunnel vs. open field: management of primocane-fruiting blackberry using pruning and tipping to increase yield and extend the fruiting season. *HortScience* 44(6), 1581–1587.

USDA (2010) Table 6: Organic berries harvested from certified and exempt organic farms: 2008. In: *Organic Production Survey (2008), 2007 Census of Agriculture*. U.S. Department of Agriculture, National Agriculture Statistical Service, Washington, DC.

USDA (2014) *Census of Agriculture, Organic Survey (2012). Volume 3, Part 4*. U.S. Department of Agriculture, National Agriculture Statistical Service, Washington, DC.

15

POSTHARVEST STORAGE AND TRANSPORT OF BLACKBERRIES

Penelope Perkins-Veazie*

North Carolina State University, Kannapolis, North Carolina, USA

INTRODUCTION

Blackberries are made up of many individual fruits (drupelets) held together by hairs (trichomes) and waxes, and attachment of drupelet bases to a central core of tissue (torus or receptacle) (Tomlik-Wyrenblewska et al., 2010). Blackberries detach from the pedicel with their central receptacle in place, which adds considerable firmness to the fruit compared to raspberries. However, blackberries are considered highly perishable because of lack of a protective barrier, high respiration rate, production of ethylene, susceptibility to decay, and lack of sugar reserves once detached from the plant.

Blackberries lose weight rapidly when moisture moves from the fruit into the surrounding drier air, as the fruit has no protective rind or cuticle. Weight loss is further aggravated by a high respiration rate, which also uses the sugars and organic acids in the fruit as a carbon source. Ethylene, a ripening hormone that often causes softening, is produced in the fruit at a low rate, and contributes to perishability, especially at temperatures above 5°C (41°F). Lastly, ripe blackberries are highly susceptible to gray mold (*Botrytis cinerea*). This fungus infects flower stamens and then appears inside the fruit following harvest (Fig. 15.1). Gray mold growth is hastened at storage temperatures above 5°C (41°F), especially if the fruit has been grown outdoors and has been subjected to dew or rainfall. Gray mold is much less of a problem if the fruit is picked from plants grown in a tunnel or a greenhouse.

As they ripen, blackberries change color from red or coffee to partly black then to solid black. The development of full black color can be very rapid, if air temperatures are warm (>30°C (86°F)), in less than a day. Blackberry color change undergoes a subtle shift as the fruit becomes overripe, with the fruit becoming more dull looking and softer (dull black stage) (Fig. 15.2). Once

---

* Corresponding author: Penelope_perkins@ncsu.edu

Fig. 15.1. Blackberries showing well established mold from *Botrytis cinerea*. Mold spreads quickly from one berry to the next (P. Perkins-Veazie).

Fig. 15.2. Shiny black (left) and dull black (right) blackberries. Note the more matte appearance of the dull black berry (P. Perkins-Veazie).

blackberries reach the dull black stage, they should be used only for local markets or for processing as further softening, juice leakage, and decay are rapid at this stage.

Steps for best blackberry postharvest life start long before the final product is ready. One of the first decisions to make is market selection. Fruits destined for commercial or long-distance markets are often different than those used for direct or local markets. Commercial marketing will require considerable thought and attention to refrigeration, storage, cold chain, and marketing, and use of cultivars with longer shelf life. In contrast, direct marketing requires attention to timing of availability of fruit, large berry size, high productivity, ease of harvesting, and good flavor.

SELECTION OF CULTIVARS FOR TARGETED MARKETS

Blackberries destined for long-distance transport to fresh markets must be firm, have some appearance of gloss, be a uniform black color, and not appear shriveled. Blackberries grown for fresh market have to be able to retain firmness during harvest, handling, and storage and have a shelf life of 14–21 days (Perkins-Veazie *et al.*, 1997, 1999a). Blackberry cultivars differ in inherent firmness and fruit must be picked when fully black as partially red fruit may develop black color but lack normal sweetness and texture (Perkins-Veazie et al., 1996). New blackberry genotypes referred to as 'crisp' have been selected in the Arkansas breeding program. These retain a very firm texture even when fully ripe (Clark and Salgado-Rojas, 2014).

Keeping fruit firm is easier when harvested under cool, low-humidity conditions. After harvest, blackberries should not be left in direct light and when conditions are warm, it is essential to frequently take harvested fruit and place them into a cool store (cold room or refrigerated walk-in cooler) for field heat to be removed. Fresh-market blackberries are harvested at a less ripe stage than fruit designated for processing, and generally retain more firmness than processing types even when overripe. Blackberries suitable for fresh-market use include cultivars from public breeding programs in the USA at the United States Department of Agriculture (USDA) in Beltsville, Maryland and Corvallis, Oregon, the University of Arkansas and North Carolina State University, at the James Hutton Institute in Scotland, Embrapa in Brazil, Niwa in New Zealand, Institute of Horticultural Research, Poland, and the Institute of Plant and Food Research in New Zealand. Driscoll's Berries, Plant Sciences, and Naturipe have also developed proprietary material in the USA. The Brazilian cultivar Tupi or Tupy, grown in Mexico, supplies most of the USA fresh blackberry demand from November to May (Nunes et al., 2014). In Europe, these cultivars are also used either directly or under license, and there are also a number of other breeding programs associated with commercial producers developing high-quality cultivars for their proprietary use. In the public domain, some of the better fresh-market cultivars grown in warmer locations are 'Navaho,' 'Ouachita,' 'Von,' and 'Prime-Ark® 45'; (Clark and Perkins-Veazie, 2011; Fernandez et al., 2013). 'Chester Thornless' and 'Obsidian' have better fruit quality when grown in cooler locations and are very important fresh-market cultivars in Oregon. 'LochNess' from Scotland was widely used in commercial fresh-market plantings in the USA and Europe in 2015.

Local or regional markets include direct (farmers market or roadside stands), U-pick (customer comes to the farm to harvest fruit), or community supported agriculture (CSA), where customers pay before the season for a specific amount of fruit delivered to them. These customers often prefer a berry high in sweetness but a less firm berry can be used because of a shorter interval from harvest to consumer. 'Triple Crown,' 'Navaho,' and 'Von' have consistently high sweetness; the berries of 'Apache,' 'Kiowa,' and 'Natchez' are very large and attract U-pick customers.

In contrast to fresh market, variety selection for processing depends on flavor, the sugar to acid balance, and good texture or low drip loss. For years, the unique flavor of 'Marion' has been the 'gold' standard for USA processing berries. This blackberry has a highly aromatic flavor that some think is a mixture of raspberry and blackberry and small seeds that are not noticeable in the processed product. Developing processing cultivars that are more productive, less cold-sensitive, thornless, and yet have this unique flavor remains a high priority in the Pacific Northwest.

Blackberries produce small amounts of ethylene as they ripen. The amount of ethylene released varies greatly with cultivar (Burdon and Sexton, 1993; Walsh et *al.*, 1983), and does not appear concomitantly with changes in respiration or anthocyanin accumulation. Although application of ethephon or ethylene to immature blackberries will hasten anthocyanin development and fruit abscission, other ripeness indicators, such as softening and sweetness, do not reach the levels of blackberries harvested mature (Burdon and Sexton, 1993, 1994). Blackberries differ greatly in storage life and relative postharvest problems primarily due to genetic differences, followed by production environment and postharvest handling. The amount of ethylene produced in the blackberry is at least partly related to postharvest shelf life (Walsh et *al.*, 1983; Perkins-Veazie et *al.*, 1996).

Color reversion, or red drupelet disorder, is the appearance of red drupelets on previously black fruit (Fig. 15.3) under cool storage or freezing. Unlike partially ripe berries, where immature drupelets are visibly red at harvest and often hard, color reversion is seen after fully black fruit are pulled out of cold storage. Color reversion appears to be a disorder rather than a disease, with red drupelets having a slightly lower pH than black drupelets, and about half as much pigment. This indicates that some physical damage may have occurred inside the drupelet. Some blackberry cultivars, such as 'Navaho,' rarely show this disorder, while others, such as 'Tupi,' can have as many as 20–30% of berries with color reversion. Various theories for the physical cause of this

Fig. 15.3. Normal black color (*left*) and red drupe or color reversion (*right*) on blackberry (P. Perkins-Veazie).

reversion include excess nitrogen during production, bruising, or cold damage of berries too close to forced air coolers. Ultimately, the breaking of cell walls or vacuoles within the drupelet allows the pigment to be exposed to more acid conditions and color change in response to the lower pH (Clark *et al.*, 2007).

HARVESTING

Fresh market

Ease of blackberry fruit abscission is critical to avoid bruising during fresh market harvest. Blackberries designated for fresh market are picked at a shiny to dull black stage, but while still firm. For fresh or direct type markets, blackberries are gently snapped from the receptacle and placed directly into packages in the field. Unlike tree fruit, fresh-market blackberries are handled only once, from the plant into the final container, and are not washed.

Because blackberries are soft, have high respiration, and produce ethylene, harvesting for fresh markets is best done when air temperatures are cool. For commercial operations in the USA, blackberries are picked quickly into plastic clamshells (Fig. 15.4), placed in master containers (cardboard cartons with holes in the sides for forced air cooling and reinforced corners for stacking) (Fig. 15.5), and then taken to a quality-control area to check for USDA standards compliance. USDA standards include few or no unripe fruit or plant debris such as stems, leaves, or calyxes, no mold or rust, no visible insects, and a generally uniformly colored and sized pack. Once passed, cartons are taken to cooling facilities, either to refrigerated trucks in place in the field or to nearby coolers (walk-in refrigerators). Some growers have found that all-terrain vehicles or golf carts can be used for quick trips to the cooler with small loads (Fig. 15.6).

Processing

Sometimes fresh-market growers will process berries that are too ripe or soft for fresh sales, using overripe fruit from picking plants clean, enabling recovery of part of the production costs. In the USA, most of the acreage grown for processing is in Oregon and is harvested by machines (see Chapter 14). Only ripe fruit at the dull black stage is easily dislodged by machines from cultivars adapted to processing and machine harvest. Proper operation of the machine, including ground speed and vibration speed of the harvesting rods, pest management, and frequency of fruit harvest are important to achieve good quality for processing. Full flats of harvested fruit must be unloaded from the machine harvester frequently enough to prevent overheating of fruit. Machine harvest is typically done at night in Oregon to maintain good fruit quality and to facilitate fruit release from the plant. Fruit are immediately transported to processing plants where further sorting and cleaning occurs.

Fig. 15.4. Clamshell boxes showing holes (*left*) or slits (*right*) on top of boxes for added ventilation and cooling, and slightly elevated top of clamshell to allow for stable stacking (P. Perkins-Veazie).

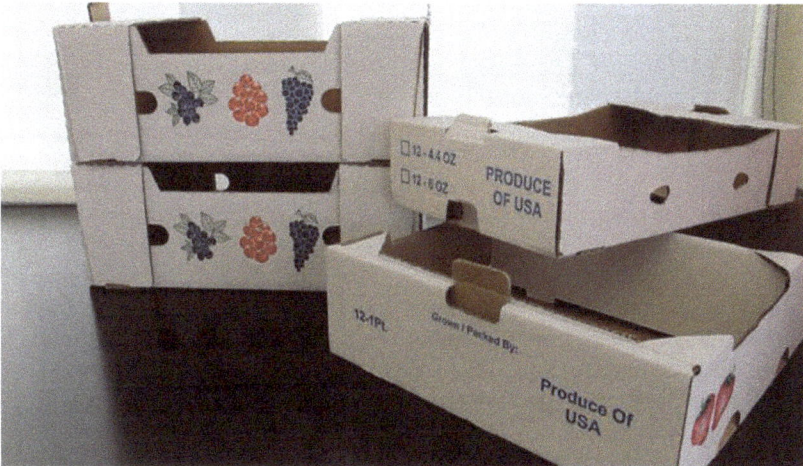

Fig. 15.5. Cardboard containers (often called 'masters' in the USA) showing reinforced corners, side vents for cooling, and interlocking tabs to help keep containers stabilized in palette stack (P. Perkins-Veazie).

Fig. 15.6. Transport systems for berries to packhouses and coolers, using all-terrain vehicles or golf carts. For larger acreages, a self-contained quality control shed is used and berries are loaded into a refrigerated truck to keep fruit cold to the packhouse (P. Perkins-Veazie).

Training for hand harvesting

When harvesting by hand for fresh, U-pick, or other markets, people have to be trained to pick the fruit at the desired stage of ripeness. Long-distance shippers desire fruit that are fully black and firm, while U-pick growers often want black-berries with a small amount of softness, yet not too soft, for highest sweetness. Picking plants clean of ripe and overripe fruit is critical in both types of opera-tions in order to decrease spread of disease and maximize returns. All pickers need to be shown how to remove fruit without breaking off vegetative shoots or harvesting partially black berries. Care must be taken not to grasp fruit too hard to avoid bruising or leakage, and injured, moldy, or discolored berries should never be placed in packs designated for fresh market, and moldy berries should not be harvested for processing. Berries are usually picked into cups or clamshells in the field, and should be placed gently in the container, not dropped. Blackberries generally are not packed more than three layers deep in a container to avoid crushing the bottom fruit. Pickers must also avoid overfilling clamshells, as snapping the hinged lid on overfull boxes will cut or crush the fruit.

Sanitation

Good agricultural practices (GAPs) enter into postharvest decisions for all types of markets, as well as in decisions for production areas. Blackberry fruit must be clean when picked as washing of fresh-market fruit will not be done. Animal entrance into fields and packing houses needs to be none to minimal, and canes may need to be trimmed or workers trained to avoid picking fruit close to the soil line. Irrigation must be with potable water, and preferably applied in a way to avoid wetting berries. Sanitary areas, including washing facilities, need to be available for customers or professional help and within easy access of plots being harvested. If drive areas in fields are dusty, then materials brought in or out of the field should be kept covered. Packing areas, box storage, and cooling facilities should be kept clean. Boxes should be covered with tarps or held in closed areas to avoid dust, birds, and insect contamination. Cool storage (refrigerated) rooms should be swept and washed down as needed to minimize debris or mildew growth. Packing sheds need to be swept to avoid dust.

A more comprehensive list of guidelines and procedures to maximize food safety can be found at www.gaps.cornell.edu and http://ucfoodsafety.ucdavis.edu/Preharvest. The introduction of the food safety modernization act in the USA highlights additional food safety issues that may need to be addressed in addition to GAPs. Some of these can be found at http://ucfoodsafety.ucdavis.edu/Postharvest_Produce. A summary list of GAPs guidelines adapted from Kader (2002) is given below.

1. Be careful when using animal or biosolids waste to avoid microbial contamination (see USDA 2011 for correct compost temperatures and application times).
2. Follow all applicable laws, regulations, and guidelines for agricultural practices.
3. Make sure all sources of water that might come into contact with blackberries at any point are safe, including irrigation.
4. Avoidance of microbial contamination is much more effective than trying to remediate contamination.
5. Blackberries can become contaminated anywhere in the food chain from field to table. The major sources of microbial contamination are associated with human or animal feces.
6. Worker hygiene and sanitation practices in all aspects of production, harvest, storage, handling, transport, and marketing is vital to minimize microbial contamination.

Packaging

Over the years, fresh-market packages for blackberries have consisted of buckets, plastic, wood, paper, or pulp (composite) boxes, and more recently clamshells. Blackberry container size is generally kept at a 170–340 g (6–12 oz) size. Occasionally, a large (454 g (16 oz)), shallow box is used for the restaurant trade. In the USA, plastic clamshells without sharp internal edges or corners, especially around ventilation holes, are the most preferred for best shelf life (Fig. 15.4). The clamshells are plastic boxes with an attached lid, with small air holes on the bottom and top of the box and are provided with an absorbent pad in the bottom to soak up any juice leaking from the berries. Clamshells have an extra advantage because of the extra rigid structure in the corners that allow several boxes to be stacked without crushing fruit. Clamshells are designed to be placed into cardboard containers (master containers) made to hold 12 or 24 clamshells (total of 1.5–3.1 kg (3.3–6.9 lb)) per master. Masters have additional rigidity and special vents for forced air or room cooling (Fig. 15.5).

Growers using fruit for farmers' markets or CSAs in the USA may harvest into 275 or 550 g (½ or 1 pint) pulp baskets, bamboo baskets, or cardboard baskets. Use of wood or composite boxes is sometimes used in direct marketing to accentuate fresh appearance. These containers present several problems for blackberries. They tend to remove water from fruit, which may be beneficial if fruit are leaky, but also leave the stains on the package. Ventilation for efficient cooling is limited, especially in wood splint boxes, and the sharp edges of wood splint boxes can easily cut the berries caught in the corners. A wrap is usually placed over the top of containers to prevent weight loss and held in place with a rubber band, requiring hand labor.

All containers should be recycled or disposed of rather than reused for blackberries or other produce. Mold spores readily enter the cardboard of masters and composite boxes, and good agricultural practices for sanitation should be followed to prevent transfer of soil, dirt, or microorganisms.

Processing fruit are packed into plastic lugs or crates that can hold 10–20 kg (22–44 lb) fruit per container. These crates are stackable, easily cleaned, and contain holes on the bottom to aid in removal of fruit heat and to facilitate good washing.

Cooling and storage

Maintaining the cold chain of rapid cooling followed by constant cold storage remains the most effective way to prolong blackberry shelf life. Blackberries can be held up to 21 days under ideal conditions. These conditions include selection of a firm cultivar, proper harvest, handling, cooling, and storage protocols, and maintenance of the cool chain through transit and distribution. The best temperature for blackberries is at −0.5°C (31°F), actually slightly below freezing. The high sugar and acid content of blackberries acts as an anti-freeze, keeping berries from solidifying. Respiration and ethylene production are greatly reduced at 0–5°C (32–41°F) (Table 15.1). Berry firmness is retained longer at a temperature closer to 0°C (32°F) (Nunes, 2008), and gray mold spore germination and mycelia growth are slowed at temperatures below 5°C (41°F). As storage temperature increases, shelf life is greatly shortened. If pre-cooling is delayed, shelf life is even shorter, regardless of subsequent storage temperature. If fruit are warmed to room temperature after proper cooling and cold chain management, the percentage of marketable blackberries are reduced as much as 50% (Perkins-Veazie et al., 1999b).

Blackberries have best shelf life if cooled to 2°C (36°F) within an hour of harvest, and if kept between −0.5° and 0°C (31° and 32°F) and 95% relative humidity during storage. Only air cooling (not water submersion, icing, or hydrocooling) is recommended for blackberries. This means that in almost all production areas, some sort of cooling system and/or cold room will be needed.

Cooling of produce depends on the density in the container, the vent and types of containers, the volume to surface area, the travel distance of cooling air (shorter distance means faster cooling), and airflow capacity (more airflow means faster cooling) (OMAFRA, 2016). Blackberries packed in clamshells and vented containers have low density, a uniform cooling surface, and a high surface area to volume (blackberries are small relative to apples), and there-fore, will take less cooling time than the same weight of apples.

Room cooling is generally done in cold rooms, and is effective if small amounts of fruit are harvested and the room is large enough to have good air circulation. Containers having well placed vents should be placed in rows with at least 30 cm (12 in) between rows, and have box fans placed to direct cold air through the boxes and tunnels, creating a serpentine cooling system (Fig. 15.7). Often efficient cooling is decreased by frequent door opening; cold

Table 15.1. Estimated blackberry shelf life, assuming 90–95% relative humidity. (Adapted from Perkins-Veazie, 2016.)

Storage temperature (°C)	Shelf life (Days)	Respiration rate (mg CO_2 kg^{-1} h^{-1})
−0.5	12–21	Not available (na)
0	10–14	18–20
1	8	na
2	6	na
3	5	na
4–5	4	31–41
10	2	62
15	1.5	75
20	1	100–130

Fig. 15.7. Palettes of berries for fresh market showing side vents in cardboard master containers, covers on top cartons, and room cooling with fans and placement of palettes (P. Perkins-Veazie).

air can quickly become replaced by warm air. Plastic strips placed over the doorway help block loss of air, as well as entrance of insects, birds, or other animals.

Cooling coils of the refrigeration system in a cold room have to be cooler than the air. If the temperature difference is large, heat transfer rate is greater and smaller cooling coils can be used (Boyette *et al.*, 1991). However, as coils cool, more water vapor from air will condense on coils as ice or moisture. This is a big concern when air coming into the room is hot and humid, forcing refrigeration to cool air and condense water vapor; sometimes the amount of ice from water vapor on the coils can stop refrigeration. Also, with blackberries, higher humidity is needed. In this case, the best system would be large evaporator coils and small temperature drops across the coils (OMAFRA, 2016), by using a forced air cooler inside a refrigerated room.

Cold rooms can be purchased from commercial contractors, built by growers, or recycled from other industries (Thompson and Spinoglio, 1996). New commercially available cold rooms are expensive; costs can be saved by buying used rooms or by grower construction. Prefabricated cold rooms are less costly, often used in restaurant industries, and can be assembled fairly easily, but must be enclosed inside another structure. The marine containers used on ships make good cold rooms, as they are well insulated and generally have refrigeration controls added. If rooms are constructed by growers, proper insulation is critical to maintain effective temperatures, with moisture resistance (styrofoam or polyurethane) plus spray-in foam giving additional insulation (Boyette *et al.*, 1991; http://storeitcold.com). All cold rooms require effective refrigeration units, with the size of unit dependent on room size and heat load (Btu or J/h).

Room cooling is not effective at rapidly removing field heat (the amount of heat in fruit when it comes in from the field), especially in warm temperatures and with many loads coming into the room over a day. Cold air from evaporator coils goes over and around the blackberries and slowly removes heat (Thompson *et al.*, 2002). Minimum airflow needs to be 100 cfm (cubic feet per minute)/ton or 0.3 m^{-3}/ton of product storage capacity (Thompson *et al.*, 2002). Since blackberries do not tolerate moisture, the most effective means of cooling is to use forced air to rapidly remove field heat.

Forced air cooling should be done inside a cold room. Once blackberries are cooled, they should be moved to another cold room to maximize space for the forced air system. Forced air cooling can be done by building a cold wall or by building home-made systems (Thompson and Spinoglio, 1996; Fraser, 2014). A very simple system is to place a box fan tightly against one end of the tunnel made of two rows of containers or palettes, so that the box fan is pulling cold air from near the room fans, place a tarp over the cartons at the other end, and push warm air out of the cartons into the room utilizing the negative pressure created by the tunnel. Exhaust air from the fans should be moved away from room cooler coils, and the pallets nearest the fan housing should be pressed into foam strips to create an air seal.

With large loads of blackberries, using a cold wall is more efficient. Here, a wall is built about 1 m (3 ft) from the cooler wall and set points are cut in the wall to place pallets of fruit. A maximum fan rate of 0.001–0.002 m^{-3}/s.kg (1–2 cfm/lb) is advised; moving air volumes above this speed increases the static pressure and energy consumption without greatly affecting cooling rates (Thompson *et al.*, 2002) (Fig. 15.8). Forced air cooling can quickly dry out fruit, especially at the top of the pallet, so a layer of cardboard is often placed over these top cartons to block airflow over the top of the fruit.

Blackberries have to be held under high humidity in order to reduce weight loss and avoid shriveling, yet have to be held below 100% relative humidity to avoid condensation or mold growth. Containers or covers that cut down on the surface area of exposed blackberries help hold in humidity once fruit are fully

Fig. 15.8. Palettes of blackberries placed for forced air cooling using a false wall plenum for pushing out air (P. Perkins-Veazie).

cooled. Effective methods can be as simple as a clean plastic tarp over pallets or boxes. Fine mist systems can be built into cold rooms to aid humidity. In some older rooms where wood boards were used for floor or wall construction, wet, clean coarse cloth, such as burlap, can be placed on the floors. Controlling air-flow during cooling and keeping cooling times within three hours also helps reduce weight loss (Fraser, 2014).

Calculating field, amount of cooling time, and refrigeration needed

The specific heat of a fruit depends on composition (amount of water), is in units of heat per weight per unit temperature, and represents the amount of heat needed to raise a unit of mass by 1°C. Blackberries are 88% water, so specific heat is usually rounded up to 1. Specific heat (kJ/kg/°C) = weight (kg) × difference in temperature (starting temperature − desired temperature (°C)) × 4.186 kJ, or 1 Btu of heat must be removed to cool one pound (454 g) of blackberries by 1°F (0.55°C).

To calculate the field heat in a load of blackberries, the following formula is used.

Field heat (kJ/h) = SH (fruit specific heat (kJ/kg/°C)) × DT (difference in field temperature (°C) − desired temperature (°C)) × W (weight (kg)).

Example:

SH: 1.0
DT: 27.8 pulp temperature of blackberry is 27.8°C (82°F) and desired tempera-ture is 0°C (32°F),
W: 171 kg weight of load is 171 kg (381 lb) for one pallet of 90 containers, each containing 12 clamshells of 160 g (half pints)
then the amount of field heat to remove is 19,899 kJ/h (1 × 27.8 × 171 kg × 4.186) or 19,050 Btu/h (1 × 50 × 381).

To cool this load in 12 hours, 1.658 kJ (19,899/12) or 1.588 tons (19,050/12) need to be removed each hour.

The amount of refrigeration capacity (RC) = Btu/h from above/refrigera-tion per hour.

One ton of refrigeration is 303,840 kJ/24 h (288,000 Btu/24 h) or 12,660 kJ/h (12,000 Btu/h). So, the amount of refrigeration capacity will be 19,050 Btu/h divided by 12,000 Btu/h ton or 1.59 ton refrigeration unit running constantly.

If more fruit is added to the cooler then the additional heat load must be calculated, plus the respiration rate of both uncooled fruit and cooled fruit must be added in (Boyette *et al.*, 1991). To reduce the amount of refrigeration needed in a room cooler, one can instead partially cool fruit with forced air then place the fruit in the cold room.

For forced air cooling of blackberries, the recommended airflow rate is 3853 cms/kg (cubic meter/s/kg) or 4 cfm/lb (cubic feet/min/lb) for one hour cooling to reach 7/8 of the final temperature, or 1204 cms/kg (1.25 cfm/lb) for 2.5 hours to reach 7/8 of the final temperature (Fraser, 2014).

COOL CHAIN

The ideal way to move blackberries to markets is to have precooled refrigerated trucks dock directly against loading docks, with doors opening into a cooled landing area with direct access to cool rooms. Pallets of cold fruit are moved by forklift into the trucks without being exposed to warmer air. Pallets need to be stacked correctly to allow for good air circulation from the top and sides of the truck. Refrigeration should be held at 2–4°C (36–40°F) or lower throughout the journey which may last 3–7 days. Trucks should be unloaded at distribution points in the same fashion, opening directly into cold rooms or refrigerated reception areas. To insure that nothing goes wrong with this system, larger producers often own and operate their own refrigerated trucks and delivery vehicles for delivery to the point of sale.

Trucks need to be adequately precooled before being loaded, and shipment temperatures need to be monitored through to the receiving agent. Refrigeration in trucks needs to be well serviced and refrigerant loading kept within design specifications; any refrigerant leaks need to be repaired without delay. Loads must be well secured and cooling needs to be adequate to accommodate delays at truck checkpoints or in dense traffic. Temperatures in trucks must not be allowed to rise upon arrival. If temperatures range from 5–8°C (41–46°F) in the freight area upon arrival, depending on placement, then care should be taken to rapidly sell fruit received at higher temperatures as quality will have been compromised.

Shipping mixed commodity loads is common in some parts of the USA, especially in areas where only a few palettes of blackberries are needed at a distribution point. In these cases, fruit or vegetables, such as tomatoes or watermelon, that need warmer temperatures should not be shipped with blackberries. Breaking the cool chain can reduce expected shelf life of black-berries by two or more days, and in extreme cases, can cause load rejection. Additionally, ethylene generating fruit, vegetables, or flowers should not be in mixed loads with blackberries, or ethylene scrubbers should be added to packs or to the truck. Vibration injury that causes fruit to move up and down or side to side will quickly cause berry leakage. Trucks should have balanced tires and minimal trailer vibration. Smooth roads, without potholes and sudden dips to the highway, help decrease transit injury, as does a lower speed on gravel, dirt, or poorly kept roads.

EXTENSION OF SHELF LIFE

Germplasm, incorporating significant fruit firmness and slow or minimal softening, has been a highly effective means of increasing shelf life (Clark et al., 2007). New blackberry varieties/cultivars should be tested for their shelf life and suitability for transportation as some components of these traits in blackberries are not immediately obvious. Other details on shelf life can be found in Clark et al. (2007).

Increased heat load shortens shelf life through increased respiration and increased rate of softening and decay. Simple methods to reduce heat load include using light-colored packaging, placing blackberries in shaded areas or providing shade, and making frequent trips to coolers. Some growers harvest at night to reduce the heat load. Rapid cooling, refrigeration, maintenance of cold chain, and suitable packaging are the most commonly used and successful postharvest methods for extending shelf life.

Use of ethylene or ethylene-promoting substances and conditions should be avoided, as it can promote undesirable softening and mold growth. Ethylene scrubbers can be added to storage units as air filters or sachets. These are usually potassium permanganate pellets that react with ethylene gas and can be monitored for loss of activity by color change from purple to brown. Activated charcoal may also be used. It will scrub carbon dioxide as well as ethylene. Large storage rooms over $425\,m^3$ (over $14,500\,ft^3$) may benefit from photo-catalytic units that use ultraviolet light to break down ethylene.

Modified atmosphere (MA) helps slow decay and ripening changes in stored blackberry. Blackberries held under 10–15% CO_2 had 15% less decay and about 10% less monomeric anthocyanin after 14 days than control fruit (Perkins-Veazie and Collins, 2002). Agar et al. (1997) reported that controlled atmosphere (CA) storage helped maintain vitamin C content in blackberries during 16 days of storage. While modified atmosphere storage offers a means to extend blackberry shelf life by a few days, especially if used during long periods of transport, it does not substitute for proper cooling and refrigeration and the extra costs associated with MA must be considered. Storage of blackberries in ozone (0.1–0.3 ppm) helped reduce decay and anthocyanin production up to 12 days (Barth et al., 1995).

SUMMARY

In spite of all of this attention and care throughout harvest and handling, consumers end up being a major cause of blackberry quality loss. Blackberries may be purchased in ideal shape then held in a hot vehicle for an hour or two, or placed on a kitchen counter overnight. Mold spores present in the blackberry wait for ideal conditions to germinate. Consumers may find their mouth-watering box of blackberries covered with gray mold within a few days.

The key for successful postharvest life in the fresh-market blackberry lies in decisions made long before harvest. Proper cultivar selection for the desired market needs to be determined, with large, fully black yet firm fruit essential for long shipping distances. Production environment is very important in determining labor and cooling needs for fresh markets. When day temperatures exceed 30°C (86°F) for 6 hours or more, blackberries can become too soft within a day. Fresh-market berries must be harvested gently by hand into the appropriate container, given rapid cooling to slow softening and color change, and held at cold temperatures throughout the shipping and marketing steps.

REFERENCES

Agar, I.T., Streif, J. and Bangerth, F. (1997) Effect of high CO_2 and controlled atmosphere (CA) on the ascorbic and dehydroascorbic acid content of some berry fruits. *Postharvest Biology and Technology* 11(1), 47–55.

Barth, M., Zhou, C., Mercier, J. and Payne, F. (1995) Ozone storage effects on anthocyanin content and fungal growth in blackberries. *Journal of Food Science* 60(6), 1286–1288.

Boyette, M.D., Wilson, L.G. and Estes, E.A. (1991) *Design of Room Cooling Facilities: Structural and Energy Requirements AG-414-2*. Raleigh: North Carolina State University. Available at: http://agris.fao.org/agris-search/search.do;jsessionid=36DD C05EBE7CF7DCC0A60F8892B972DA?request_locale=es&recordID=US201301 743511&sourceQuery=&query=&sortField=&sortOrder=&agrovocString=&adv Query=&centerString=&enableField= (accessed September 20, 2017).

Burdon, J.N. and Sexton, R. (1993) Fruit abscission and ethylene production for blackberry cultivars (*Rubus* spp.). *Annals Applied Biology* 123, 121–132.

Burdon, J.N. and Sexton, R. (1994) Practical implications of differences in the ethylene production of *Rubus* fruits. *Acta Horticulturae* 368, 884–892.

Clark, J.R. and Perkins-Veazie, P. (2011) 'APF-45' primocane-fruiting blackberry. *HortScience* 46, 670–673.

Clark, J.R. and Salgado-Rojas, A. (2014) Determination and inheritance of firmness and texture of the 'crispy' trait in the Arkansas blackberry breeding program. *Southern Region Small Fruits Research Consortium Final Report*.

Clark, J.R., Stafne, E.T., Hall, H.K. and Finn, C.E. (2007) Blackberry breeding and genetics. *Plant Breeding Reviews* 29, 139–144.

Fernandez, G., Ballington, J. and Perkins-Veazie, P. (2013) 'Von' thornless blackberry. *Hortscience* 48(5), 654–656.

Fraser, H. (2014) Forced-air cooling systems for fresh Ontario fruits and vegetables. Factsheet 14-039. Ontario Ministry of Agriculture, Food and Rural Affairs (OMAFRA). Available at: www.omafra.gov.on.ca/english/engineer/facts/14-039.htm (accessed May 11, 2017).

Kader, A.A. (2002) Postharvest biology and technology: an overview. In: Kader, A.A. (ed.) *Postharvest Technology of Horticultural Crops*, 3rd edn. Publication 3311. University of California, Oakland, California, pp. 39–47.

Nunes, M.C.N. (2008) Blackberry. In: Nunes, M.C.N., *Quality of Fruits and Vegetables*. Blackwell Publishing, New York, pp. 139–146.

Nunes, M.C.N, Nicometo, M., Emond, J.P., Melis, R.B. and Uysal, I. (2014) Improvement in fresh fruit and vegetable logistics quality: berry logistics field studies. *Philosophical Transactions of the Royal Society A*, 372. Available at: http://rsta.royalsocietypublishing.org (accessed May 11, 2017).

OMAFRA (2016) Forced-air cooling systems for fresh Ontario fruits and vegetables. Ontario Ministry of Agriculture, Food and Rural Affairs. Available at: http://www.omafra.gov.on.ca/english/engineer/facts/14-039.htm (accessed 29 August 2017).

Perkins-Veazie, P.M. (2016) Blackberry. USDA Handbook 66, p. 237. Available at: https://www.ars.usda.gov/ARSUserFiles/oc/np/CommercialStorage/CommercialStorage.pdf.

Perkins-Veazie, P.M. and Collins, J.K. (2002) Quality of erect-type blackberry fruit after short intervals of controlled atmosphere storage. *Postharvest Biology and Technology* 25(2), 235–239.

Perkins-Veazie, P. M., Collins, J.K. and Clark, J.R. (1996) Cultivar and maturity affect postharvest quality of fruit from erect blackberries. *HortScience* 31(2), 258–261.

Perkins-Veazie, P.M., Collins, J.K., Clark, J.R. and Risse, L. (1997) Air shipment of 'Navaho' blackberry fruit to Europe is feasible. *HortScience* 32(1), 132.

Perkins-Veazie, P.M., Collins, J.K. and Clark, J.R. (1999a) Cultivar and temperature effects on the shelflife of blackberry fruit. *Fruit Varieties Journal* 53(4), 201–208.

Perkins-Veazie, P.M., Collins, J.K. and Clark, J.R. (1999b) Shelflife and quality of 'Navaho' and 'Shawnee' blackberry fruit stored under retail storage conditions. *Journal of Food Quality* 22(5), 535–544.

Thompson, J. and Spinoglio, M. (1996) *Small-scale Cold Rooms for Perishable Commodities. Publication 21449*, University of California, Davis, California, pp. 1–9. Available at: http://ucce.ucdavis.edu/files/datastore/234-701.pdf (accessed May 11, 2017).

Thompson, J.F., Mitchell, F.G. and Kasmire, R.F. (2002) Cooling horticultural commodities. In: Kader, A.A. (ed.) *Postharvest Technology of Horticultural Crops*, 3rd edn. Publication 3311. University of California, Oakland, California, pp. 97–112.

Tomlik-Wyremblewska, A., Zielinski, J. and Guzicka, M. (2010) Morphology and anatomy of blackberry pyrenes (*Rubus* L., Rosaceae). Elementary studies of the European representatives of the genus *Rubus* L. *Flora* 205(6), 370–375.

USDA (2011) *Guidance: Processed Animal Manures in Organic Crop Production*. National Organic Program 5066, pp. 1–3. Available at: www.ams.usda.gov/sites/default/files/media/5006.pdf (accessed May 11, 2017).

Walsh, C.S., Popenoe, J. and Solomos, T. (1983) Thornless blackberry is a climacteric fruit. *HortScience* 18(3), 482–483.

16

MARKETING OF BLACKBERRIES

Gail Nonnecke,[1,*] Michael Duffy,[1] and Richard C. Funt[2]

[1]Iowa State University, Ames, Iowa, USA; [2]The Ohio State University, Columbus, Ohio, USA

INTRODUCTION

Successful business components of a sustainable blackberry enterprise include the production and management of the planting and marketing (selling) of items (blackberry fruit or fruit products) for a profit. Today, the value chain of blackberries from production through sales includes the critical step of marketing. The major details of marketing include placing the fruit into a bulk or retail container, labeling, and delivering to local, retail or wholesale markets. For blackberry production to be profitable, selling of quality fruit at fair prices creates a sustainable enterprise, and markets provide the opportunities for sales. Recommendations indicate the importance of marketing in advance by stating that blackberries should be sold before production is started and by when fruit is first harvested. Determining and developing appropriate markets are critical for sales of highly perishable blackberries. It is vital to a successful blackberry operation to determine how the crop will be marketed.

Diverse market channels for blackberries provide unique opportunities for distribution and sales. In the USA, blackberries are produced for the fresh or processing markets. Fresh markets include on-farm sales that are either pick-your-own (PYO) or pre-picked (hand-picked) and pre-picked berries for community farmer markets, community supported agriculture, roadside markets, or for wholesale (large stores). Processed markets include berries that are immediately frozen or made into jam, jelly, and juice or fermented into wine. A majority of processed berries in the USA are mechanically harvested. Blackberries may be processed into many delicious products and offer the potential sales of 'value-added' products by the farm enterprise through their specialty markets. PYO berries often are processed by the customer, immediately after harvest, into desserts, jam or jelly, juice or wine, and/or frozen for consumption

---

* Corresponding author: nonnecke@iastate.edu

at special events or holidays. Blackberries from home gardens typically are eaten fresh or processed.

Variables that are controlled by the grower include the value of the fruit (price), quality, form, and packaging associated with the fruit and its sales (product), publicity to sell the product (promotion and advertisement), and location of the distribution and sales (place). To determine price, growers need to consider their costs associated with producing the blackberries and the amount needed to obtain a profit (Funt, 1990) (see Chapter 17 for costs of producing blackberries). Growers must keep accurate records of variable and fixed production costs to be used in determining appropriate prices for sale of the blackberries. Additional factors include knowing the amount customers are willing to pay and any competitor's prices. Selling blackberries below the costs of production to undercut competitors does not contribute to a profitable and sustainable operation and needs to be avoided (Bushway *et al.*, 2008). A grower's cost of producing and marketing blackberries in the future can be estimated. The most important costs to consider are labor and fixed costs and/ or overhead costs. Consideration must be given for making a return to a grower's management time for the planting and labor and costs of capital. Also, the cost per unit (pint, crate, master tray) is dependent on how many units are actually sold. One general estimate is that 80% of the expected yield will be sold in either direct market or wholesale markets and that 20% could be lost to birds, mold, or spoilage.

FRUIT QUALITY AND MARKETING EXPERIENCE

A high-quality product contributes to successful marketing of blackberries, and the statement that 'quality sells' is as accurate for blackberry fruit and its products as for any consumer product. Quality of fresh blackberries includes fruit that receives top grade, such as U.S. No. 1, which is based on the fruit being 'firm, well colored, well developed (not misshapen) and not overripe' (USDA, 2016b). They are also 'free from caps (calyxes), mold and decay, and from damage caused by dirt or other foreign matter, shriveling, moisture, disease, insects, mechanical or other means' (USDA, 2016b). Tolerances to allow for variation in quality are usually provided in percentages for the different grades, such as 'not more than 10 percent, by volume, of the berries in any lot may fail to meet the requirements of the grade' (USDA, 2004, 2016b). Blackberry producers should check with their local agencies regarding inspections of fruit quality. High-quality blackberries are evenly and fully colored black, firm and hold their shape. Fruit that are shiny indicate freshness and will draw repeat customers (see Chapter 15, Fig. 15.2).

Factors of quality associated with marketing and sales of blackberries include some aspects that are distinct from the fruit, itself. If a grower sells fruit through agricultural tourism experiences, the experience also must be of high quality. Attractive, clean, and welcoming farm markets and PYO farm conditions can showcase blackberries and blackberry products and provide a pleasant and high-quality entertainment or tourist experience.

Promotion of blackberries is facilitated through a relationship with the public and customers. Its goal is to encourage demand for and sales of blackberries and blackberry products. Promotion strategies are unique to each grower and seller and include any aspects related to public relations, advertising through mass media, direct mailings, the internet and social media, and sponsorship of grower or non-grower events. Determining who will buy the blackberry fruit allows for targeted promotion. Customers can range from those visiting a farm for a PYO or an agricultural tourism experience to large-scale processors purchasing large volumes of blackberry fruit.

The blackberry market is the location of the fruit's distribution. For customers to purchase blackberries, the fruit must be available in the right condition at the right place at the right time. Efficient and effective locations are important for growers to meet their goals for sales. In the planning process for direct marketing, growers should determine how many units of land are already in the market, the population within the area, and the impact your supply of berries will have on the market (Courter and How, 1990).

Markets that are used for blackberries and blackberry products, include selling directly to the consumer or through wholesale channels. Direct-to-consumer markets have the potential to provide good returns to the grower, since the grower is the distributor of the fruit and products and no other intermediary is involved. Direct markets include selling pre-picked fruit at community farmers' markets, on-farm markets, off-farm markets, and through markets of community-supported agriculture (CSA) enterprises. PYO operations sell fruit directly to the consumer, but the customers harvest the blackberries at the grower's farm.

Fresh pre-picked blackberries, that are offered directly to the consumer or as wholesale, are sold in smaller quantities, typically in ½-pint (about 125 g) or 1-pint containers that are made of biodegradable pulp, vented plastic (hinged containers that are called 'clamshells'), meshed plastic, or wood. Shallow or smaller containers are better because the weight of the top fruits can damage the quality of the lower fruits. Wooden containers are usually placed into 24–36-quart crates and clamshells are placed into 6-quart or 12-pint masters (see Chapter 15, Fig. 15.5). In the USA, masters are made of cardboard, which are easily folded together to hold clamshells and may be recycled. In New Zealand and elsewhere in the world, blackberries are sold in punnets (125 g, ~1/4 lb) where 2 punnets equal the weight of 1 pint. Blackberries to be processed are taken to the processor in shallow containers, called lugs, which are used in the field during either hand or mechanical harvest (see Chapter 5, Fig. 15.6). Lugs are food-grade polyethylene, reusable, and stackable.

DIRECT-TO-CONSUMER MARKETS FOR BLACKBERRIES

Community farmers' markets bring customers to a central location for the convenience of the shoppers, such as in an urban setting of a town or city, where farmers offer products for sale (Sabota et al., 1980). Most farmers' markets are organized whereby growers benefit from having customers come to one location (marketplace), and the sponsoring location also benefits from its support of the farmers' market by increasing customers' activities in the town or city. Major considerations for a farmers' market include location, facilities, parking, operational procedures, including days and hours of operation, stall fees, publicity, food safety considerations, and legal requirements, such as permits, items to be sold, and the methods of sales (Sabota et al., 1980). Many local farmers' markets are part of associations that facilitate the development and organization of community farmers' markets and training of market managers. Commercial growers often participate in several farmers' markets across multiple communities within a week, so that they can sell blackberry fruit daily or at least several times in a week, since fresh blackberries decline in quality after storage of several days.

On-farm markets are retail operations located on the grower's farm that offer blackberries and blackberry products for sale. Highly perishable blackberries benefit from on-farm market sales where fruit can be precooled and stored in refrigeration on the farm (see Chapter 15). Markets should include an attractive sales area that is clean and staffed by competent and welcoming workers, a place for parking of cars, proximity to major roadways, and convenient hours of operation to increase on-farm market sales (Bushway et al., 2008).

Off-farm markets are operated by the grower but located at a site different from the farm. The off-farm market may provide a better location for customers than the farm site, such as being located along major roadways or intersections. These markets also require an attractive sales area with sufficient space for parking and convenient hours. The off-farm market structures may be permanent facilities, typical of an on-farm market, or temporary. If off-farm facilities are temporary, such as a sales shed, refrigerated storage still should be available, because blackberries are highly perishable.

Community supported agriculture (CSA) is a partnership between the grower and the community, who work together to establish a local food system (Gradwell et al., 1999). In a CSA, community members agree to a full-season price before production occurs, sharing with the grower the risks and benefits of the production season. The CSA model includes the grower estimating production costs and how many members can be supported from the production; from that information, the price of each membership or share is calculated. The produce harvested over the season is divided into each share's amount and distributed among all of the shareholder members. Each CSA is designed to meet the needs of the grower and community, and some have more than one producer and often include a variety of produce or products. CSAs can be broad or specific in the products that are available to members. Some CSAs focus on fruit crops and include blackberries; some have a CSA model that is comprised of multiple growers, with a grower providing blackberries to shareholders while other growers provide different produce or products to members. Spreading out fresh blackberry production as long as possible, such as using both primocane- and floricane-fruiting cultivars, field and high tunnel production, and processed products, creates a steady supply of blackberries throughout the growing season for increased value and satisfaction to CSA members, creating repeat members in following years.

PYO marketing of blackberries includes customers traveling to the grower's farm and harvesting fruit themselves. Growers save in harvest labor, but staff members are still needed to manage visitors to the farm and facilitate their purchasing of the blackberries. PYO market success can be highly dependent on the farm's location, with the best location within 20 miles of a densely populated area (Bushway *et al.*, 2008; Courter, 1979, 1982). Any competition of PYO blackberries within the market area should be considered. Because customers harvest fewer blackberries than other PYO crops, such as strawberries, 350 PYO customers are needed to harvest 1 acre of blackberries (Bushway *et al.*, 2008). The logistics of operating a PYO enterprise for selling blackberries include determining the best parking arrangements, check-in details, safe transportation of customers to the blackberry fields, field supervision, containers to pick into, and promotion and communication strategies to inform customers and promote the blackberries. An attractive farm offering the necessary facilities, including parking, transportation, and restrooms, and with welcoming and competent staff is important in PYO enterprises so that the customers have a pleasant and satisfying experience and become repeat customers (Courter, 1979).

Direct-to-consumer markets bring the general public to a grower's farm to purchase blackberries or harvest and purchase PYO blackberries. It is essential for a grower to consider all risks and liability associated with the possibility of a customer being injured on their property. Even when the best possible care and precautions are taken, it is important to have adequate insurance (liability) protection (Uchtmann, 1979).

WHOLESALE MARKETS FOR BLACKBERRIES

Wholesale markets include selling blackberries in fresh and/or processed form, and typically include intermediary persons involved in the sales. Examples of wholesale markets include processors, supermarket chains, restaurant chains, terminal markets, auctions, and institutions, such as schools or hospitals. A major wholesale market includes processors of blackberries because the perishable nature of blackberries necessitates processing to get the product to the consumer throughout the year. The majority of blackberries grown in Oregon, a major state of production in the USA, are sold for processing. In 2015 in Oregon, over 50 million pounds (lb) were produced; 94.7% were processed, and 5.3% were used for fresh markets (USDA, 2015, 2016a).

Wholesale market sales of fresh fruit to supermarkets, restaurants, terminal markets, auctions, and institutions include postharvest and packaging requirements to keep the shelf life of blackberries as long as possible (see Chapter 15). In the short run, price changes for hand-picked berries in supermarkets are due to a change in supply from the local area. Roadside market price fluctuates with supermarket prices. In the USA in 2013, the average retail price of one pound of fresh blackberries was US$5.77 and US$3.39 for frozen blackberries (USDA, 2013).

Seasonal prices of USA blackberries are affected by both local and imported blackberries. Mexico is the largest supplier of fresh blackberries by volume imported into the USA with over 95% of the total share. The top three countries of origin for frozen blackberries imported into the USA include Chile (59.1% share), Mexico (29.9% share), and Serbia (7.6% share) (USDA, 2016a). New primocane cultivars offer blackberries out of the normal summer season (see Chapter 5) and technology, such as high tunnels (see Chapter 12), can extend the normal season into late autumn (Geisler, 2015).

Additional marketing opportunities are from the production and sale of organic blackberries. Investigations in Oregon researched production systems in certified organic blackberries (Dixon and Strik, 2016; Dixon et al., 2015, 2016a,b; Fernandez-Salvador et al., 2015a,b,c; Harkins et al., 2013, 2014; Strik, 2016). Certified organic acreage of blackberries in Oregon grew 96% from 2008 to 2014 (USDA, 2014). The climate in Oregon offers advantages for organic production, including relatively mild winters and dry, warm summers that reduce weed- and disease-management costs and provide optimal conditions for fruit ripening (Strik, 2016). Cultivars differ in their adaptation to organic production systems, likely due to cultivar susceptibility to disease and insect pests (Dixon et al., 2015; Fernandez-Salvador et al., 2015c) (see Chapter 14). In field trials, yield of processed and fresh blackberry cultivars grown in the best organic production systems was similar to what would be typical for conventional production (Dixon et al., 2015; Fernandez-Salvador et al., 2015b). Organic blackberry production is expected to increase in Oregon (Strik, 2016).

Cook (2011) summarized the importance of shippers in the market chain for fresh produce: 'streamlining' the supply chain of wholesale fresh fruit has forced non-value-adding costs to be removed and increased the importance of shippers in key production regions. Many shippers now act as the marketing agents for growers and are the first handler in the market chain. Larger-scale shippers also may source blackberries from many production regions to offer fresh fruit throughout the year, including shipping from several states and also other countries. Many of the shippers are specialized grower-shippers, who control a sizeable portion of fresh sales at the 'first-handler level.'

The costs for processing facilities are not available in the public domain. These costs will vary by individual location. The best way to determine costs for a facility would be to use an economic engineering procedure. With this procedure, the individual plant is constructed first on paper and the costs at each step are determined through interviews with existing plants operators and manufactures.

Approximately 90% of Oregon blackberry production is specifically for the processed markets, including individually quick frozen (IQF), bulk frozen, freeze dried, puree, juice, or concentrate. Most of the processed fruit in this region consists of high-quality, aromatic, small-seeded, trailing cultivars (see Chapter 5).

The Washington Red Raspberry Commission describes the current processed raspberry products, available on the world market, and the raspberry information is similar for processed blackberries (WRRC, 2013).

IQF blackberries are the choicest whole frozen blackberries. Individual blackberries are frozen in a quick-freeze tunnel or on trays at temperatures between $-20°C$ to $-23°C$ ($-4°F$ to $-10°F$). This 'quick freezing' seals in juices and maintains the original shape of each berry. IQF blackberries are packed in polyethylene bags and sealed in corrugated fiber cartons. This ensures that each IQF blackberry is 'fresh frozen' and protected from damage or shipping shock. IQF blackberry packs range from 336 g (12 oz), 448 g (16 oz), and 869 g (32 oz) polyethylene bags for retail sale to 13.5 kg (30 lb) cartons. Restaurant, institutional, and food service packs of six 2.25 kg (5 lb) IQF packages to a case are also offered (WRRC, 2013). Frozen blackberries should be stored at $-20°C$ ($-4°F$) and may be kept at these temperatures up to two years (Tree Top, 2016).

Blackberry purees, concentrates, and juices are specified in 'Brix.' Degrees brix are the approximate percent of sugar or soluble solids. Blackberry puree typically is sold with a minimum of $9°$ Brix (soluble solids) (Tree Top, 2016). Passing cleaned and sorted berries through a sieve to achieve a consistent particle size produces frozen blackberry puree. Screen meshes from 0.75 mm to 3.1 mm (0.03–0.125 in) determine the fineness of puree and the amount of seed removed. Blackberry puree may be processed at ambient temperatures or heated for pasteurization. Puree of blackberries is frozen at $-20°C$ to $-23°C$ ($-4°F$ to $-10°F$) for storage and marketing. The most common puree packs are 2.9 kg (6.5 lb) and 12.6 kg (28 lb) containers and 180 kg (400 lb) drums. Puree may be custom packed in quart and gallon equivalents. Puree is also available in concentrated form at $30°$ Brix (soluble solids) (WRRC, 2013; Tree Top, 2016).

Blackberry concentrate is an intense capture of both blackberry essence and form. Blackberry juice is first extracted from the fruit. This juice is filtered and heated. High temperature allows the flavor and aroma (essence) to be distilled from the concentrate. The essence is captured in liquid form and may be packaged separately or mixed back into the concentrate (recapture), and concentrated juice is available at 70° Brix (soluble solids). Blackberry concentrates are packed in 190 l (50 gal) drums and 19 l (5 gal) enamel-lined cans or 1.8, 2.25, and 2.7 kg (4, 5, and 6 lb) polyethylene-lined pails with essence packed separately or recaptured (WRRC, 2013).

Traceability or the trace-back and trace-forward process is an important component of good agricultural practices (GAP) that is intended to reduce liability and prevent the occurrence of food safety problems in the marketplace (USFDA, 1998). By having a traceability system in place, blackberry growers can market fruit knowing the exact planting/field of the fruit's origin, dates of harvesting and shipping, the persons involved in harvesting and handling of the fruit, and the intended market to which the blackberries were distributed. Additional aspects may include the date and location of postharvest storage. Recently, the purpose of traceability systems has come from food safety concerns in the produce and food industries (Golan et al., 2004) and some markets, such as wholesale produce, may require traceability. If contaminated blackberries occur, the contaminated product can be identified, allowing non-contaminated blackberries to be marketed (Golan et al., 2004). Information that typically is required for wholesale markets includes harvest date, packing date, harvesting and handling personnel, including the staff member completing packing and quality control, and even the picker, shipping date, field identification, and customer records. While traceability systems are used in wholesale fresh and processed fruit marketing, their use also is relevant and may be used effectively in direct-to-consumer sales.

Traceability systems are often linked to computer software that is designed for tracing the fresh or processed fruit back to the blackberry grower or forward to the customer. Traceability codes are computerized, and there is a goal in the produce industry to have an adoption of electronic traceability (PTI, 2012). Barcodes are often printed on labels that are placed on the flat or container of blackberries to be marketed. Information contained in the barcode includes any aspect about the blackberries that the grower wishes to document and retain. Most traceability systems use barcodes that are read by electronic scanning machines. The electronic barcode system can be set up to print automatically with all information stored in a computer.

Successful marketing of blackberry fruits requires intentional activities that contribute to the profitability and sustainability of a blackberry farm. Sales can include fresh or processed forms of blackberries through direct-to-consumer or wholesale markets. The market channel that will be used for quality blackberries determines the final product and its packaging, price, place of sale, and promotion strategies.

REFERENCES

Bushway, L, Pritts, M. and Handley, D (2008) *Raspberry and Blackberry Production Guide*. NRAES-35. Cooperative Extension, Natural Resource, Agriculture, and Engineering Service, Ithaca, New York.

Cook, Roberta L. (2011) Fundamental forces affect U.S. fresh produce growers and marketers. Agricultural and Applied Economics Association, Milwaukee, Wisconsin. *Choices* 26(4). Available at: www.choicesmagazine.org/UserFiles/file/cmsarticle_202.pdf (accessed May 15, 2017).

Courter, J.W. (1979) *Pick-your-own Marketing of Fruits and Vegetables*. Cooperative Extension Service, Horticulture Facts HM-1-79. University of Illinois, Urbana-Champaign, Illinois.

Courter, J.W (1982) Establishing the trade area and potential sales for a pick-your-own strawberry farm. Cooperative Extension Service, Horticulture Facts HM-6-82. University of Illinois, Urbana-Champaign, IL.

Courter, J.W. and How, R.B. (1990) Marketing small fruits. In: Galletta, G.J. and Himelrick, D.G. (eds.) *Small Fruit Crop Management*. Prentice-Hall, Upper Saddle River, New Jersey, Ch. 13.

Dixon, E.K. and Strik, B.C. (2016) Weed control increases growth, cumulative yield, and economic returns of machine-harvested organic trailing blackberry. Proceedings XI International *Rubus Ribes* Symposium. *Acta Horticulturae*, 1133, 323–328.

Dixon, E.K., Strik, B.C., Valenzuela-Estrada, L.R. and Bryla, D.R. (2015) Weed management, training, and irrigation practices for organic production of trailing blackberry. I. Mature plant growth and fruit production. *HortScience*, 50(8), 1165–1177.

Dixon, E.K., Strik, B.C. and Bryla, D.R. (2016a) Weed management, training, and irrigation practices for organic production of trailing blackberry. II. Soil and aboveground plant nutrient concentrations. *HortScience* 51(1), 36–50.

Dixon, E.K., Strik, B.C. and Bryla, D.R. (2016b) Weed management, training, and irrigation practices for organic production of trailing blackberry. III. Accumulation and removal of aboveground biomass, carbon, and nutrients. *HortScience* 51(1), 51–66.

Fernandez-Salvador, J.A., Strik, B.C. and Bryla, D.R. (2015a) Liquid corn and fish fertilizers are good options for fertigation in blackberry cultivars grown in an organic production system. *HortScience* 50(2), 225–233.

Fernandez-Salvador, J.A., Strik, B.C. and Bryla, D.R. (2015b) Response of blackberry cultivars to fertilizer source in an organic fresh market production system. *Hort-Technology* 25(3), 277–292.

Fernandez-Salvador, J., Strik, B.C., Zhao, Y. and Finn, C.E. (2015c) Trailing blackberry genotypes differ in yield and post-harvest fruit quality during establishment in an organic production system. *HortScience* 50(2), 240–246.

Funt, R.C. (1990) Economics of small fruit production. In: Galletta, G.J. and Himelrick, D.G. (eds.) *Small Fruit Crop Management*. Prentice-Hall, Upper Saddle River, New Jersey, Ch. 14.

Geisler, M. (2015) Blackberries. Updated by Marzolo, G. Agricultural Marketing Resource Center. Available at: www.agmrc.org/commodities-products/fruits/blackberries (accessed November 2, 2016).

Golan, E.H., Krissoff, B., Kuchler, F., Calvin, L., Nelson, K. and Price, G. (2004) Traceability in the U.S. food supply: economic theory and industry studies. *Agricultural Economic Report, Number 830*. United States Department of Agriculture, Economic Research Service, Washington, DC.

Gradwell, S., De Witt., J., Mayerfield, D., Salvador, R. and Libbey, J. (1999) *Community Supported Agriculture. Local Food Systems for Iowa*. Bul. 1692. Iowa State University Cooperative Extension Service, Ames, Iowa.

Harkins, R.H., Strik, B.C. and Bryla, D.R. (2013) Weed management practices for organic production of trailing blackberry. I. Plant growth and early fruit production. *HortScience* 48(9), 1139–1144.

Harkins, R.H., Strik, B.C. and Bryla, D.R. (2014) Weed management practices for organic production of trailing blackberry. II. Accumulation and loss of biomass and nutrients. *HortScience* 49(1), 35–43.

PTI (2012) *PTI Vision*. Produce Traceability Initiative. Available at: www.producetraceability.org (accessed November 2, 2016).

Sabota, C.M., Courter, J.W. and Archer, R. (1980) *Establishing a Community Farmers' Market*. Cooperative Extension Service, Horticulture Facts HM-4-80. University of Illinois, Urbana-Champaign, Illinois.

Strik, B.C. (2016) A review of optimal systems for organic production of blueberry and blackberry for fresh and processed markets in the northwestern United States. *Scientia Horticulturae* 208, 92–103. Available at: www.sciencedirect.com/science/article/pii/S0304423815303186 (accessed May 15, 2017).

Tree Top (2016) Fruit ingredients. Available at: http://foodingredients.treetop.com/fruit-ingredients (accessed November 3, 2016).

Uchtmann, D. (1979) *Liability and Insurance for U-pick Operations*. Cooperative Extension Service, Horticulture Facts HM-2-79. University of Illinois, Urbana-Champaign, Illinois.

USDA (2004) Strawberries and other berries. Shipping point and market inspection instructions. Fruit and Vegetable Programs. United States Department of Agriculture, Agricultural Marketing Service. Available at: www.ams.usda.gov/grades-standards/fresh-dewberries-and-blackberries-grades-and-standards (accessed November 2, 2016).

USDA (2013) Blackberries – average retail price per pound and per cup equivalent, 2013. United States Department of Agriculture, Economic Research Service. Available at: https://www.google.co.uk/search?source=hp&q=Blackberries+%E2%80%93+average+retail+price+per+pound+and+per+cup+equivalent%2C&oq=Blackberries+%E2%80%93+average+retail+price+per+pound+and+per+cup+equivalent%2C&gs_l=psy-ab.12...2741.2741.0.3982.3.2.0.0.0.0.71.71.1.2.0....0...1..64.psy-ab..1.1.75.6..35i39k1.75.fq90xdKtu8Y (accessed September 20, 2017).

USDA (2014) *Census of Agriculture, Organic Survey* (2012) Volume 3, Part 4. United States Department of Agriculture, National Agriculture Statistical Service, Washington, DC.

USDA (2015) Blackberries. United States Department of Agriculture, National Agricultural Statistics Services. Available at www.nass.usda.gov/Statistics_by_Subject/index.php (accessed November 3, 2016).

USDA (2016a) Fruit and tree nut data. 2016. Yearbook Tables. 1980 to date. Blackberries: Commercial acreage, yield per acre, production, and season-average grower price, Oregon. United States Department of Agriculture, Economic Research Service. Available at: www.ers.usda.gov/data-products/fruit-and-tree-nut-data/yearbook-tables/#Berries (accessed November 3, 2016).

USDA (2016b) United States standards for grades of dewberries and blackberries. Specialty Crops Program. Effective Sept. 6, 2016. United States Department of Agriculture, Agricultural Marketing Service. Available at: www.ams.usda.gov/grades-standards/fresh-dewberries-and-blackberries-grades-and-standards (accessed November 2, 2016).

USFDA (1998) Traceback. In: *Guidance for Industry: Guide to Minimize Microbial Food Safety Hazards for Fresh Fruits and Vegetables*, Section 9. United States Food and Drug Administration. Available at: www.fda.gov/food/guidanceregulation/guidancedocumentsregulatoryinformation/ucm064574.htm (accessed November 4, 2016).

WRRC (2013) Washington Red Raspberry Commission. Available at: https://www.redraspberry.org (accessed November 4, 2016).

17

BLACKBERRY FARM MANAGEMENT AND ECONOMICS

Richard C. Funt*

The Ohio State University, Columbus, Ohio, USA

INTRODUCTION

Blackberry farm management refers to the decisions that are made by the farm owner (grower) or farm manager. These decisions will be made with regards to the allocation of resources available to the grower, such as land, labor, water, and capital. In many of the previous chapters, biological and technical information have been given with regards to making wise decisions about soil and site selection, water sources for irrigation, and infrastructure for the farm. The economics of farm management refers to the fixed and variable costs and the potential returns on investment. For blackberries, high initial costs for medium priced land, irrigation, plants, farm equipment, and overhead costs are invested in years 0, 1, and 2 while returns begin in years 2 or 3. The ultimate factors for a successful berry farm business are whether there will be sufficient capital to establish a crop and receipt of sufficient revenue to cover the establishment costs several years after planting, and in the final analysis, receiving a positive return on the investment over the life of the planting. This chapter is written to emphasize decisions that are necessary and critical for a successful farm business. Generally, growers are less advanced in handling business decisions than with the production of fruit (Nicholson, 1984). Therefore, an understanding of blackberry farm management from a business plan to a positive rate of return over the life of the planting is necessary for a satisfying experience.

FARM MANAGEMENT

Commercial blackberry production is a business and is considered a high-risk investment. It is also considered to be a high-value crop due to high potential returns. Blackberries are well suited for small-scale agriculture, particularly

* Corresponding author: richardfunt@sbcglobal.net

where sufficient and efficient local labor is available. Blackberries are harvested by hand or machine and are native to many countries (Strik et *al.*, 2007). They are produced for fresh, frozen, juice, dried, or pureed markets; and can be grown from spring to autumn depending whether they are grown in the field under protective cover, as row covers, high tunnels, and/or greenhouses. Blackberries are a labor-intensive crop.

Blackberry farm managers are faced with a complex, multifaceted, and dynamic industry. New cultivars and production systems, insect- and disease-management strategies, and climate change (global warming) need to be addressed (Strik et *al.*, 2007). In the 21st century, blackberry growers need reliable information from many sources in order to create a business plan that leads to success (see Chapters 3, 6, and 13). Overall, the 21st-century grower lives in a global environment, global market, and global economy (see Glossary 2).

Many growers have plantings of only blackberries, while some growers have diverse crops, as other berries, vegetables, and/or tree fruits. Enterprises, that have a large number of acres and several crops that are spread over the seasons, have an economical advantage when the same type of equipment, such as sprayers, mechanical harvesters, and cultivation equipment, is used. Thus, in large enterprises there is a higher level of efficiency and a lower cost per hour or per hectare (acre) for equipment. Small farms may have less equipment and most of the labor is accomplished by family members but generally have higher overhead cost than larger farms.

When the land is first purchased, it is wise to consider cash flow in the choice of crops. Vegetable crops, such as sweet corn or pumpkins, and other crops such as wheat or rye, are recommended biologically to prepare the land before planting blackberries. Further, the annual cash flow in the business needs to be considered when different crops are being harvested on the farm. In the eastern USA, a whole farm plan should also consider annual crops for cash flow, because blackberries will take at least 2–3 years before an optimal yield is produced (Table 17.1). Selecting land on a suitable site is a very important decision prior to planting. Information on soil type, site selection, and preparation required for a successful business is provided in Chapters 5 and 9. There needs to be sufficient land for buildings, roads, parking for customers, and storage of equipment.

Biologically, the land must be internally well drained and fertile, unless the grower is producing in substrate culture. There should be an adequate quantity of safe, clean water available for all crops produced. The land should be located so that it can be irrigated to produce optimal yields consistently from year to year. Economically, certain high-quality soils have a greater value and provide greater profit margins than sites or soils of lower quality. In many cases, farms that sell locally grown fruits and vegetables are near metropolitan areas, where, for tax purposes, the land is zoned for agricultural production. Thus, economically, these high-value crops are considered as being produced on medium-priced land, and therefore, have less risk than those on high-priced

Table 17.1. Example of expense and return by year from preplant to first harvest.

Year	Expense	Return
−2	Removal of old plants and weeds, plus overhead costs	0
−1	Grow cover or green manure crop, plus overhead costs	0
−0	Apply compost, fertilizer and lime, install irrigation, install high tunnel, plus overhead costs, return from grain crop	0
+1	Plant blackberries; apply weed control, irrigate, install trellis, plus overhead costs	0
+2	Prune, irrigate, control weeds, harvest costs, spray application equipment, plus overhead	Less than annual cost
+3	Pruning, irrigation and weed control costs, spray application equipment, harvest costs (large crop), and container costs (prorate if used multiple years), plus overhead costs	Larger than annual cost
+4 to +20	Same as year 3	Equal to or greater than year 3

land (land that could be sold for houses or industrial buildings). If the land and water for irrigation are unsuitable for crops, it should not be purchased.

WHOLE FARM MANAGEMENT

The whole farm budget must allocate resources that account for different enterprises and for all expenses and receipts from the different enterprises. Whole farm budgets are useful in determining the appropriate mix of enterprises to maximize returns (Castaldi and Lord, 1989). A budget leads to a plan that will utilize general equipment, refrigeration, utilities, labor, and transport to market effectively and efficiently. In the short term, some of the enterprises will be replaced by another enterprise and/or replanted. Thus, at the completion of a budget, a plan (or planning horizon) in the short and long term is necessary. This plan also needs to include the hours of labor needed and seasonality of labor requirements.

A long-term planting scheme needs to consider a successful mix of enterprises that maintain a high level of returns early in the planning process. For example, some crops should not be replanted or rotated to the same crop land. Monoculture is not a good soil practice and may lead to low yields. An alternative is to use fumigation, where allowed, if the grower wishes to replant to the same crop. However, this practice may not be allowed, due to environmental laws. The use of tunnels or hoop houses and greenhouses (glasshouses), combined with substrate culture and hydroponics, can make a long-term blackberry production system viable without significant environmental issues, especially if nutrient-rich eluate can be recycled. Utilizing the biological and technological aspects with the economical potential is necessary under these conditions. Further, the selection of crops always requires an estimate of labor required by workers and managers. A budget and plan per enterprise is necessary for expansion or contraction of production or marketing (see Glossary 2).

Pritts (2016) in New York did a comparison of field and high tunnel blackberry production. Blackberries produced in the high tunnel generally were higher in yield and marketable percentage. Further, cash flow and expenses, involved in growing and harvesting 'Triple Crown' blackberries in a $9 \times 30\,m$ single bay tunnel with no crop failures, indicated expenses of nearly US$12,000 in year 0 with tunnel construction of US$9600. The cover was replaced every three years at a cost of US$360. Expenses were US$7000 per year from years 5 to 10. With a price of US$2.50 per ½ pint (170 g) for the fresh market, a positive cash flow was achieved in year 5. Yield in years 5–10 was equal to $40\,T.ha^{-1}$ (16 t/ac)

ENTERPRISE BUDGET

An enterprise budget accounts for one specific enterprise, such as pick-your-own (PYO), autumn-bearing blackberries, or mechanically harvested berries for processing. Enterprise budgets provide the initial process (blueprint) for farm planning and analysis of the enterprise. This is particularly important when adding a new blackberry operation to the existing farm site. An example would be, mechanically harvested blackberries for the processing market as a new enterprise being planned with an existing fresh PYO blackberry enterprise.

Enterprise budgets have fixed and variable costs. Fixed costs are those costs of ownership, such as machinery, capital (principal on a loan), and interest expenses. These costs have to be paid whether or not a blackberry crop is produced. Fixed costs vary from farm to farm, due to farm size and choice of equipment and marketing, irrigation system, taxes, and computers for record keeping. Variable costs vary directly with production and machinery usage. Variable costs include harvest supplies (containers) and activities, picking labor, mechanical harvester, transportation, and field supervision of labor. It is useful to put all variable costs on a cost-per-hour basis. Knowing the fixed and variable costs in an enterprise can be useful in deciding whether to buy, lease, or rent equipment during the non-productive years (Castaldi and Lord, 1989).

RISK MANAGEMENT

Blackberry production, along with other crops and enterprises, is a high-risk industry. Risk arises from the lack of certainty. Sources of uncertainty (risk) are related to change, such as a change in the weather from extremely wet to dry (drought) to floods and incursions of pests or disease, which can cause lower yields (Schwab et al., 1989). Other risks, in this 21st-century global economy, are changes in the world supply and demand for oil and energy sources, changes in the supply and demand for blackberries, declining land values in certain areas, high interest rates, and labor costs (see Chapter 18). Berry growers also need to assess changes in political, social (local consumers of food and entertainment and/or health benefits), economic, and environmental circumstances where they operate.

Uncertainty can be a situation where a number of outcomes are possible; the greater the uncertainty, the greater the risk. Berry growers generally have crops that have a great number of years of life, and therefore, must consider the future increase/decrease in interest rates, cost of labor, equipment, and supplies. Creating budgets for the life of the planting and using a net present value (NPV) or internal rate of return (IRR) analysis (NPV and IRR are explained later in this chapter) can assist in managing long-term risk. Risk can be reduced when a system of culture allows the establishment costs to be covered early in the life of the planting as compared to a system that covers the establishment costs later in the life of the planting. Successful management depends on taking risks consistent with the goals and financial position of the business.

Risk management strategies for production and financial risk include:

1. diversification, i.e. growing different crops, such as grains and berries (Ross and Auchter, 1930);
2. spatial dispersion (spreading crops over different climatic areas);
3. enterprise selection, using different production and marketing schemes (PYO as well as hand or machine harvest);
4. purchase of crop, fire and/or liability insurance; and
5. sufficient reserves to provide liquidity and cash flow for 12 months.

In Oregon, Julian et al. (2009) evaluated risk management of two systems known as every year (EY) and alternate year (AY) for 'Marion' blackberry production where a 'typical' producer of 'Marion' in the Willamette Valley has 8 ha (20 acres) of 'Marion' on a 40 ha (100 acres) farm. All berries are mechanically harvested and taken to the processed market, general labor and equipment labor is US$13.50 and US$19.50 (includes social security, workmen's compensation, etc.), respectively. Irrigation costs are US$150/ha (US$60/acre), with a trellis cost of US$800/ha (US$320/acre) and operating interest at 8.5%. Management, family labor, and state and federal income taxes were not included. The EY system had an annual deficit of US$3243/ha (US$1297/acre) (includes an amortized establishment cost of US$5778/ha (US$2331/acre)) and the AY system had an annual deficit of US$6053/ha (US$2421/acre) (includes an amortized establishment cost of US$20,321/ha (US$8128 per acre)), when both were in full production. The projected returns for these plantings do not cover all cash costs of establishment and production over 21 years. The EY system had an estimated 1271 kg/ha (7000 lb/acre) average each year while the AY system had 1998 kg/ha (11,000 lb/acre) in alternate years. For these systems to be productive, a sensitivity analysis of price or yield indicated that profitability could be achieved by:

1. increasing price by 80% from US$1.43–2.57/kg (US$0.65–1.17/lb) or increasing production by 87% for the EY system; or
2. increasing the price by 98% from US$1.43–2.83/kg (US$0.63–1.29/lb) or increasing the yield by 107%.

These breakeven yields exceed the production capacity of 'Marion.'

Managers need to consider their own well-being by considering health, work stress, and safety in the work place. Reducing stress and monitoring fatigue can reduce accidents, injury, and lost work time. If these are not considered, managers could lose time on the job. The health and safety of non-hired (generally family) and hired workers is paramount to the business. Potential health hazards include dehydration, hearing loss (from loud engine noise or other sources), skin cancer, respiratory diseases, and exposure to and inhalation of pesticides. Owners and managers should understand the need for protective measures, and insist that workers use protective clothing, including gloves and long-sleeved shirts, for protection from the sun and pesticide applications.

The health of workers is also important as the transmission of communicable disease is of risk to both people on site and to those that consume the fruit. Provision of clean, hygienic toilets, with hand-washing stations, soap and clean water, or alcohol spray/hand sanitizer for hand cleansing is essential. Training and monitoring of staff hygiene is important. Hand-cleansing stations should be outside toilets so that they are clearly visible when staff sanitize hands after using the facilities. For large production units, the provision of mobile toilets for in-field use is very important, both for convenience and for reducing the cost of down-time and loss of productivity. Reduction or elimination of birds in the production area is also important, as bird droppings are a potential hazard for human health, especially when berries are sold in bulk packs for processing.

Waiting periods after application of fungicides and insecticides should be clearly understood and observed, with areas that have been sprayed being signposted to warn against entry until the waiting period has elapsed, especially when members of the public are permitted on site, as in a PYO operation. A careful management system for the collection of harvested fruit and transport to packing and cool stores is essential. Care should be taken to insure that fruit are never left in direct sunlight and that field heat is removed as soon as possible after harvest. Failure to manage this well may result in the spoilage of fruit, an increase in the grade (pack out) of poorer-quality fruit, and a reduction of storage and shelf life. Investment in high-quality packing and storage facilities is essential, especially for the registration of facilities for Hazard Analysis Critical Control Point (HACCP), Europe Gap, or GLOBALGAP and protecting the health of the consumer, which is essential for a large production unit (see Glossary 1).

Managers should take inventory of worksite hazards that could be eliminated to reduce loss of work time from the job, particularly during the stressful times at planting, pruning, or harvest. An inventory should include checking on power-take-off (PTO) shaft coverings, locking pesticide storage areas to keep children and other unauthorized persons away, minimizing slippery areas, and discarding unsafe ladders from which people could fall and be injured. Keep tools safe and sharp to reduce accidents and fatigue; when power tools are used, safety equipment should be used. (For example, the use of metal gloves with attached cut-off circuits is advisable when power shears or hedge trimmers are used.) It is advisable for management to insist that staff take refreshment breaks during the work day or create a shorter work day, if possible. Many times a shorter work day can be as productive as a longer one. It is often worthwhile for workers to begin early in the morning, especially in hot climates, when daytime temperatures are expected to be high. This also improves the quality of harvested fruit. Safety is a habit!

MACHINES AND EQUIPMENT

Besides land and packing and storage facilities, farm machinery (equipment) is the largest investment for berry growers. Growers substitute machines for manual labor, especially where machines can be more efficient and cost less per hour than labor. In fact, one of the highest machine costs is the farm truck (pickup).

Each operation has its own unique needs and aspirations. It is important to focus on equipment cost per unit of land (hectare or acre) and include the cost of the farm truck in overhead costs (cost spread over the entire business). In the 21st century, having new equipment with appropriate technology is important in maintaining a high level of productivity, fuel efficiency and effectiveness, as in pesticide delivery and coverage, and in completing the tasks on time. Growers should know their machinery costs and compare them with those of other growers with similar enterprises and/or with different types of equipment that may benefit the business. Berry growers who wish to maximize long-term returns will compare buying versus leasing, custom hire versus owning, investment in new versus used (pre-owned), or upgrading old, low technology to new technology. Any of these choices can either increase the cost per unit of land or reduce the cost of operating the equipment over the entire enterprise, such as more units of berries harvested and shipped per hour of use.

LABOR

Blackberries are labor-intensive. There are 'peaks and valleys' for labor needs; during the year the amount of labor can vary within a season and among years (a variable cost). Thus, a large amount of labor is needed during certain seasons. For example, pruning is performed in the spring, planting in the spring or autumn, and harvesting in late spring, summer, or autumn. Examples of how technology has made a reduction in labor possible are use of herbicides or plastic weed barriers instead of hand weeding or cultivation, using mechanical pruners that can partially reduce the amount of hand pruning, and using mechanical harvesters as a substitution for hand harvesting berries. Generally, on small farms, the manager and/or members of the family can accomplish the pruning, irrigation, and chemical sprays without hired labor. However, small blackberry growers may need to hire labor for fresh, hand-harvested berries for roadside or wholesale markets, or for additional supervisory labor for PYO (on-farm sales) harvest.

Where there are other crops on the farm, the manager must increase or decrease the number of workers to achieve the completion of tasks in a particularly short period of time. For example, Mason and Cross (1993) found in a survey of Oregon raspberry, blackberry, and Boysen growers that the harvest season was 31 days. The season in this area required more than 17,500 workers to either hand harvest, machine harvest, or accomplish a mix of hand or machine-harvest operations.

Recruitment and retention of workers during the year is strongly influenced by the wage received (hourly or piece work). Also, recruitment and retention of workers is influenced by the local labor supply, the cultivar being picked, whether housing is being supplied, whether the crop is good, and a myriad of other reasons. Further, growers who use a mix of hand and machine harvest recognize that it takes fewer workers to harvest a unit of land than hand harvest alone.

In 1992, growers were able to manage the workforce more efficiently than in 1990 (Mason and Cross, 1993). In 1992, growers also increased the piece rate for fresh raspberry/blackberry hand-harvest workers near the end of the season when berries were fewer and more scattered in the row. This increase in the piece rate retained workers, keeping them from leaving the raspberry/blackberry crop and moving on to other crops. Workers were likely to leave at harvest when unharvestable, moldy berries and poor-quality fruit had to be sorted from the quality berries. A large proportion of unharvested berries can be due to poor weather conditions, not enough workers to pick the entire planting at the proper time, and/or the lack of fungicide applications. In 1992, raspberry/blackberry growers also hired workers for non-harvest work. Growers, who provide off-season work, are more likely to develop long-term relationships with workers. These workers can also be employed in one location and have their children attend the same school year after year (Mason and Cross, 1993).

An economic evaluation of improved mechanical harvesting systems for eastern thornless blackberries in 1999 indicated that harvesting was the single largest annual expenditure and could require up to 900 h/ha (360 h/ac) for hand harvest. With the scarcity and the cost of hired labor for the harvest of eastern thornless blackberries, a feasibility study of mechanical harvest for fresh market and/or processed market was conducted. (Harper *et al.*, 1999). Based on field data of 4.7 ha (11.7 acres) at the United State Department of Agriculture (USDA) Appalachian Fruit Research Center, Kearneysville, West Virginia, USA, an economic analysis of machine harvesting V, Y, and rotatable Y trellis systems indicated that all are profitable under a broad range of fresh market prices and yields. The total preproduction cost one year after establishment (year 2) was US$13,758, US$16,266, and US$21,640 per ha (US$5503, US$6506, and US$8656 per acre) for V, Y, and rotatable Y trellis, respectively. The ownership and operating cost for the mechanical harvester was US$11,039, US$12,786, and US$15,417 per ha (US$4416, US$5114, US$6168 per acre) per year for the V, Y, and rotatable Y trellis, respectively (Harper *et al.*, 1999).

PREPLANT TO MARKETING

Blackberry growers should first develop enterprise budgets for the preplant years (start-up costs), the establishment years, and then for the years of production and marketing. Worldwide, blackberry plantings have a life of 5–20 years. One blackberry grower will have different costs from another grower, particularly the fixed costs. Failure to include start-up, labor, fixed, and marketing costs will greatly overestimate potential profits. Yields will also vary from farm to farm and among different systems of growing blackberries. As an example of a system, there would be a trellis with drip irrigation designed for hand-harvested, fresh-market berries. There are many different plant densities as well as different types of blackberries (semi-erect, erect, primocane-fruiting, floricane-fruiting, and trailing). Also, production systems can have designs for PYO, hand harvest or mechanical harvest, with or without irrigation.

From an economic perspective, the preplant year is year 0, the planting year is year 1 and the next year is year 2, etc. (Table 17.1). Some growers may prepare the site two years before planting; this could be called year minus 1 (−1). A commercial yield is expected in year 3. In the productive years, the cost of labor, materials, equipment, irrigation, etc., do not vary greatly from year to year. Yields also vary from farm to farm, from one system to another, and from different in-row and between-row spacing as well as from different cultivars. In West Virginia, data from test plots of eastern thornless blackberries were recorded for plant densities on different trellis systems ranging from 1280 to

1861 plants/ha (512–744 plants/acre), with yields ranging from an average of 16,500–24,750 kg/ha (14,520–21,780 lb/acre) for years 5–15. Trellis labor ranges from 210 to 299 h/ha (84–120 h/acre) worldwide, semi-erect blackberries can have an in-row spacing of 1.0–1.5 m (3–4.5 ft) with 2.5–3 m (7.5–9 ft) between rows, erect blackberries 0.8–1.2 m (2.5–3.5 ft) in-row spacing with rows 3 m (9 ft) apart, and trailing blackberries with an in-row spacing of 0.9–1.8 m (2.75–5 ft) and 3 m (9 ft) between rows (Strik *et al.*, 2007).

ESTABLISHMENT AND HARVEST LABOR

Preplant costs in 2008 were estimated at US$1500/ha (US$600/acre) and included a herbicide and its application. Costs in the planting year include plants, installing irrigation, mulch, establishing sod middles, management of weeds and pests, and an installed trellis for a total of about US$3200/ha (US$1280/acre) in the USA (Pritts, 2008).

Labor costs need to be part of the budget even if the grower or his family does most of the work. Generally, 8–10 hours are used in preplant activities, 40–50 hours in establishment (year 1), and 15–25 hours per 1 ha (2.5 acres) in the second year, when there is no harvest. Labor costs were US$10.80 per hour (including benefits) in 2008 (Pritts, 2008). Some blackberry systems require tying of canes, which would increase the amount of labor. In the year of harvest, harvest labor may average 3.6–7.2 kg (8–16 lbs) of berries per hour (Funt, personal communication). Harvest labor costs could be 40–50% of the total hours of labor per year. However, in PYO systems where the customer picks the berries, hand-harvest labor costs are eliminated, but labor is still involved with managing and exchanging money with the customers. Generally, one person will interact with 12–14 people per hour in a pick-your-own business.

Multiple-year budgeting includes a measure of the total investment and returns over the life of the planting. This could be from preplant to the removal of the planting (10–12 years in eastern or midwestern USA) or up to 50 years (average 20 years) for production in Oregon. The preplant cost is then expected to be recovered midway through the life of the planting.

Overhead costs are considered the cost of doing business over the entire farm. Farm managers should add these costs per year in multiple-year budgeting. These costs include electricity, refrigeration, irrigation ponds, buildings (shop and machine shed), insurance (fire and liability), truck(s), taxes, advertising, etc. On a large farm, 16–32 ha (40–80 acres), estimated overhead costs can be US$750–1250/ha (US$300–500/acre) and on small farms, 4–8 ha (10–20 acres), overhead costs can exceed US$1500/ha (Funt, personal communications).

NET PRESENT VALUE AND INTERNAL RATE OF RETURN

A USA dollar of profit today is usually worth more than a U.S. dollar of profit tomorrow, due to inflation. This is known in economic terms as the 'time value of money.' The sum of adjusted USA dollars over a period of years is called *net present value* (NPV). The NPV can be a useful tool for growers to be able to pay a rate of interest on the initial investment and allowance for risk. Generally, a discounting rate of 6–8% is used in a NPV analysis. A NPV analysis is used for a single, long-term enterprise budget. To compare two enterprises' long-term budgets that have different cost and return streams, an internal rate of return (IRR) analysis is appropriate. For example, when two crops are compared, one with a high and early return and another with a low and later return on investment, a decision on a long-term investment can be more accurate with an IRR analysis. Software packages are available for both of these analyses. In summary, long-term investments that utilize all labor and fixed and variable costs are essential in the management process to determine potential profits.

Failure to provide a return to equity capital in the long term for one crop means that the owner/manager did not receive a return on invested capital equal to that from alternative investment (8% on berries versus 10% on bonds). Thus, it would be more profitable for the owner/manager to invest funds elsewhere than growing blackberries.

Profitability is total revenue minus fixed and variable costs. Since revenues may not occur until the second or third year (Table 17.1) after planting, growers may want to compare the payback period between crops. The payback period is the estimated time required to recover the initial costs (establishment costs or initial investment) of preplant and establishment costs. Thus, if one crop has a payback period of five years and another of seven years, then the five-year payback period is preferred. See Glossary 2 for economic term definitions.

In 1997, Funt, in Ohio, conducted several economic analyses for raspberries and blackberries. First, he compared different blackberry systems (PYO, hand-harvested, and machine-harvested thornless trellised blackberries) with different farm sizes of 8.1, 16.2, and 32.4 ha (20, 40, and 80 acres). While the different farm sizes had different overhead costs and equipment costs spread over different acreages, farm size did not influence the internal rate of return as much as other variables. In the long term, the higher the labor requirement and start-up costs, the lower the rate of return. These systems were either hand harvested or PYO for fresh market or machine harvested for processing. The PYO system had the highest rate of return over a 12-year period, generally due to a lower number of hours of labor. However, the hand-harvested system required US$1.00 more per 0.45 kg (1 lb) to make a rate of return equal to the PYO system. The hand-harvest system was not considered to be profitable at an average annual yield of 1818 kg/ha (4000 lb/acre) receiving US$1.90 per 0.45 kg. The production of 2272 kg/ha (5000 lb/acre) receiving US$1.90 per

0.45/kg (1 lb) was more desirable than 1818 kg (4000 lb) receiving US$2.00 per 0.45 kg (1 lb). Generally, there was a 3% increase in the rate of return over the life of the planting for every US$0.10 per 0.45 kg (1 lb) increase in selling price (Funt et al., 1999a).

COST OF MACHINE HARVEST

For machine harvest of a trellised blackberry system, lower yields and only a few hours of machine usage per year increased the cost per kilogram harvested. Under these conditions, there was a negative rate of return over 12 years. When a blackberry harvester is used more than 120 hours per year and harvests 1364 kg/0.45 ha (3000 lb/acre), the machine harvest cost per unit is less than the hand harvest cost. However, nearly all machine harvest berries are considered for the processing market that generally receives a lower price per unit than fresh hand-picked berries (Funt et al., 1999b). In the report of Harper et al. (1999), the estimated annual cost to own and operate an improved mechanical harvester for both fresh and processed markets for a single cultivar planting of 4.7 ha (11.6 acres) was US$2359–3294 for different trellis systems. The machine cost for fresh berries ranged from US$0.53–0.77/kg (US$0.24–0.35/lb) and hand harvest ranged from US$1.48 to US$1.52/kg (US$0.67–0.69/lb). Growers who have the management skills to produce high yields and can use a machine 80 or more hours per year are likely to lower their harvest costs by using a mechanical blackberry harvester. Berry quality, the development of suitable markets for mechanically harvested fruit, and rates of return per unit of berries sold to fresh or processing markets, respectively, must be considered before growers can make informed decisions concerning the suitability of mechanical berry harvest for their own operations. The rule of thumb is that a grower needs 8–12 ha (20–30 acres) before machines become economically preferable to hand harvest.

PRODUCTIVITY

Productivity is a key to profitability whether it is expressed as kilograms per hour of labor or kilograms per hectare. Managers will be rewarded in the long term when many factors are considered. Productivity is correlated to the quantity of quality berries harvested. Berries that are moldy or spoiled are not usable, and, as explained earlier in this chapter, the harvest cost increases when moldy berries are among the clean, fresh, unblemished berries. Systems that allow for few moldy or spoiled berries are preferred over those that have greater levels of spoilage. Systems that present berries at a comfortable reach for hand-picked or pick-your-own berries most likely have more harvested berries per hectare (acre).

If hand harvested or mechanically harvested berries are measured by weight, then a large berry size (4 g/berry as compared to 8 g/berry) is an important economic factor in the decision-making process. Therefore, workers, who are paid at a piece rate and pick a large number of kilograms (pounds) per unit of land, can make more dollars (income) per hour than those who harvest fewer pounds per hour. Further, vigorous blackberry plants can produce a greater number of fruit per unit of land if the plants are in an ideal reproductive:vegetative (fruit:foliage) balance. Therefore, berry size and number of berries that are either hand harvested or mechanically harvested while maintaining quality are major economic factors leading towards a high level of productivity (high amounts per hour per person). In the domestic and world markets, productivity needs to increase as costs increase.

SUMMARY

The commercial blackberry grower in the 21st century will need a greater amount of information to make good decisions than was needed in the 20th century. Blackberries require a larger investment in preproduction costs than other berry crops. The risk is much larger because the global market moves much faster and input costs will rise much faster than ever before. Information is needed to put many factors into a realistic plan and budget. Growers or groups of growers may require consultants on production techniques, as well consultants for the economic and business aspects. Clearly, a greater level of satisfaction will occur when the biological, technological, and economic factors are considered together in one package.

REFERENCES

Castaldi, M. and Lord, W. (1989) Bramble crop budgeting. In: Pritts, M. and Handley, D. (eds.) *Bramble Production Guide.* NRAES-35. Cornell University, Ithaca, New York, pp. 131–152.

Funt, R.C., Ellis, M.A., Williams, R., Doohan, D., Scheerens, J.C. and Welty, C. (1999a) *Brambles – Production Management and Marketing.* Ohio State University Extension Bulletin 782. Ohio State University, Columbus, Ohio.

Funt, R.C., Wall, T.E. and Scheerens, J.C. (1999b) *Yield, Berry Quality and Economics of Mechanical Berry Harvest in Ohio. Fruit Crops: A Summary of Research 1998.* Research Circular 299. Ohio Agricultural Research and Development Center, Ohio State University, Wooster, Ohio, pp. 62–81.

Harper, J.K., Takeda, F. and Peterson, D.L. (1999) Economic evaluation of improved mechanical harvesting systems for eastern thornless blackberries. *Applied Engineering in Agriculture, American Society of Agricultural Engineers* 15(6), 597–603.

Julian, J.W., Seavert, C.F., Strik, B.C. and Kaufman, D. (2009) Establishing and producing Marion blackberries in the Willamette Valley. EM 8773-revised. Oregon State University Extension Service, Oregon State University, Corvallis, Oregon.

Mason, R. and Cross, T. (1993) *Labor Demand, Recruitment and Worker Retention of the 1992 Caneberry Harvest Workforce*. Special Report 929. Oregon State University Agricultural Experiment Station, Corvallis, Oregon.

Nicholson, J.A.H. (1984) Management by objective. *Acta Horticulturae* 155, 403–408.

Pritts, M.A. (2008) Budgeting. In: Bushway, L., Pitts, M. and Handley, D. (eds.) (2008) *Raspberry and Blackberry Production Guide*. NRAES-35. Cornell University, Ithaca, New York, pp. 148–153.

Pritts, M.A. (2016) Growing blackberries in a cold climate using high tunnels. *Acta Horticulturae 1133*, Proceedings XI International *Rubus and Ribes* Symposium, pp. 263–267.

Ross, H. and Auchter, E.C. (1930) *A Production and Economic Survey of the Black Raspberry Industry of Washington County, Maryland*. Bulletin No. 322. University of Maryland Agricultural Experiment Station, College Park, Maryland, p. 208.

Schwab, G., Barnaby, G.A. and Black, J.R. (1989) Strategies for risk management. In: Smith, D.T. (ed.) *Yearbook of Agriculture*. United States Department of Agriculture, U.S. Government Printing Office, Washington, DC, pp. 151–155.

Strik, B.C., Clark, J.R. and Banados, M.P. (2007) Worldwide blackberry production. *HortTechnology* 17(2), 205–213.

18

WORLD BLACKBERRY PRODUCTION

Harvey K. Hall*

Shekinah Berries, Ltd, Tauranga, New Zealand

Harvest of blackberries for food and for preserving has been practiced for millennia, both in the Old World and in the New World with European and North American blackberries, respectively. The Greek writer Theophrastus mentioned the consumption of blackberries in 370 BC, and since this time the fruit have been gathered for fresh consumption, drying, and processing. From the time of these earliest records to the 1800s, fruit were harvested from wild stands of blackberries, and only in the 1800s was there a beginning of intentional plantings of superior types for production of fruit. From these early beginnings, varieties of blackberries were adopted into small, and then later into large-scale plantings, for commercial production.

The earliest varieties displayed many faults, including growth being difficult to manage, thorniness, extreme vigor, poor fruit set, low production, and poor quality. Intentional breeding for improved varieties began in the 1800s and the breeding efforts were formalized by government-sponsored programs in the early 1900s, both in the United Kingdom and in North America.

(i) By the mid-1900s there was significant commercial production of blackberries, primarily for processing in California, Oregon and the eastern USA, Europe, and temperate locations in the southern hemisphere. Commercial production of blackberries has expanded significantly in California and Oregon since that time, although significant production regions have been become urbanized, especially near Los Angeles. Four types of blackberries for commercial production have been reported: i) trailing blackberries or dewberries, derived from *Rubus (R.) trivialis*, in southeastern USA; (ii) upright types, derived from *R. allegheniensis*, and other species in eastern USA; (iii) semi-erect types, derived from eastern USA upright types and from European blackberry types, including *R. ulmifolius, R thyrsiger, R. procerus (R. armeniacus)*; and (iv) western trailing blackberries and their hybrids, derived from *R. ursinus*, and other species in western USA (Darrow, 1937; Strik, 1992).

* Corresponding author: shekinahberries@icloud.com

The 1974, USA blackberry production was reported to be approximately 12,500 tonnes (T) (13,750 tons (t)) from 2374 hectares (ha) (5864 acres), with 72% from Oregon ('Marion' and 'Thornless Evergreen'), 14% from California ('Olallie'), and 5% each from Washington and Texas. Loganberries and Boysenberries were planted in Washington State, but production was not reported and no information was given on the production of these cultivars in Oregon or California (Moore, 1980).

In 1978, the blackberry plantings and production in the USA were reported as 11,500T (12,650t) produced on almost 2200ha (5434 acres), about 95% of which was produced in Oregon (75%), California (15%), and Washington (5%). Another 3% was produced in the USA in Texas, Oklahoma, and Arkansas (Skirvin and Hellman, 1984). As in the publication by Moore (1980), Boysenberries (CA 173 ha (427 acres) and 1019T (1121t), OR 304 ha (751 acres) and 907T (998t), WA 0.4 ha (1 acre) and 0.4T (0.4t)) and Loganberries (OR 109 ha (269 acres) and 533T (586t)) are included separately and not included in the blackberry totals. In addition, a further 142 ha (351 acres) and 470T (517t) production are recorded separately for 'Olallie' in California.

In the UK in the 1980s, commercial interest in the production of blackberries was based on the hybrid cultivars, 'Loganberry' and 'Bedford Giant,' and the new cultivars, 'Tayberry' and 'Tummelberry,' were also finding a place in commerce (Jennings, 1988).

By 1991, the worldwide blackberry industry, growing blackberries and hybrid blackberries, was focused on the Pacific Northwest (western Oregon and western Washington in the USA and southwestern British Columbia in Canada), parts of coastal USA in California and Arkansas, and New Zealand. New production regions were becoming important in Colombia and Brazil, South America (Jennings et al., 1991; Strik, 1992). Until that time, a small amount of sales was to the fresh market, especially to pick-your-own (PYO) and on-farm sales (gate sales), but the bulk of the fruit in each country was used for processing (see Chapter 16), as the handling ability and shelf life was not sufficient for market requirements. In the 1990s, fresh fruit were not found on grocery store shelves in eastern and, only rarely, in western USA (Strik et al. 2008).

In the USA Pacific Northwest, production in 1990 on 3184 ha (7864 acres) was listed as mostly trailing blackberries (over 98%), primarily in Oregon. However, a closer examination of the information presented by Strik (1992) showed that production was 71.3% blackberries and hybrid blackberries, derived from *R. ursinus*, and 28.7% tetraploid blackberries, 27% of which was from 'Thornless Evergreen,' 0.8% from semi-erect types, and 1% from erect types. In Oregon, over 85% of production was machine harvested, but only 10% was machine harvested in California. In Oregon and Washington, more than 90% of the fruit from trailing blackberries was marketed for processing, but 50–80% of the erect and semi-erect crop was marketed fresh. In California,

95% of the erect and semi-erect crop was marketed fresh with the remainder as PYO.

In 1990, there were 1205 ha (2976 acres) in blackberry production east of the USA Rocky Mountains (Clark, 1992). This was divided into 12 ha (30 acres) of trailing (43% Boysenberry and thornless Boysenberry, 19% 'Gem,' 7% 'Lucretia,' 2% 'Youngberry,' and 29% other), 411 ha of semi-erect (23% 'Black Satin,' 21% 'Chester,' 21% 'Hull,' 19% 'Dirksen Thornless,' 10% 'Thornfree,' and 5% other), and 782 ha (1932 acres) of upright types (35% 'Shawnee,' 14% 'Rosborough,' 13% 'Brazos,' 13% 'Cheyenne,' 8% 'Darrow,' 6% 'Cherokee,' 4% 'Womack,' 3% 'Comanche,' 1% 'Navaho,' 1% 'Choctaw,' and 3% other). All harvesting east of the USA Rocky Mountains was done by hand. The marketing for these berries was 62% PYO, 36% pre-picked, fresh market, and 2% processing.

Another survey of blackberry production was conducted in 2005 (Strik et al. 2007; Strik et al. 2008). This survey was expanded from examining USA production to looking at production around the world (Table 18.1). The ongoing delineation of four types of blackberries was reported as per Darrow (1937, 1967) and Strik (1992). However, a differentiation of thornless types and new types, specifically suited for fresh market, was also evident across the western USA trailing types and eastern USA erect types and the semi-erect types. Additionally, this period marked the beginning of commercial development of primocane-fruiting and dual-cropping types among the eastern USA erect types. By this time, the southeastern USA trailing types or dewberries had almost disappeared, except for the contribution of genetic background to other types through Brazos.

In Europe in 2005, 69% of the production area (5300 ha (13,091 acres)) was in Serbia (25,000T (27,500t)), the largest area of production in the world and fourth highest production, with the predominant cultivars being 'Thornfree,' 'Dirksen Thornless,' 'Smoothstem,' and a new cultivar, 'Čačanska

Table 18.1. World blackberry production in 2005 (Strik et al. 2008).

Region	Area planted (ha)*	Production (T)**
Europe	7692	43,000
North America	7159	59,123
Central America	1640	1590
South America	1597	6380
Asia	1550	26,350
Oceania	297	3650
Africa	100	200
Total	20,035	140,292

*To convert hectares (ha) to acres, multiply ha×2.5.
**To convert tonnes (T) to tons (t), multiply T×1.1.

Bestrna,' This was followed by Hungary with 21% of the area (1600 ha (3952 acres)), dominated by 'Loch Ness' with 75% of the planted area and 90% of the production. The other 10% of the production in Europe was spread over the remaining countries with the largest producers being Croatia (180 ha (445 acres)), Germany (110 ha (272 acres)) and the UK, Romania, and Poland with 100 ha (247 acres) each. In Germany and Romania, 'Loch Ness' was the dominant cultivar, and in Poland, the cold hardy cultivar 'Gazda' was the main variety. In the UK, 'Loch Ness' was also grown along with a wide range of other cultivars, including trailing and semi-erect types.

In North America in 2005, the main production region was in the USA, comprising 67% of the area (4818 ha (11,900 acres)) (Strik *et al.* 2008). Almost two-thirds of the planted area in the USA was in Oregon (3138 ha (7,751 acres)), of which 95% of the production (22,848 t (25,133 t)) was machine harvested and sold for processing. Cultivars derived from *R. ursinus* comprised 83%: 'Marion' (61%), Boysenberry (15%), 'Silvan' (7%), and tetraploid blackberries of European and/or North American descent (17%) ('Thornless Evergreen' (11%), 'Chester' (3%), 'Cherokee' (0.6%), 'Navaho' (0.3%), and others (0.7%)).

The second largest production in the USA in 2005 was based in California with 283 ha (699 acres) and 2359 T (25,949 t) production (Strik *et al.* 2008). Since the survey conducted in 1992, the focus of Californian production had shifted so that Boysenberry then occupied only 40 ha (100 acres) and the main focus of blackberry production became blackberry types for fresh market, with 'Chester' occupying 113 ha (279 acres). In 2005, fresh market production in California was predominantly proprietary cultivars, comprising 60% of the trailing types and 85% of the erect types; Texas had 275 ha (679 acres) and 726 T (799 t) production and AR 243 ha (600 acres) and 1400 T (1540 t) production with upright cultivars being the only cultivars grown. In Texas, 'Kiowa,' 'Brazos,' and 'Roseborough' comprised 85% of the planted area; in AR the cultivars grown included 'Arapaho,' 'Navaho,' 'Ouachita,' 'Apache,' 'Chickasaw,' and 'Kiowa.' Georgia (GA) USA was also included with 127 ha (314 acres) planted.

In 2005, Mexico accounted for 32% of the planted area in North America, (2300 ha (5681 acres)), 93% of that in the state of Michoacán. The main cultivars grown were 'Brazos' and 'Tupy'; the remainder were comprised of a small amount of proprietary cultivars. Production systems in Mexico were modified to extend the harvest season, so that cropping in the region covered much of the year.

In Central America, there were 1640 ha (4051 acres) of blackberries in 2005, with a production of 1590 T (1749 t), (mainly 'Brazos' and *R. glaucus*). The majority were in Costa Rica (1550 ha (3829 acres)) with the remainder in Guatemala.

In South America, there were 1597 ha (3945 acres) of blackberries reported in 2005 and 6380 T (7018 t) produced (Strik *et al.* 2008). Ecuador accounted for 53% of the planted area (850 ha (2100 acres)), producing 'Brazos' and *R. glaucus*, which is also produced in Costa Rica. Chile had 450 ha (1112 acres) of commercial blackberries, with a production of 3879 T (4267 t). This was supplemented by 5800 T (6380 t) of fruit harvested from wild *R. ulmi-folius*. Production from commercial plantings was from *R. ursinus* derivatives, primarily Boysenberry, as well as from semi-erect and erect cultivars, most of which were exported for the processing industry. In Brazil, plantings of 250 ha (618 acres) of 'Tupy' and 'Guarani' were producing 780 T (858 t), mostly for local processing; 15% were exported. Other southern hemisphere plantings were reported in NZ (221 ha (546 acres)), South Africa (100 ha (247 acres)), and Australia (38 ha (94 acres)).

In 2005, all the production in Asia was reported from China, with 26,350 T (28,985 t) from 1550 ha (3829 acres), mostly in seedlings of 'Hull Thornless' and 'Chester Thornless'; the remaining plantings were in 'Shawnee,' Boysenberry, 'Marion,' and 'Siskiyou.' Most production was in the Jiangsu province, but plantings were also underway in Liaoning, Shandong and Hebai provinces.

Since 2005, there have been significant developments in blackberry production and in the cultivars available for planting. The breeding program in Arkansas has provided much of the impetus for the change with the continued development of primocane-fruiting cultivars. This has resulted in the USA release of the home garden cultivars, 'Prime Jim®,' 'Prime Jan®' (thorny), and 'Prime-Ark® Freedom' (thornless), and the fresh-market cultivars, 'Prime-Ark® 45' (thorny) and 'Prime-Ark® Traveller' (thornless). Sales of genetics from the University of Arkansas to proprietary marketing companies and nurseries has also led to the development of breeding programs for primocane fruiting blackberries in Australia, California, Chile, and the UK. A program in Switzerland also has been developed, using the cultivars 'Prime Jim®' and 'Prime Jan®' as starting material.

The Hargreaves Plants Breeding Program in the UK has released 'Reuben,' a productive variety suited to conditions in England. It is likely to find an ongoing place in home gardens in Europe. However, it does not have the fruit qualities required for fresh marketing through supermarket outlets.

In Australia, the program initiated by Costa's Berry Exchange has produced new selections that appear suitable for fresh marketing (Bardon, personal communication). In Switzerland, an early, thorny (spiny) variety, Direttissima® Montblanc®, derived from the Arkansas primocane-fruiting material, has also been released. New varieties of primocane-fruiting blackberries, including 'Amara,' have been released in Chile from this same source of genetics.

Driscoll's Strawberry Associates (DSA) in the USA has invested signifi-cantly into blackberry cultivar development. The release of the primocane-fruiting, 'Elvira,' and the floricane-fruiting cultivar, 'Victoria,' has created significant expansion of fresh-market blackberry production for sales in the

Driscoll brand in Europe, Australia, and the USA. 'Victoria' has outstanding flavor, fruit size, fruit firmness, and shelf life.

In the commercial market of berries around the world, there has been significant increase in the supply of strawberries and raspberries resulting in a saturated market and a drop in prices, especially at some times of the year. Marketers and producers have seen blackberries as the next step in profitable growing and marketing. Mexico has led the way with expansion of 'Tupy'. It is the major contributor to increased production, reaching almost 11,000 ha (27,170 acres) in 2015 (Ahumada, 2016). 'Tupy' has provided growers with a high yield of berries with good fresh-eating quality and the ability to be picked, handled, stored, and shipped to distant markets, primarily in the USA. Flavor is mild and sweet; the texture of fruit is good for fresh eating, without objectionable seediness, and production can be tailored to supply at defined marketing windows, allowing retail sales to be made throughout the year.

In 2016, the main region for process blackberry growing in Europe remained in Serbia with the hand-picked production of 'Čačanska Bestrna.' The development of cold hardy blackberries in Poland continued and production for processing was likely to increase significantly for the future. In the USA, Oregon continued to be the main producer of blackberries for processing with 'Marion' the dominant cultivar for machine harvesting. However, there had been a number of new cultivar releases in Oregon in recent years and the plantings of 'Columbia Star' were expanding rapidly. 'Columbia Star' appeared poised to eclipse 'Marion' before 2020.

In 2016, DSA fresh-market production was unable to meet the market demand for fruit of 'Victoria' in Europe. Plantings of this variety and 'Elvira' were expanding rapidly (Van Der Most, 2015). In the USA, the number of blackberry growers and planted area had increased significantly since 2005, (grower numbers expanded 76% between 2008 and 2015), with the majority of the increased production being for fresh-market production, for supply to markets, and for PYO and grower sales.

Expansion throughout the world is continuing, primarily dependent on the supply of cultivars with adaption to the growing regions and the quality of fruit being satisfactory for the market. The future of blackberries is promising, for this healthy, nutritious fruit, which is becoming an expanded part of the diet in countries around the world.

REFERENCES

Ahumada, M. (2016) Blackberry production in Mexico. Available at: http://cesanluisobispo.ucanr.edu/files/239571.pdf (accessed November 20, 2016).

Clark, J. R. (1992) Blackberry production and cultivars in North America east of the Rocky Mountains. *Fruit Varieties Journal* 46(4), 217–222.

Darrow, G.M. (1937) Blackberry and raspberry improvement. *USDA Yearbook of Agriculture, Yearbook 1937*. U.S. Department of Agriculture. Government Printing Office, Washington DC, pp. 496–533.

Darrow, G.M. (1967) The cultivated raspberry and blackberry in North America – breeding and improvement. *American Horticultural Magazine* 46(4), 203–218.

Jennings, D.L. (1988) *Raspberries and Blackberries: Their Breeding, Diseases and Growth*. Academic Press, London.

Jennings, D.L., Daubeny, H.A. and Moore, J.N. (1991) Blackberries and raspberries (*Rubus*). *Acta Horticulturae* 290, 331–392. DOI:10.17660/ActaHortic.1991.290.8

Moore, J.N. (1980) Blackberry production and cultivar situation in North America. *Fruit Varieties Journal* 34(4), 36–41.

Skirvin, R.M. and Hellman, E.W. (1984) Blackberry products and production regions. *HortScience* 19(2), 195–197.

Strik, B.C. (1992) Blackberry cultivars and production trends in the Pacific Northwest. *Fruit Varieties Journal* 46(4), 202–206.

Strik, B.C., Clark, J.R., Finn, C.E. and Bañados, M.P. (2007) Worldwide blackberry production. *HortTechnology* 17(2), 205–213.

Strik, B.C., Clark, J.R., Finn, C.E. and Bañados, M.P. (2008) Worldwide production of blackberries. *Acta Horticulturae* 777, 209–217.

Van Der Most, B. (2015) High demand for blackberries leads to increased production in Europe. Available at: www.freshplaza.com/article/147267/High-demand-for-blackberries-leads-to-increased-production-in-Europe (accessed November 20, 2016).

Appendix 1: Windbreaks

Michele Stanton*

University of Cincinnati, Cincinnati, Ohio, USA

Wind is possibly the most important source of environmental mechanical stress in blackberry production (Casierra-Posada and Aguilar-Avendano, 2008). When choosing a specific site for blackberry production, it is important to evaluate the amount of wind to which the plants might be exposed. In production areas such as Colombia, where blackberries are grown across different elevations, microclimates can vary over short distances with changes in altitude (Fischer, 2005).

Blackberry plants are fairly sensitive to wind damage. In Colombia, air movement at speeds of $1.7\,\text{m·s}^{-1}$ (3.8 mph) was beneficial, increasing water movement throughout the plants and cooling the leaves in warm weather; but wind at speeds of $2\,\text{m·s}^{-1}$ and $4\,\text{m·s}^{-1}$ caused stress that resulted in reduced fruit yields (Casierra-Posada and Aguilar-Avendano, 2008). In Scotland, raspberries were found to respond favorably to wind protection where average wind speed was $1.7\,\text{m·s}^{-1}$ (3.8 mph) (Waister, 1970). If protection is desired, it is recommended that it be in place during establishment of the orchard (Prive and Allain, 2000).

Initially, excessive wind tears, abrades, or shakes unprotected plants, causing the edges of one leaf or cane to rub or strike another. This type of injury presents initially as a brown mottling (which eventually turns to purple) on upper leaf surfaces, canes, and petioles (Newenhouse, 1991). Torn leaves, blossom or fruit drop, 'sandblasting' by airborne soil particles, and plant lodging may also occur. Wind also indirectly damages plants by increasing evapotranspiration rates so that available soil moisture is rapidly depleted. Desiccation in winter can be especially problematic in areas where soil freezes. Winds may reduce the current season's crop; in strawberries, wind also reduces the next season's yields (Waister, 1973). Chronic wind exposure reduces mature plant size, causes plants to put more energy into leaf and cane structure, and reduces fruit yield (Prive and Allain, 2000).

---

* Corresponding author: michele.stanton@uky.edu

Acute damage can occur through infrequent high winds, such as found in passing storms. As little as 30 seconds' exposure to wind gusts of $10–20\,\text{m·s}^{-1}$ ($22–44\,\text{mph}$) or higher is sufficient to injure plants. In either case, plants benefit from protection against wind (Hodges and Brandle, 1996).

Two main methods of protection are utilized by blackberry growers: trellises, which help stabilize plants (Vanden Heuval et *al.*, 2000), and windbreaks or shelterbelts. A windbreak is a group of trees, shrubs, or other plants that is installed and maintained so as to reduce and redirect wind away from a specific area. Fences composed of alternating slats or synthetic materials are also used. A windbreak or shelterbelt is designed to slow or filter wind, but not stop it entirely.

Windbreaks create a better environment for the orchard so that plant growth and yield are higher than in unprotected areas. In some cases, windbreaks make possible crop production where it is otherwise not feasible. Both air and soil temperatures in protected areas can be a few degrees warmer, especially early in the season, which may give growers a slightly earlier harvest than in adjacent fields (Brandle et *al.*, 2004). Windbreaks can also reduce coastal salt sprays, drift of chemical sprays, lessen dust and soil erosion, reduce infiltration of windborne disease inoculum, provide a more favorable habitat for pollinating and predatory insects (Vaughn and Black, 2008), and sequester more carbon in the landscape (Brandle et *al.*, 2004). In regions where winter snowfall is a significant source of soil moisture, windbreaks help capture that moisture by decreasing drift (Brandle et *al.*, 2004). They lessen energy needs for protected dwellings and help create an easier environment in which to work.

Windbreaks have associated costs. In addition to the expense of installation and maintenance, the grower must consider the cost of land lost to production. The cost/benefit analysis should include land costs, crop prices, yield differential, irrigation, years to plant maturity, increased pollinator presence, decreased pesticide requirements, and decreased erosion.

The height of a windbreak is its most important design factor. A windbreak protects crops on the windward side for a distance of two to five times its height, expressed as 2–5H. Plants on the leeward side (away from the wind) are protected for a distance of up to 20–30H; protection decreases as you move away from the windbreak, with maximum protection obtained within 10H. A single row of trees 10 m (33 ft) in height, for example, would afford protection to plants up to 200–300 m (650–980 ft) away, with the greatest degree of protection for the first 100 m on the leeward side. This single row should be designed without large gaps, since wind can funnel through openings at a greater intensity. Large plantings may need rows of windbreaks which should be planted every 10H, if space allows (Fig. A1.1).

The next design aspect is density or porosity. Medium density windbreaks (40–60%), such as provided by a single row of native evergreen trees, are recommended for most field crops. Their density affords adequate protection while their porosity allows for dissipation of turbulence. Medium density windbreaks also give excellent soil erosion control (Brandle and Finch, 1991). Density can be visually estimated.

Fig. A1.1. Layout of a blackberry planting, where the shelter belts are placed at regular intervals at ten times the height of the mature shelter belt.

The windbreak should be perpendicular to the direction of oncoming winds (Fig. A1.2). Its continuous length should be at least ten times the mature height of the tallest trees. In the case above, a row of trees 10 m (33 ft) tall would suggest that the windbreak be a minimum of 100 m (330 ft) in length. In many areas, winds may come from one direction in winter and from a different direction another part of the year. In such cases, windbreaks should be composed of multiple lines, each line at least ten times the tree height.

A single species of native coniferous or broadleaf evergreen trees is efficient; however, a mixed-species row may give insurance against diseases or pests that might afflict a singles-species planting. Species should be selected for excellent adaptability to existing climate and soils, with preference given to those which will not compete aggressively with adjacent blackberries for water and nutrients.

Growers may also wish to place a second or third row of plants, often deciduous perennials or annuals, on the leeward side of the initial row of evergreen trees; these can be chosen to provide pollinator habitat, or as an additional crop for extra income. Species diversity is often helpful in providing

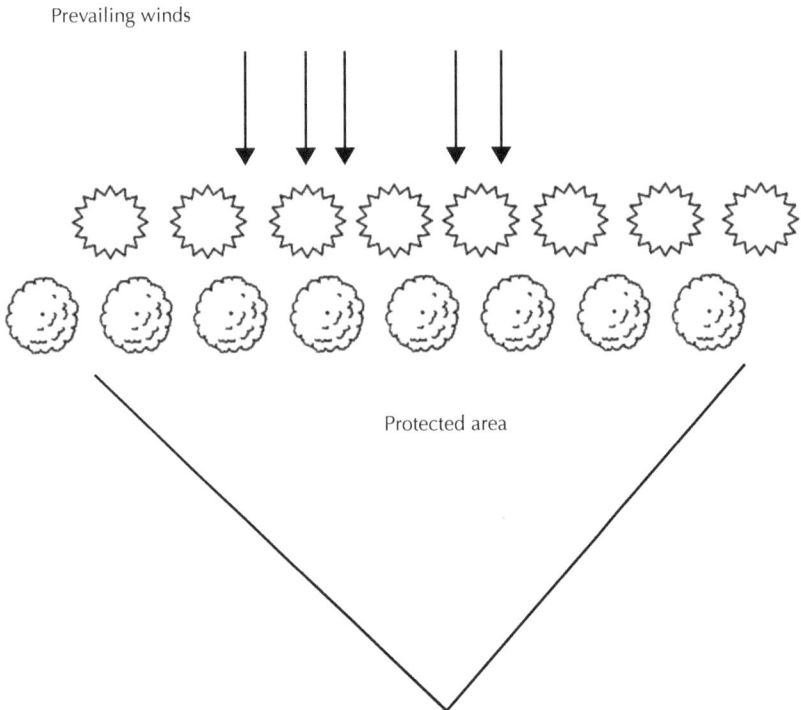

Fig. A1.2. A shelter belt placed perpendicular to the prevailing winds offers optimum protection.

biological control of pests or diseases (Wilkinson and Elevitch, 2000; Hodges and Brandle, 2006).

REFERENCES

Brandle, J.R. and Finch, S. (1991) *How Windbreaks Work*. University of Nebraska-Lincoln Extension publication EC91-1763-B. University of Nebraska-Lincoln, Lincoln, Nebraska.

Brandle, J.R., Hodges, L. and Zhou, X.H. (2004) Windbreaks in North American agricultural systems. *Agroforestry Systems* 61(1), 65–78.

Casierra-Posada, F. and Aguilar-Avendano, O.E. (2008) Respuestas fisiologicas y morfologicas de plantas de mora (*Rubus* sp.) sometidas a estres por viento inducido. *Revista Colombiana de Ciencias Horticolas* 2(1), 43–53.

Fischer, G. (2005) Aspectos de la fisiología aplicada de los frutales promisorios en cultivo y poscosecha. *Revista Comalfi* 32(1), 22–34.

Hodges, L. and Brandle, J.R. (1996) Windbreaks: an important component in plasticulture systems. *HortTechnology* 6(3), 177–181.

Hodges, L. and Brandle, J. (2006) *Windbreaks for Fruit and Vegetable Crops*. University of Nebraska-Lincoln Extension. Publication EC1779. Available at: http://extensionpublications/unL.edu/assets/pdf/ec1779.pdf

Newenhouse, A. (1991) How to recognize wind damage to leaves of fruit crops. *Hort-Technology* 1(1), 89–91.

Prive, J.P. and Allain, N. (2000) Wind reduces growth and yield but not net leaf photosynthesis of primocane fruiting red raspberries (*Rubus idaeus* L.) in the establishment years. *Canadian Journal of Plant Science* 80(4), 841–847.

Vanden Heuval, J.E., Sullivan, J.A. and Proctor, J.T.A. (2000) Cane stabilization improves yield of red raspberry (*Rubus idaeus* L.) *Journal of Horticultural Science* 35(2), 181–183.

Vaughn, M. and Black, S.H. (2008) Native pollinators: how to protect and enhance habitat for native bees. *Native Plants Journal* 9(2), 80–91.

Waister, P.D. (1970) Effects of shelter from wind on the growth and yield of raspberries. *Journal of Horticultural Science* 45(4), 435–445.

Waister, P.D. (1973) Climatic limitations on horticultural production with particular reference to Scottish conditions. *International Journal of Biometeorology* 17(4), 379–383.

Wilkinson, K. and Elevitch, C. (2000) *Multipurpose Windbreaks: Design and Species for Pacific Islands*. Permanent Agriculture Resources, Holualoa, Hawaii, Ch. 8, 203–232.

Appendix 2: Fertigation and Drip Irrigation Primer

David S. Ross[1]* and Richard C. Funt[2]

[1]University of Maryland, College Park, Maryland, USA; [2]Ohio State University, Columbus, Ohio, USA

The topic of fertigation involves water quality, making and using stock solutions, selecting and using delivery equipment, calibration of delivery equipment, irrigation system considerations, and crop considerations. All of these will be discussed in this Appendix.

Fertigation is an important part of crop production and is the process of applying one or more agricultural plant nutrients through an irrigation system to the crop's root zone to meet a portion of a crop's fertilizer needs.

A concentrated solution of nutrients known as a stock solution is used. Through this liquid feed method, nutrients may be applied as needed for maintaining good crop growth, as long as irrigation water can be applied. A well-designed drip irrigation system with uniform water application can provide an excellent partner for utilizing fertigation with a blackberry production system in the field, a high tunnel, or a greenhouse (see Chapter 10, Fig. 10.2)

Nitrogen- (N) and potassium- (K) based fertilizers are the most commonly applied nutrients by fertigation. Some formulations of phosphorus (P) and micronutrients, such as zinc (Zn), copper (Cu), iron (Fe), and manganese (Mn) can also be applied if compatible with the irrigation water, particularly a pH of less than 6.5. Also, because of precipitation problems, be careful not to mix K fertilizers with calcium (Ca) nitrate and Fe. For calcium nitrate and iron chelate, use stock tanks separate from those used for other fertilizer materials. Applying all P before planting, based on a soil test, is recommended (Rosen et al., 2004).

Fertigation is used over the growing season during several irrigation events to apply one-half to two-thirds of the total plant needs for the season. The frequency of application is not as important as the total rate applied. Crop nutrient requirements may vary over the season and application should match

* Corresponding author: dsross@umd.edu

those seasonal needs as a supplement to dry applied nutrients. Soil tests and leaf analyses are recommended to determine the amounts of nutrients to apply. Many local factors can affect the amount of fertilizer and water to apply. Irrigation moisture sensors should be used to determine when to apply water, and for how long, with the goal to maintain moisture at 50% or more of field capacity. Excessive water application will leach the nutrients away from the root zone (Rosen et *al.*, 2004).

Nutrients in the irrigation pipeline can support the growth of algae in drip irrigation lines, so treating the water with chlorine or other material may be needed to keep the lines clean. Fe in the water can also be a problem. Maintaining a low level of active chlorine (1–3 parts per million (ppm) at end of lines) with line flushing, if necessary, can reduce problems.

If the cropping is done in containers with substrates or using hydroponics, then the salt level should be monitored regularly for electrical conductivity (EC) levels so that salt levels do not climb too high. As needed, the containers should to be flushed to lower EC values.

Soil acidity and alkalinity may need to be adjusted for the crop, but fertigation is not expected to cause a problem in most cases. Iron formulation used is subject to pH levels but no toxity of elements is to be expected (Kafkalfi and Tarchitzky, 2011).

WATER QUALITY AND NUTRIENT ISSUES

Water quality and fertilizer choice are the two most important considerations per Katkafi and Tarchitzky (2011). Saline irrigation water is common in arid and semi-arid climatic regions and is found in coastal areas. Sensitivity of plants to solution salinity varies between plant species and cultivars, so growers in these regions should be aware of their EC levels and plant sensitivity. Presence of nitrates ($70–140\,$g N m^{-3}) and calcium ($200–400\,$g Ca m^{-3}) in solution can reduce the salinity hazard to irrigated crops (Kafkalfi and Tarchitzky, 2011).

Specific ion toxicity can be a problem. Sodium (Na) in large quantities hampers root elongation and competes with Ca^{2+} on specific adsorption sites in the cell walls of the elongation zone. Chloride (Cl) is present in large quantities in saline water and can cause scorching and complete leaf death, but potassium nitrate or calcium nitrate in the solution can reduce Cl uptake (Kafkalfi and Tarchitzky, 2011).

Treated waste water is being used in different parts of the world in increasing amounts as demand for potable water increases. It can add to the salt content of soils in a negative manner. However, the nutrient levels of N, P, K, and micronutrients are typically higher than in fresh water. The form of the nutrient depends on any treatment process used for the TWW. The nutrient content can vary with season, source, and water treatment process and the availability, amount, and timing of nutrients in the waste water can vary with time. Growers who have the opportunity to use TWW should investigate the source and the treatment process, and use continuous monitoring of water quality (Kafkalfi and Tarchitzky, 2011).

For those using a micro-irrigation (drip) system, Fe, particularly ferrous iron (Fe^{2+}) with concentrations as low as 0.15–0.22 g Fe m^{-3}, is considered potentially hazardous for clogging drip emitters. Any water source with concentrations of more than 0.5 g Fe m^{-3} should not be used for irrigation without treatment (Kafkalfi and Tarchitzky, 2011). Waters with levels of Fe above 4.0 g Fe m^{-3} are considered useless for irrigation. Fe bacteria acts on the Fe slime to oxidize it to form ferric iron (Fe^{+3}), which is insoluble. The ferric iron surrounded by the filamentous bacterial colonies creates the sticky Fe slime gel that clogs emitters. Cl gas is a good treatment if good mixing of the water and gas is achieved and sand filters followed by backup filters are used to remove precipitate (Kafkalfi and Tarchitzky, 2011).

High concentrations of Ca, magnesium (Mg), and bicarbonate (HCO_3) mean high total hardness and increase the hazard of clogging, especially when P fertilizers are introduced. Precipitation of calcium carbonate to form scale deposits is common in alkaline water that is rich in Ca and HCO_3 (Kafkalfi and Tarchitzky, 2011).

Preparing the stock solution and applying it at the correct or desired rate is important. A basic description of a dosimeter (fertigation controller) that meters a concentrated nutrient solution into irrigation water going to the crop is covered here. Maintenance of the dosimeter is critical to success as is the regular maintenance of the equipment. Regular calibration should also be done each season to ensure proper application of fertilizer occurs (Weiler and Sailus, 1996).

DOSIMETER

The purpose of a dosimeter is to meter the concentrated fertilizer solution (stock solution, described in a later section) into the irrigation water. An injection ratio of 1:100 or 1:200 is typically used. This means that the dosimeter set at a 1:100 ratio will put one l (1 qt) of the concentrated stock solution into 99 liters (25 gal) of irrigation water to make 100 liters (25 gal) of dilute, nutrient rich irrigation water.

If the dilute, nutrient rich irrigation water is to contain 125 parts per million (ppm) nitrogen, then the stock solution (concentrate) will contain 125 ppm × 100 (proportioner ratio) = 12,500 ppm nitrogen (for a 1:100 dilution). The stock solution is made 100 times more concentrated than the final application rate, so a smaller container is required and enough concentrate can be made to cover a large amount of the crop. The total amount of nutrients is dissolved into a quantity of water that matches the requirements of the dilution ratio. Care must be taken to have 100% solubility with no interaction between nutrient ingredients. The stock solution is then metered into the irrigation water at the specified ratio to make the dilute nutrient solution desired for the crop.

Different types of dosimeters are available:

1. One type is an *electric motor driven pump* that has a metering valve for adjusting the percentage of total flow rate that can pass through. It pumps the stock solution into the irrigation pipe after the irrigation pump to achieve the desired dilution ratio.

2. Another type has an *irrigation water powered pump that drives a second injection pump*. This self-contained unit causes injection to occur when irrigation water flows through the pump; the injection rate is adjusted by fixed ratio settings to control the amount of concentrate added. This type of injector is available as a portable unit that can be moved around and connected into the irrigation line in the field. It is available in many small sizes for small systems and is easy to use. It takes its power from the flowing water.

3. A third type of dosimeter is a *venturi device* that requires a large pressure drop across it to create a suction to pull the fertilizer solution into the irrigation line. The venturi device is placed in parallel to the main irrigation line where a valve restricts the irrigation water flow to cause a pressure drop (pressure loss) to occur across the valve. A high velocity water flow through a small diameter section of the venturi causes a suction on a pipe coming from the stock solution to draw in the nutrients. The venturi device may appear to be the lowest cost system, but it may require a booster pump to create the higher pressure needed to make it work properly. The device is inexpensive, but the larger pump and energy cost to operate it must be considered. The pressure must be constant to get uniform flow.

Large operators may design their irrigation systems to use a specific steady flow rate of water, say 600 liters per minute (lpm), for all irrigation zones. The dosimeter can be an electric motor driven piston or diaphragm pump that moves a certain lpm of stock solution, say 3.1 lpm at a 100% flow rate setting, into the pipeline. The dosimeter's injection rate can be adjusted between 1 and 100% or 0 and 3.1 gallons per minute (gpm) of stock solution. Thus, the dosimeter injection ratio can be varied. The actual ratio at any setting is roughly the lpm pumped by the dosimeter divided by the lpm pumped by the irrigation pump. A calibration will give the ratio.

Small operators who have irrigation zones of varying sizes may choose to use irrigation water driven dosimeters sized for their range of water flow rates. The irrigation water flow rate through dosimeter drives the pump injecting the fertilizer so no other power source is required. The dosimeter pump speeds up or slows down with the irrigation water flow rate and so does the rate of injection. Therefore, a constant injection ratio can be maintained over a range of water flow rates. These dosimeters are purchased in a size that matches the irrigation water flow rate. Or, they can be used in a parallel circuit that carries part of the irrigation water (see Chapter 10).

CALIBRATION

Calibration is a process of checking the fertilizer injection system to see whether it is functioning correctly. The goal is to verify that the desired amount of nutrients is being delivered by the irrigation system. Calibration should be done at least once per year or when any changes are made to the irrigation system or settings. A good manager will check the calibration frequently to insure correct and economical operation. Incorrect fertigation may cause crop damage or delay in growth, which will affect crop sales.

Preparation for calibration

Several things should be checked before beginning the calibration. The equipment must be serviced and made ready in good condition. Check the suction tube strainer to see that it is clean and fluids are moving through it freely. Clean the stock tank of all settled material. Install the suction tube at least 5 cm (2 in) from the bottom of the stock tank so any solids are not pulled in. Inspect the injector and service all O-rings, if any, in the injector. Check pipe connections and any valves to see that they are properly connected and valves are open or closed as appropriate. The owner's or operator's manual for the injector should have specific instructions for maintenance and should be followed.

Calibration by volume

One method of checking the injection system is to verify that the correct amount/volume of stock solution is being delivered into the irrigation water. This method requires a means of verifying the irrigation water flow rate. An accumulating water meter is good, or for low flow rates, a 20 l (5 gal) bucket may be used. Also needed is a means of accurately measuring the amount of stock solution being injected. Sizes of these containers can be varied to fit the situation.

A dosimeter ratio is determined by dividing the amount of stock solution used in liters (gallons) by the amount of irrigation water used and reduce the ratio.

liters (gallons) stock solution
liters (gallons) irrigation water

If the dosimeter is set for 1:100 dilution ratio, the answer should reflect that.

E.g.
3 liters stock solution
300 liters irrigation water

The following is a simple procedure for calibrating a small dosimeter system to determine the ratio:

1. Get a measuring cup marked for 1000 ml (34 oz), a 20 l (5 gal) bucket, paper and pencil, and a calculator.
2. Measure a small quantity of stock solution into a bucket or other container from which the liquid can be removed by the dosimeter. A suggested amount would be 1000 ml (34 oz) in a measuring cup marked in ml/oz. During the test, the system will be run until the 20 l (5 gal) bucket is filled with irrigation water. If necessary, several buckets of water can be collected and dumped, just keep track of the total volume of water collected.
3. Run the injector and irrigation system for a few minutes to purge air from the injector and irrigation line.
4. Place the suction tube into the 1000 ml (34 oz) measuring cup filled to the 950 ml (32 oz) mark. Place the discharge irrigation line into a 20 l (5 gal) bucket to catch the dilute fertilizer and irrigation water solution from the injector.
5. Run the injector and irrigation system until the 20 l (5 gal) bucket is filled (or more than one bucket if necessary). Stop the injector and irrigation system at that point and remove the suction tube from the measuring cup.
6. Measure accurately the amount of stock solution remaining in the measuring cup. Starting with 950 ml (32 oz) and ending with 650 ml (22 oz), for example. Subtract the remaining amount from the initial amount. In this example, 300 ml (10 oz) were injected.
7. Determine the injection ratio by comparing the amount of stock solution injected to the amount of water collected on the discharge side.

Injection ratio = amount injected in ml (oz): amount of dilute irrigation solution discharged in ml (oz).

Injection rate (Metric units) = 300 ml: 19,200 ml (collected from nearly full 20 liter (5.3 gal) container)

$$= 300 \text{ ml}: 19{,}200 \text{ ml}$$
$$= 1{:}64, \text{ which is obtained by dividing both sides by 300.}$$

Injection ratio (United States (US) units) = 284 g (10 oz):19 l (5 gal)) × 3.81 (128 oz/gal)

$$= 284 \text{ g} (10 \text{ oz}){:}18.9 \text{ l} (640 \text{ oz})$$
$$= 1{:}64, \text{ which is obtained by dividing both}$$
$$\text{sides by 10.}$$

8. Check the injection ratio set on the injector, if it has settings. If the injector is working properly, it is set to inject at about 1:64 ratio. If the injector does not have ratio settings, then use the 1:64 ratio to create the stock solution to give you the amount of nutrient in ppm that you wish to inject.

9. Just to review this calculation, the stock solution was made 64 times more concentrated than the dilute solution to be applied through irrigation so the injector is mixing one part concentrate into each 64 parts of clear irrigation water. This results in the desired application rate of the fertilizer. In this case the injection ratio was learned by doing the calibration. Common ratios are 1:100, 1:150, 1:200, etc.

10. If an error occurs, check the setting of the dosimeter and perform a maintenance check on the equipment. Do the test in the same manner as normal watering so there are no differences in water flow rate or time of application. The equipment operator's manual should give instructions on making adjustments to correct the ratio.

11. Repeat this calibration procedure several times over the season.

12. This calibration should be repeated two or three times to verify the ratio value. Volumes of water and stock solution can be adjusted to better fit the size of equipment used and amount or flow rate of irrigation water used. The main idea is to measure both the concentrate solution and the dilute solution so the ratio can be calculated and compared to the ratio setting on the injector, if it has them. Some injectors may have a percentage total flow scale so relate the ratio to the percentage setting. If the injector does not have an injection ratio, then the ratio that is determined can be used to calculate the stock solution if sufficient information is known (refer to calibration example in the next section).

Calibration by nutrient analysis

Another way to verify proper operation is to test the nutrient enriched irrigation water delivered to the plants. In this case, water samples are taken in the field and the samples are sent to a water analysis lab to test total N (nitrate NO_3 plus ammonium NH_4), P (ortho P and total Kjeldahl P), and K. Consider the nutrient value of the irrigation water before fertilizer is added.

For this process, knowing when the nutrients are at the sampling location is important. One way to verify this is to use an electrical conductivity (EC) meter. EC is reported in millisiemens per centimeter (mS/cm) which equals 1 mmho (millimo)/cm. Be aware, the EC will vary for different fertilizer products, but the EC reading will show when the nutrients have reached the sampling location. This caution is for large systems where it takes several minutes for the nutrient solution to reach the far ends of a system. Sample in the middle part of the injection process to get an accurate result.

The calibration procedure is as follows:

1. Start the irrigation system with clear water until the pipes are filled and the system is working.

2. Start to inject nutrients (stock solution) at the desired setting.

3. At the water sampling point in the system, monitor the EC readings each 1 or 2 minutes until the reading has peaked and three steady readings have occurred.

4. Take one or several water samples at 1-minute intervals. More than one sample helps to show any variability and allows a representative average to be calculated.

5. A water sample of the clear irrigation water should be taken to establish a baseline value of the nutrients in the clear water.

6. Have a good water testing lab analyze the water samples. Test the irrigation water before fertilizer is included to have the base level.

7. Check test results against the planned values.

UNDERSTANDING PARTS PER MILLION

Nutrient concentration is often described in terms of parts per million (ppm) and this section discusses what this means (Barrett, 2000). Creating a stock solution means, for example, taking 'chemical A' that contains 4% active ingredient or nutrient and calculating the amount required to make 20 l at 1000 ppm concentration. Metric units are usually much easier to work with than American units and conversion to American units can be done later.

The concept of ppm can be difficult to understand, but it is a unit used to express a low concentration of a chemical in water. At a higher concentration, the amount of chemical is usually expressed as a percentage.

For example one ppm is one red raspberry in a pile of 999,999 black raspberries (1/1,000,000) to give a total of 1,000,000 berries. The concentration of red raspberries is 1 ppm. If there were 600 red raspberries and 999,400 black raspberries, the concentration of red raspberries would be 600 ppm. Parts per million is an expression of concentration and not an actual measure. The 600 ppm would be the same if there were 300 red raspberries and 499,700 black raspberries. One percent (1%) would be 1/100 or 10,000/1,000,000 equal to 10,000 ppm.

STOCK SOLUTIONS

A stock solution is a concentrated mixture of soluble dry fertilizer plus water or a purchased liquid fertilizer concentrate. A specified weight of the soluble dry fertilizer is mixed into solution in a known quantity of water to prepare a stock solution. The first thing that needs to be verified is the recipe for making the stock solution. N is a primary ingredient but two or more fertilizers may go into a recipe to supply a given mix of N, P, K, and minor nutrients. Manufacturers usually supply the recipe for mixing a stock solution.

When using a dry fertilizer to make a stock solution, it is best to use a Solution Grade fertilizer so there are not conditioners and other insoluble materials in the dry fertilizer that might clog the injection equipment. Also, there is a limit on the amount of dry fertilizer one can dissolve in a quantity of water to make a concentrated solution. It is best to learn the solubility (g/100 ml) of the fertilizer. Read the label on the fertilizer bag to get instructions for using and mixing that fertilizer with water.

Errors may be introduced over time in several ways to cause the recipe to become inaccurate so verify a recipe before use. Changes may include changing suppliers, product content changing, or bag weights changing. A new dosimeter may be installed with different flow rates. People may change and recipes may be understood differently. Thus, an old recipe may become a different mix by accident.

A series of mathematical calculations are required to convert grams of fertilizer or pounds of fertilizer into ppm of N, P, or K in a stock solution. Weiler and Sailus (1996) in NRAES-56 explain how to calculate in American units. The mathematical calculations are easier in metric units. When using American units of measure, one must be careful to write out all units used for conversions to be sure that the correct conversion (78 is a conversion for ounces per gallon to get ppm) is made. Units must cancel to give the final units needed. An example of calculations will be given here.

Dry (soluble) or liquid fertilizer is dissolved in water to make the stock solution. Ammonium nitrate (34-0-0) has a solubility of 18.3 g/100 ml. At 34% N, 18.3 g of the fertilizer yields 6.2 g of N. This 6.2 g of N in 100 ml is $6.2/100 \times 10,000/10,000 = 62,000/1,000,000$ or 62,000 ppm concentration. This would be a maximum concentration at 0°C (32°F). A known amount of fertilizer has been dissolved in a known amount of water, the amount of N in the solution is then calculated and then its concentration in parts per million is calculated. To apply the nutrient at some concentration, like 100 ppm or 200 ppm, the concentrate needs to be diluted by mixing (injecting) a small amount of it with a larger amount of irrigation water. This rate is typically the injection ratio of the dosimeter. In reverse, to make a stock solution to work with a fixed injection ratio, the amount of fertilizer needed to be dissolved in a known amount of water to make the concentrated ppm needs to be determined. The fertilizer bag generally gives instructions for making the stock solution (Burt *et al.*, 1995).

The math can be tricky, so the instructions on the fertilizer bag help. To apply 200 ppm N of ammonium nitrate and it is known that, for example, 64.5 milligrams (mg) N per 100 ml solution is to be applied, then for a dosimeter with an injection ratio of 1:100, the stock solution must be 100 times more concentrated than the solution being applied. If 64.5 mg N is the applied solution concentration, then the stock solution must be 100×64.5 mg N/100 ml or 6450 mg N/100 mg or 6.45 g N/100 ml. Note from above, that this is about one-third the maximum concentration that can be made.

One can work from recommendations of kilograms (kg) per hectare (ha) and knowing the amount of water applied in irrigation in liters (l) to get a recommendation of an amount of fertilizer to apply in a volume of water. Since applications of fertilizer would be over time, the weekly amount could be used. This could be used to work out a concentration of g/l.

MAKING DILUTIONS AND CONCENTRATIONS

Barrett (2000) explains dilutions and concentrations. Fertigation involves making a concentrated stock solution or making a dilution of a concentrated fertilizer solution by using the dosimeter to mix the concentrate into the irrigation water. The equation below is handy for calculating the amount of a concentrate that is needed to make the amount of dilute solution to be applied through the irrigation system.

$$V_1 = C_2 \times V_2/C_1$$

Where:

V_1 is the volume or amount of original chemical needed.
C_1 is the concentration of the original chemical
C_2 is the concentration desired for the solution being made.
V_2 is the volume of the second solution being made.

In using this equation, the concentrations for C_1 and C_2 must be expressed either in ppm or as a percentage. Likewise, the units for volume must be the same: liters, quarts, gallons, etc.

Note that the concentration of 'chemical A' is 4%, while the concentration of the final solution is given as 1000 ppm. Therefore, converting the 4% to ppm gives 40,000 ppm, from 1% = 10,000 ppm.

Next, the amount of 'chemical A' needed to make 20 l at a concentration of 1000 ppm is to be determined. Given that C_1 is 40,000 ppm; C_2 is 1000 ppm and V_2 is 20 l:

$$V_1 = C_2 \times V_2/C_1$$
$$V_1 = 1000 \times 20/40,000$$
$$V_1 = 0.5 \text{ liters}$$

When using a dosimeter, a dilution ratio of 1000 ppm/40,000 ppm would be used. This would be 1/40 setting. One unit of concentrate (at 40,000 ppm) would be applied in 40 units of irrigation water to deliver 1000 ppm.

Calculations *by weight* can also be done using the equation where weight is exchanged for volume:

$$W_1 = C_2 \times W_2/C_1$$

Taking a dry 'chemical B' that contains 20% of a particular active ingredient or element to make 50 kg with a concentration of 0.5% (C_1 is 20 percent (200,000 ppm), C_2 is 0.5% (5000 ppm) and W_2 is 50 kg.):

$$W_1 = C_2 \times W_2/C_1$$
$$W_1 = 0.5 \times 50/20$$
$$W_1 = 1.25 \text{ kg}$$

In other words, to make the lower concentration, use 1.25 kg of the active 20% material in 50 kg of final mix for a 0.5% (5000 ppm) concentration.

Manufacturers may have charts or tables for their product to assist one in making the stock solutions and for calculating dilutions.

CALCULATIONS FOR MAKING A STOCK SOLUTION

Boyle (2003) explains the calculation of a stock solution. A stock solution is a concentrated solution of fertilizer made by dissolving fertilizer in water that is intended to be injected into the irrigation water by a dosimeter. By mixing with the irrigation water, the concentrated solution is diluted to the rate at which the nutrient is to be applied. Thus, knowledge of the dilution ratio (injector ratio) of the dosimeter must be known. Also, one must know the application rate of the fertilizer in terms of ppm of N, P, or K. The rate might be 100 ppm N or 200 ppm N, for example. An injection ratio of 1:200 means that the stock solution will be made 200 times more concentrated than the rate at which the nutrient is applied. A stock solution for application of 150 ppm N by an injector with a dilution rate of 1:200 would mean the stock solution would be $200 \times 150 = 30,000$ ppm N. The goal is to determine the weight of fertilizer to dissolve in a volume of water to make the stock solution.

The basic equation for this calculation is:

$$\frac{\text{Amount of fertilizer to make}}{\text{1 volume of stock solution}} = \frac{\text{Desired concentration in parts per million} \times \text{Dilution factor}}{\text{\% of element in fertilizer} \times \text{Constant, C}}$$

where the dilution factor is the larger number of the fertigation controller injection ratio and the conversion constant C is determined by the units desired:

Unit	Conversion Constant, C
Ounces per U.S. gallon	75
Pounds per U.S. gallon	1200
Grams per liter	10

Example 1. A grower has a 1:200 fertigation controller and a fertilizer with an analysis of 15-16-17 (%N-%P$_2$O$_5$-%K$_2$O). He wants to apply a 100 ppm solution of N at each watering. How many grams of fertilizer would he have to weigh out to make 1 liter of concentrate?

List all variables:

1. Desired concentration in parts per million (ppm) at application = 100.
2. Dosimeter ratio = 1:200; dilution factor = 200
3. Fertilizer analysis = 15-16-17 (15% N).
4. Grams of fertilizer to make 1 liter of concentrate = X (unknown). Use 10 as the conversion factor C.

Set up and solve the problem using the equation:

$$X = \frac{100 \text{ ppm N} \times 200}{15\% \text{ N} (15) \times 10} = \frac{20,000}{150} = 133.33 \text{ g} = 0.13 \text{ kg} (0.13 \text{ kg/l})$$

Add 133 g/l of water to create this stock solution. Larger quantities can be made by increasing the amount of fertilizer times the number of liters of water.

Equations can be used to make various calculations based on what one knows and what one wishes to find. Also, many fertilizer companies have charts that give the amount of fertilizer to use for each of several injection ratios.

UNDERSTANDING N-P$_2$O$_5$-K$_2$O VERSUS NPK

If the calculation is for mixing actual P or K then an adjustment is needed when using the fertilizer analysis. Government regulations in the USA require content to be listed as %N: % P$_2$O$_5$: %K$_2$O. However, plants respond to N, P, K.

P$_2$O$_5$ = 43.7% P
K$_2$O = 83% K

Actual P available is only 43.7% of the amount of P$_2$O$_5$ in the fertilizer and actual K is 83% of the K$_2$O in the fertilizer. An additional calculation would be required if actual P or K is to be determined.

Recommendations are often given in terms of N, but one can consider the amount of P and K being delivered by a fertilizer product.

WATER PIPE FLOW RATES (VELOCITIES) (AND TIME FOR CHEMICAL TO REACH END OF THE IRRIGATION LINE)

Understanding water flow rates in the main line distribution system and in manifolds or submains is important when doing a calibration test, particularly when collecting samples to test the EC at the end of an irrigation line. It may take some time for water to reach a given location in the system and samples should not be drawn until the nutrient solution has reached the sample site.

The water flow rate, Q, is explained by the following relationship:

$$Q = A \times V \ (\times \text{conversion units})$$

Where

Q = water flow rate, gpm or lpm
A = cross-sectional area of pipe, sq. ft. or sq. cm.
V = water velocity in pipe, fps or cm per s.

Basically, this relationship says that for a given flow rate, if the pipe size (A) changes, the water velocity (V) must change in the opposite direction. If the pipe gets smaller, the water velocity must get larger, in order to get the same amount of water through the pipe. Also, if the flow rate reduces because water is being discharged from the manifold into many laterals and the manifold stays the same size, then the water velocity must decrease past each lateral.

The equation above and its relationships can be read on a friction loss chart that shows the water velocity and friction loss (an energy loss in pressure) for various flow rates in different size pipes. A dealer can provide this chart for all types and sizes of pipe and explain the relationships. In designing a system, maintaining the pressure to the end of the system requires careful selection of all pipe sizes. A dealer uses the friction loss chart for this purpose and it can be useful to the grower also.

The travel time for a chemical from the injection point to the final application point can be calculated through each pipe segment using the following equation and then adding all the times for all segments. For one pipe segment:

$$\text{Time} = \frac{\text{Distance}}{\text{Velocity}} = \text{Distance} \times \frac{\text{Inside area of pipe segment}}{\text{Flow rate through the segment}}$$

Where:

Time = minutes
Distance = pipe segment length (feet)
Inside area = square feet
Flow rate = gallons per minute (gpm)

BACKFLOW PREVENTION TO PROTECT DRINKING WATER

When a potable (drinking) water source is used for irrigation, particularly where fertilizers or other chemicals are added to the water, it is very important to avoid a cross-connection that might allow the chemical to flow back into the potable water source. This would be a backflow of the water to contaminate the source. A backflow could happen if there is a siphoning effect due to a low pressure on the source side of the connection. A hose connected to a household faucet with the other end in a container would be a cross-connection. To prevent backflow, an air gap device that opens at low pressure can prevent that suction.

Irrigation systems are typically under pressure, so check valves (combined with an air gap device) that close if the water tries to flow in the reverse direction are used to prevent backflow. There are different types of and different levels of protection provided by different backflow preventers.

At a minimum, an atmospheric vacuum breaker (AVB) or pressure vacuum breaker (PVB) is required. Both of these must be located at a certain point above the highest point in the irrigation system. Both only protect against back siphonage. The AVB opens to the atmosphere to create an air gap to break the siphonage, while the PVB opens to dump the flow of water that was under pressure. A double-check-valve assembly (DCA) protects against both back pressure and back siphonage. A DCA is allowed in some USA states. The reduced pressure (RP) type backflow preventer is most expensive and protects against both back pressure and back siphonage. It will dump large quantities of water when tripped (Vinchesi, 1999). An agricultural manufacturer is using an atmospheric vacuum breaker and a check valve in a low-cost backflow preventer for low hazard situations to protect irrigation wells (non-potable) and ponds.

BLACKBERRY NUTRIENT RECOMMENDATIONS

As indicated at the beginning of this Appendix, drip irrigation is not just for applying water to blackberries. Using a drip irrigation system to deliver nutrients directly to the roots can be a very economical method of improving yields while reducing material and labor costs. In general, drip irrigation is used in the humid eastern USA from mid-spring through the summer months, ending in early summer for floricane-fruiting raspberries. Primocane-fruiting blackberries will flower and fruit in mid-summer to autumn until the first frost. Applying N and P to blackberries at critical growing stages can improve primocane and floricane growth and improve the number of flower buds, firmness, and size of berries.

Fertigation is a term for a process for injecting one or more nutrients through an irrigation system using a dosimeter to deliver a quantity of nutrients only to the root zone of plants. Many studies indicate that drip irrigation, especially when water is applied only to the root zone and not the row middles, can reduce the amount of water needed per ha (acre) by up to 50%, that N can be reduced up to 50% when applied directly to 50% of root zone, especially when N is fed over time so that little is leached as compared to broadcasting dry fertilizer. Typically, drip irrigation applies water to a limited surface area above or within the root zone so the amount of applied water is much less than is applied by overhead broadcast application. Therefore, there are lower costs for water, fertilizer, labor, and farm equipment (tractor and fertilizer spreader) as compared to conventional systems. For general recommendations for fertigation for conventional production, see Chapters 10 and 11.

SUPPLEMENTAL FERTIGATION IN FIELDS

Blackberry fertilization can follow the instructions given here but nutrient testing (soil or leaf (tissue) testing) is recommended to account for different soils and plant responses. Some nutrients are incorporated into the soil before planting (see Chapters 9 and 11) and fertigation is used to supplement plant requirements over the growing season. In the planting year (year 1-first leaf), transplants are set in soil which has compost or manure and nutrients such as Ca, P, K, and minor elements that were applied in the preplant year (year 0). At planting transplants are fertilized in early spring with a transplant solution and followed by 5.5 kg to 11 kg/ha (5 lb/acre to 10 lb/acre) of actual N, 2–4 weeks after planting. Beginning 8 weeks after planting and for the next 8 weeks, 0.5 to 1.0 kg/ha (1 lb/acre to 2 lb/acre) of actual N can be injected per week for an additional 8 weeks. Take leaf samples during week 3 and adjust the rate of N. More N may be necessary in sandy soils and/or soils having less than 2% organic matter.

In the second and subsequent fruiting years, after plants are 5 cm to 10 cm (2–4 in) tall in the spring, begin injection for eight weeks at the rate of 2–3 kg/ha (4–6 lb/acre) of actual N and approximately the same amount of K, if K levels are medium to low. Potassium nitrate can be a good source of N and may be supplemented with calcium nitrate in another stock solution (caution: do not mix in same stock solution). In regions where P is low, phosphoric materials may be injected alone at the beginning of the season and then dropped out of the stock solution during the growing season. The same amount of N should be continued for 8 more weeks for a total of 16 weeks. Do not over-water or fertilize floricane blackberry plants in late summer to allow acclimation to occur normally in cold regions. For autumn-bearing primocane production, plants should receive the higher rate of N beginning three to four weeks before the first harvest (10% of total crop). Further, maintain optimal soil moisture three to four weeks before harvest and during the first six weeks of harvest for optimal fruit size. Monitor soil moisture with moisture sensors.

FERTIGATION IN HIGH TUNNELS

In high tunnel production, the soil should be well prepared before planting by incorporating compost, P, K, N (just before or after planting), and minor elements in the optimal range for raspberries based on soil tests, including pH testing. Maintaining optimal soil moisture will be most critical for the entire season for root and cane development since plants will not receive water from rainfall. Therefore, the injection of nutrients may be more beneficial on a daily basis rather than a weekly basis. The primary rationale is to supply the amount of actual N over the season. Since plants will be in a warmer temperature regime plant growth and development will be more rapid inside than outside of the high tunnel. Secondly, for autumn-fruiting primocanes that ripen in late autumn the injection of N may be continued up to and through the second harvest (20% of the entire crop). When N is supplied by injection, levels in the soil and plant should remain high for 3 weeks after the injection of N is stopped.

For organic production in high tunnels, synthetic fertilizers are not allowed. Composted chicken or dairy manure is preferred for their N levels. Grass or leaf composts have also shown some good results with grass being high in N and leaves having a high level of Ca. These are particularly good incorporated 10–15 cm (4–6 in) into the soil in the row before planting on either flat or raised beds. For poultry or dairy composted manure, the N levels should be assessed to determine the amount of manure required to supply N equal to 132–220 kg/ha (60–100 lb/acre); these are placed on top of the soil around the plants. Composted manures are preferred since they have very low weed seed germination as compared to fresh manures.

FERTIGATION FOR GREENHOUSE BLACKBERRIES IN CONTAINERS

Blackberry plants (tissue culture virus indexed transplants are best) are planted into 12 l pots (3 gal), set into soilless medium (standard peat, vermiculite, and sand in a 2-2-1 ratio) with calcium carbonate (lime), P, and N fertilizer plus micronutrients and set outdoors after the first frost for the first growing season. Plants are irrigated daily (more often when hot weather exists) and 100 ppm of N is injected once per week. In early fall, water and fertilizer are reduced significantly, but pots should not be allowed to become excessively dry. This acclimation process of 1200 hours at −2°C to +6°C (28–43°F) is necessary for flowering and fruiting. Plants are moved into a cooler for 8 weeks so that dormancy is completed by early winter. Another method is to buy dormant plants in early winter, pot them, and set pots immediately into the greenhouse (Demchak, 2008).

Summer-grown potted plants may be moved into the greenhouse and given 50 ppm N per week plus a soluble fertilizer of 5-11-26 (% of N-P-K) (separate injection tank). Once moved into the greenhouse, flowering and fruiting begins over an eight-week period. Fertigation may be regulated to lower amounts if first fruits become soft. Adding additional K (as potassium nitrate) may increase fruit color and firmness. After reaching frost-free days, the plants are then moved outdoors and the cycle is repeated. Placing moisture sensors in a few pots will be useful to maintain optimal moisture and oxygen levels in the root zone.

REFERENCES

Barrett, J. (2000) Getting comfortable with PPM, percentages and metrics. *Greenhouse Product News* (GPN), May, 20.

Boyle, T.H. (2003) *Fertilizer Calculations for Greenhouse Crops.* University of Massachusetts Extension Greenhouse Crops & Floriculture Program. University of Massachusetts, Amherst, Massachusetts. Available at: http://extension.umass.edu/floriculture/fact-sheets/fertilizer-calculations-greenhouse-crops (accessed May 16, 2017).

Burt, C., O'Connor, K. and Ruehr, T. (1995) *Fertigation.* California Polytechnic State University Irrigation Training & Research Center, San Luis Obispo, California, Ch. 9.

Demchak, K. (2008) Production methods in raspberry & blackberry production guide In: Bushway, L., Pitts, M. and Handley, D. (eds.) *NRAES-35*, pp. 28–38.

Kafkalfi, U. and Tarchitzky, J. (2011) *Fertigation – A Tool for Efficient Fertilizer and Water Management.* International Fertilizer Industry Association. International Fertilizer Industry Association, Paris, Ch. 9.

Rosen, C., Wright, J., Nennich, T. and Wildung, D. (2004) Fertility and fertigation management – high tunnel production. *Minnesota High Tunnel Production Manual for Commercial Growers.* Section 8 of M1218. University of Minnesota, St. Paul, Minnesota.

Vinchesi, B.E. (1999) Backflow basics. *Landscape & Irrigation.* February, pp. 48, 49, 52.

Weiler, T.C. and Sailus, M. (ed.) (1996) *Water and Nutrient Management for Greenhouses.* NRAES-56. Cooperative Extension. Natural Resource, Agriculture and Engineering Service, Ithaca, New York.

GLOSSARY 1: BIOLOGICAL TERMS

Richard C. Funt and Harvey K. Hall

Acclimation: the natural process of plants adapting to a climate; hardening; the phase under declining hours of sunlight and temperatures in late autumn, when shoots stop elongation and tissues acquire increased cold hardiness for winter.

Adventive: term used as to a plant that is introduced but not fully naturalized.

Airblast sprayer: a machine using a pump, specially designed nozzles, and a large fan to deliver mixtures of pesticides and/or nutrients to a plant canopy in order to reduce pest damage and to improve plant growth, fruit size, and fruit quality.

Antioxidant: a substance that opposes oxidation or inhibits reactions promoted by oxygen or peroxides; raspberries may contain antioxidants, as calcium, potassium, magnesium, or selenium and other compounds; may be expressed in total amount in an ORAC analysis.

Anthocyanin: water-soluble pigments that may appear red, purple, or blue according to the pH. They are flavonoids and are odorless and nearly flavorless.

Apomistic: a plant that reproduces or is reproduced by apomixis, which is the development of an embryo without the occurrence of fertilization; asexual seed production by reproduction budding in the ovary without fertilization.

Axil: the angle between a petiole and the stem to which it is attached.

Axillary buds: buds that develop in the leaf axils of a primocane, which may break bud and begin growing vegetatively if the growing apex of the plant has been removed.

Blackberry breeding program: program for the development of new cultivars of blackberries by the use of hybridization, growing of seedlings, and selection of elite lines for commercial development.

Botrytis cinerea: the fungus responsible for gray mold of fruit, one of the most serious diseases of raspberry throughout the world.

Brix: a measure of a fluid's total soluble solids concentration; expressed in degrees or as a percentage of weight of sugar in the solution.

Calyx: the five green petal-like projections beneath a flower; usually seen as a cap attached to the top of the fruit just beneath the stem.

Cane: a main stem of a small fruit plant; also a woody, mature shoot after leaf fall.

Cation: generally a positively charged ion.

Cation Exchange Capacity (CEC): the maximum quantity of total cations, of any class, that a soil is capable of holding, at a given pH value, for exchanging with the soil solution. CEC is used as a measure of fertility and nutrient-retention capacity.

Certified stock of raspberries: plant material certified as free of pests and disease; it is sold as Foundation grade plants to nurseries.

Chilling requirement: number of hours at or below a certain maximum temperature that is necessary to produce internal changes in a plant that result in uniform bud break and the normal sequence of growth following winter dormancy; plant needs to be exposed to temperatures between 0°C and 7°C (32°F and 45°F) once it is dormant.

Chilling unit: period of time (usually 1 hour) at or below a specific threshold temperature that has the maximum effect toward fulfilling the chilling requirement of a given plant.

Clay: smallest soil particle class size, less than 0.002 mm in diameter; so small they cannot be seen except with an electron microscope.

Climacteric fruits: fruits which have high respiration rate during fruit ripening, a process mediated by the phytohormone, ethylene, which dramatically increases up to 1000 times of the basal ethylene level as ripening occurs. Raspberries and blackberries are non-climacteric fruits.

Cloche: bell-shaped transparent plant cover used outdoors for protection against cold.

Clone (adjective: clonal): a group of genetically identical individuals, all having been vegetatively (asexually) propagated and having been ultimately derived from a single individual.

Crop coefficient (Kc): Represents the percentage of water used against a standard and relates to amount of crop and ground cover canopy on the area; it increases with increased ground cover and crop canopy. The crop water use or crop evapotranspiration (Etc) can be determined by multiplying the potential evapotransporation (ETo) by a crop coefficient (Kc) which takes into account the ground cover and crop canopy or the amount of leaf area available to transpire soil moisture. $Etc = ETo \times Kc$.

Note: It represents, while having no units, the percentage (as a decimal) of the maximum evapotranspiration possible against a standard (ETo) and depends on the amount of leaf area available. Bare soil and early spring crop leaf area (small, few leaves) means a low Kc. The value of Kc can change over a season. Irrigation may or may not be designed to be able to meet the maximum Kc for the area to meet peak water demand.

Dormancy: a plant's physiological stage of rest; usually environmentally induced as nights lengthen in fall and temperatures drop below 7°C (45°F).

Dosimeter: in agriculture, an instrument that injects fertilizer or pesticide into irrigation water at a set ratio or proportion; also referred to as a fertigation controller, injector, and proportioner.

Drip irrigation: refers to the application of water to crops in small amounts, under low pressure and applied directly to the root zone rather than broadcast water (as done by overhead irrigation); also referred to as micro-irrigation or trickle irrigation.

Drupe: fruit derived entirely from an ovary, one seeded with a pericarp, fleshy mesocarp, and stony endocarp.

Drupelet: any of the small individual drupes forming a fleshy aggregate fruit, such as a blackberry or raspberry.

Effective neutralizing value (ENV): a measurement of the effectiveness of the particular quality of lime for agricultural use. The EVN of a lime is the ability for a unit mass of lime to change soil pH.

ELISA (enzyme-linked immunosorbent assay): a common serological test for particular and unique antigens or antibodies that have been produced from viruses or microbes attacking a plant. Detection of a response to an antibody produced to a specific virus is a clear indication of the presence of a virus.

Eluate (verb: elute): 1. A liquid solution resulting from eluting; 2. A solution of solvent and dissolved matter resulting from elution; 3. Elute – to wash out a substance by the action of a solvent, as in chromatography.

ET: abbreviation for evapotranspiration.

Evapotranspiration: the amount of water lost from the soil surrounding plants plus the water lost by transpiration from plants (crop and ground cover).

Explant cultivar: technique used for the isolation of cells from a piece or pieces of tissue. Tissue harvested in this manner is called an explant.

Fertigation: the application of dissolved mineral nutrients via irrigation water.

Fertilizer: a substance, such as manure or chemical mixture (e.g. N-P-K), used to make soil more fertile and with a balanced approach improves plant growth, fruit size, and fruit quality.

Field capacity: the amount of water held in a soil after it has been saturated and then allowed to drain away the water not held by soil particles.

Flame thrower, Flamer: a torch used for spot application of heat to kill weeds in lieu of chemical herbicides, usually propane-fueled.

Floricane: The second-year canes (stems) which bear fruit and then die.

Floricane-fruiting (FF): cultivars of raspberries that fruit in the summer of their second year.

GLOBALGAP: a private-sector body that sets voluntary standards for the certification of production processes of agricultural (including aquaculture) products around the globe. The GLOBALGAP standard is primarily designed to reassure consumers about how food is produced on the farm by minimizing detrimental environmental impacts. GLOBALGAP serves as a practical manual for Good Agricultural Practice (GAP) anywhere in the world.

Hardening or **hardening off:** refers to the ability of the plant to withstand fluctuating stress such as wet to dry, or cold to heat, and/or sunlight to darkness; hardening off is a process to regain strength of plants by placing plants into an environment of sufficient heat, sunlight, and water after they have been subjected to stress of low moisture, low sunlight, and low temperature, as in shipping of plants in a box over several days.

Hazard Analysis Critical Control Points (HACCP): a management system in which food safety is addressed through the analysis and control of biological, chemical, and physical hazards from raw material production, procurement, and handling, to manufacturing, distribution, and consumption of the finished product.

Heading: the cutting (pruning) of a single main cane (stem) (central leader) for the purpose of encouraging lateral branching; also referred to as tipping or pinching on primocanes of black raspberries and blackberries or on floricanes and primocanes of red raspberries and blackberries.

Herbaceous host indicators: species and selections of herbaceous plants, including beans, cucumbers, tobacco, and amaranths that are particularly sensitive to viruses. When young plants of these herbaceous plants are inoculated with plant material of raspberry they will react in a visible and predicted manner to the presence of known viruses.

High tunnels: unheated, plastic-covered structures that provide an intermediate level of environmental protection and control compared to open field conditions or heated greenhouses. They are also known as hoophouses or plastic tunnels.

***Idaeobatus*:** subsection of the genus *Rubus* that contains raspberry species, including *Rubus idaeus*, *Rubus strigosus*, and about 200 other species that are spread around the world in the northern hemisphere and into Africa, Australia, and Oceania.

Introgressed: introduction into the gene pool of genes from one species.

Irrigation: the watering or application of water to farm land by means of ditches (furrows), trickle (drip) systems, and/or pipes, either overhead, big guns or large traveling applicators.

Leaf light interception: the amount of light received by leaves; it is measured with light meters and quantified as photosynthetic photon flux (PPF) or photosynthetically active radiation (PAR).

Liquid feed: refers to nutrients being supplied to plants in a liquid form as compared to a dry form.

Loam: soil that is composed of balanced amounts of the three particle sizes; typically this is 40% sand, 40% silt, and 20% clay.

Long day plant: a plant that flowers only after receiving illumination longer than a critical photoperiod which generally is 11–14 hours depending on the species.

Manure tea: the liquid obtained when manures are soaked in a large volume of water; used as a source of dissolved mineral nutrients.

Marketable yield: the amount of product (fruit) that is sent to market; the total yield minus damaged, bruised, or moldy berries equals marketable yield.

Mediterranean climate: climate that is characterized by warm to hot, dry summers and mild to cool, wet winters. Average temperatures are above 10°C (50°F) in their warmest months, and in the coldest months between 20°C (64°F) and −3°C (26.6°F).

Micropropagation: practice of rapidly multiplying stock plant material to produce a large number of progeny plants using tissue culture.

Mosaic virus of raspberries: disease caused by a virus complex (more than one virus involved). Viruses of the mosaic complex cause the greatest reduction in growth, vigor, fruit yield, and quality of any of the bramble viruses.

Municipal solid waste: garbage collected by community sanitation services. The biodegradable portion of this waste is sometimes composted and made available for farms and gardens.

Murashige and Skoog (MS) medium: the most commonly used plant growth medium used for plant cell culture. MS medium was invented by plant scientists Toshio Murashige and Folke K. Skoog in 1962.

Mycelium (plural: mycelia; adjective: mycelial): mass of hyphae constituting the body (thallus) of a fungus.

Nematodes: microscopic, unsegmented round-worms. They are found worldwide in water, soil, decaying organic matter, plants, and animals.

Nutraceutical: foodstuff, as a fortified food or a dietary supplement, that provides health or medical benefits in addition to its basic nutritional value.

Nutrient: a nutritive substance or ingredient; nutriment or something that promotes growth and repairs the natural wastage of organic life.

ORAC: Oxygen Radical Absorbency Capacity; a method of measuring antioxidant capacities in biological samples *in vivo.*

Permanent wilting point: the point at which remaining soil water is not available, as it is held too tightly for the plant to extract it.

Phenology: the scientific study of periodic biological phenomena, such as growth and flowering in relation to climatic conditions.

Photosynthetically active radiation (PAR): the amount of light available for photosynthesis, which is light in the 400–700 nanometer wavelength range. PAR changes seasonally and varies depending on the latitude and time of day.

Phytonutrient: nutrient derived from plant material that has been shown to be necessary for sustaining human life.

***Phytophthora*:** any of the *Phytophthora* spp. water molds that cause root rots in wet soils. The main species of concern for raspberry growers is *Phytophthora rubi* (formerly *P. fragariae* var. *rubi*).

Phytophthora root rot: a disease caused by *Phytophthora fragariae* var. *rubi*, otherwise known as *Phytophthora rubi* (Wilcox and Duncan) Man in 't Veld, that limits production and longevity in raspberries in many areas in the world where raspberries are cultivated.

Phytoplasma: specialized small bacteria that are obligate parasites that cause plant disease by infecting plant phloem tissue and are transmitted by homopteran insect vectors; also called mycoplasma-like organisms.

Plant available water (PAW): the amount of water available in a soil between field capacity and the permanent wilting point.

Potential Evapotranspiration (ETo): Potential evapotranspiration (water loss from soil surrounding plants) of a reference crop is used as a standard. The reference crop used is a cool season grass grown on deep soil and under well-watered conditions.

Precipitate: a substance separated from a solution or suspension by chemical or physical change.

Precipitation: in a mineral, the process of forming a precipitate; a deposit on the earth, as hail, mist, rain, sleet, or snow.

Pre-emergent: an herbicide that kills weed seeds as they germinate; usually applied to the soil surface just before the weeds would normally appear.

Primocane: new cane (first-year stem) on a raspberry plant that will flower and fruit the following year (floricane-fruiting type), or fruit from the tip basipetally in the current year (primocane-fruiting).

Primocane-fruiting (PF): fruiting on the first-year canes, also called autumn-fruiting, ever-bearing, fall-bearing, or remontant.

Primocane suppression: the practice of removing some or all of the primocanes during the early portion of the growing season.

Raspberry Bushy Dwarf Virus (RBDV): a pollen- and seed-borne virus that is commonly found in red and black raspberry.

Recalcitrant: difficult to manage or operate; not responsive to treatment.

Receptacle: modified or expanded portion of an axis that bears the organs of a single flower or the florets of a flower head. The receptacle of a blackberry fruit remains inside the blackberry when pulled from the stem; while the receptacle of a raspberry fruit remains on the stem and is not a portion of the fruit.

Relative humidity (RH): The amount of water vapor in the air relative to the greatest amount that could be held at a given temperature. A 50% RH at 20°C is half (50%) the amount of water possible at that temperature.

Rogue: in agriculture, this is a selective removal of canes (stems) to the ground, usually in reference to those that emerge outside of the bed or row.

Rotating Cross-Arm Trellis (RCA): is a plant trellis (support) system that allows blackberries to be positioned horizontally and close to the ground (source of heat) and covered in early winter to prevent low winter temperature injury to blackberries. Further, in early spring the RCA is moved to a near-vertical position during flowering and early fruit development to allow more efficient harvest from one side of the trellis.

Rubus allegheniensis: a widely distributed blackberry ranging from Minnesota to Nova Scotia; also occurs in the mountains of North Carolina and Tennessee; one of the most erect growing of blackberries; large, long-shaped fruits.

***Rubus coreanus*:** a black raspberry species found in Japan, Korea, and northeastern China. This species is very spiny and released cultivars have not been widely adopted as they are not very grower-friendly.

***Rubus crataegifolius*:** a widespread red raspberry species found in Japan, Korea, China, and the Eastern part of the Russian Federation; the species is cultivated in its own right and harvested extensively from the wild in northeastern China.

***Rubus idaeus*:** naturally occurring European wild species of red raspberry that is endemic or naturalized from Scotland through Europe and the Russian Federation into Asia; also known as or *Rubus idaeus* var *vulgatus* or *Rubus idaeus* ssp. *vulgatus*.

***Rubus neglectus* Peck:** naturally occurring or artificial hybrids between red and black raspberries.

***Rubus occidentalis* L.:** the black raspberry; a species native to the eastern United States (USA) and Canada; grown and cultivated in North America and in South Korea.

***Rubus strigosus*:** naturally occurring North American wild species of red raspberry; endemic or naturalized from Alaska through Canada from west to east and through the USA to northern Mexico; also known as *Rubus idaeus* var. *strigosus* or *Rubus idaeus* ssp. *strigosus*.

***Rubus ursinus* Cham. et Schlecht:** western American blackberry which is geographically different from the eastern American blackberries; the common California dewberry (an octoploid trifoliate species) of the coast and foot hills of southern Oregon to southern California.

Sand: the largest-sized soil particles, easily visible to the naked eye. These measure 0.05–2.0 mm in diameter.

Scald (in raspberries): results from overheating of fruit, even in the shade, and results in cell death; fruit turns a muddy red color and collapses.

Sclerotium (plural: sclerotia): a compact mass of hardened mycelium stored with reserved food material that in some higher fungi functions as a resistant form.

Shelterbelt: similar to a windbreak; a row or multiple rows of shrubs and trees that decrease the force of oncoming winds. Shelterbelts are also used to reduce soil erosion, provide pollinator habitat, and reduce drift of chemicals.

Silt: medium-sized soil particles, measuring 0.002–0.05 mm in diameter; roughly analogous to particles of white flour.

Soil drench: a pesticide applied in liquid form directly to the soil.

Sunburn (in raspberries): causes white drupelet disorder; the drupelet contents are overheated and the drupelet dies. This only occurs in direct sunlight. Similar symptoms may occur in some regions due to insect feeding.

Taxon (plural: taxa): in biology a taxon is a group of one or more populations of an organism or organisms seen by taxonomists to form a unit as a species or genus.

Tetraploid: an individual with four sets of chromosomes.

Tipping: practice of removing the end portion of a raspberry cane (stem) to promote lateral branching; also referred to as pinching, particularly in black raspberries. See **Heading**.

Torus: the receptacle of a flower. In the raspberry fruit, the torus remains on the plant when the fruit is removed. In blackberry, it remains with the fruit after harvest, becoming an edible portion of the fruit.

Trueness to type: a plant that has been examined and shown to be free of abnormal or atypical morphology. Trueness to type may be shown by DNA fingerprinting, where another genotype will quickly be distinguished from the desired clone.

Venturi: a flow device having a center section with a constricted/reduced diameter to create suction. The constricted section causes high velocity flow which creates suction (vacuum) on a small tube entering the constricted section. A venturi is used to inject chemicals or fertilizers into irrigation water. It is named for physicist G.B. Venturi.

Verticillium wilt: one of the most serious diseases of raspberries; caused by a soil-borne fungus and causes wilting, stunting, and eventually the death of a fruiting cane or the entire plant. The disease is usually more severe in black than in red raspberries.

Water stress: physiological stress that occurs when water is not easily accessible to roots.

White drupelet disorder: sunburn; caused by direct sunlight where the drupelet contents are overheated and the drupelet dies; similar symptoms occur in some regions due to insects, such as the stink bug, feeding on the fruit.

Wick and wiper (Wick wiper): a hand-held device used for spot treatment of herbicide, usually glyphosate.

Windbreak: a barrier composed of trees and shrubs or artificial materials to lessen the force of oncoming winds.

Yield: total amount of product (fruit) that is produced on a unit of land, quantified in kg/ha or lb/acre.

GLOSSARY 2: BUSINESS TERMS

Richard C. Funt and Harvey K. Hall

Agribusiness: business activities engaged by agriculture and farm production.

Agriculture: the production of livestock and/or crops; farming.

Agritourism: agricultural operations which bring visitors to the farm for educational or leisure activities, i.e. hayrides, corn mazes, pick-your-own berries, apple picking, horseback riding, and wine tasting.

Asset to debt ratio: total assets divided by total liabilities, shows the proportion of a firm's assets which are financed through debt.

Assets: tangible or intangible economic resources stated on the balance sheet showing the value, i.e. cash on hand or in the bank, growing crops, crops in storage, livestock, etc.

Business: the activity of providing goods and services recorded by financial transactions.

Capital: the total finance required in order to operate a farm business.

Cash: money in any form, as coins, notes, checks, money order, credit, which is immediately available.

Cash crop: crop produced to be sold for profit.

Cash flow: total amount of money in-flows and out-goings from a business.

Collective farm: a large farm run by a group of people (families) working together under government supervision for prices received and quantities produced. Generally, each family is given a house and plot of land for their own consumption.

Commercial: agricultural business with the intended purpose to make a profit.

Community (village-organized) farming: arable fields, pastures, and woods held in common, where each family has the right to produce food for the family and others in the village and were a major part of Europe and Russia in feudal times in the 19th and early 20th centuries.

Community Supported Agriculture (CSA): an important social invention in industrialized countries starting in the USA in the 1980s and defined as where non-farm public 'members' pay for food in advance and receive a share of the food produced by the farm at harvest.

Cooperative: a jointly owned enterprise that produces and distributes goods and services; is run for the benefit of its owners.

Cooperative credit: credit unions or banks formed by groups of people for the benefit of those using it; these have been shown to benefit families with low incomes on small tracts of land.

Credit: amount of money available for an individual or business account to borrow from a bank.

Currency: the type of money used in any particular country.

Debt: an amount that is owed or due.

DIRTI 5: a method used to determine fixed costs of machinery, irrigation, etc., which includes depreciation, interest, repairs (includes fuel, maintenance), taxes, and insurance (for equipment, shelter) or DIRTI.

Economics: a social science concerned chiefly with the description and analysis of production, distribution, and consumption of goods and services.

Economies of size: related to the efficiencies of larger machines (technology) and the fully employed operator/laborer; average cost of production per unit declines as the size of operation grows.

Enterprise: a business organization, company.

Family farm: refers to farms owned and operated by a family, generally passed down from generation to generation.

Farm: a tract of land devoted to agricultural purposes, such as crop production or raising livestock animals.

Farmer: a person who cultivates land to produce crops or raise livestock.

Farmers market: customers come to a centralized location where farmers offer their farm products for sale.

Farm management: the decision-making process of a farm business, in which the resources are utilized wisely or an applied science dealing with the biology, economics, technology, and the social aspects to achieve a return on investment to land, management, and labor, including farm markets and community-supported agriculture, where the public interacts and connects with the land.

Farmstead: a farm and the adjacent building area, which may include a farm house for the owner or tenant, barns, housing for livestock, with a garden and orchard near the house.

Finance: to raise or provide funds or capital.

Financial risk: any financial situation which jeopardizes variability of the farm and involves the proportion of debt and equity in the entire farm firm or current assets and current liabilities.

Garden: a plot of land where herbs, fruits, flowers, or vegetables are cultivated.

Gross farm income: the total cash income from the sale of farm products, such as lumber, livestock, crops, and/or machinery.

Human capital: skills, knowledge, and experience possessed by an individual allowing the capacity to do valuable work to produce goods and services and interact in society.

Income: financial earnings or benefit, measured in money received from capital or labor.

Increasing risk principle: the greater the percent of borrowed assets in total farm assets (leverage), the greater the risk of becoming insolvent or illiquid.

Internal rate of return: a discounted cash flow (DCF) method used to compare two or more investments with different costs and revenue streams to calculate a rate of return of cash flow across time. It takes into account the 'time value of money' economic principle.

Labor: expenditure of physical work or mental effort by an individual, which provides goods and services in an economy.

Labor efficiency: the amount of input (labor hours) required to produce an amount of output (kilograms or pounds harvested, etc.), expressed as a ratio, i.e. kilograms or pounds of blackberries harvested per hour.

Law of diminishing return: the economic principle in a production function, which states that while additional units of input increase output, it occurs at a declining incremental (marginal) rate.

Liquidity: the ability of the farm manager to meet short-term debt as it becomes due; debt-paying ability.

Long run: the time horizon beyond one production season, as for grain, but can be 5–15 years for fruit crops; where many costs are considered variable.

Land costs should be entered as a cost (either lease, rent, or purchased) or where family labor may increase or decrease.

Manual labor: physical work done with the hands; labor accomplished by people manually, such as in manual pruning, manual planting, or harvesting by hand harvest.

Market: a meeting together of people for the purpose of trade by private purpose and sale (not an auction).

Marketable: fit to be offered for sale in a market; saleable.

Marketable yield: only those items that are most saleable or berries that are not bruised, moldy, or of poor color.

Mechanical: of or relating to machinery or tools, such as mechanical pruning, mechanical harvest, or mechanical planting.

Net farm income: net cash income adjusted for inventory increases or decreases and depreciation.

Net Present Value: a discounted cash flow (DCF) method which is the sum of adjusted dollars over a period of years (cash flow); can indicate the ability to pay interest on the investment over time from future returns and also assess risk. An interest rate is selected for this procedure.

Net return: income less expenses; the remaining funds when subtracting costs from returns.

Net Worth Statement: the listing of all assets (values of land, buildings, equipment, etc.) minus all liabilities (debt such as loans, mortgage, and payments due, etc.).

Peasant: one of many persons tilling the soil, as small land owners or as laborers; generally those who are in the lower socio-economic class in Europe.

Pick-your-own (PYO): the public who are allowed to enter a commercial farm to pick (harvest) fruit or vegetable crops and pay for the amount (volume or weight) that is harvested.

Profit: the excess of returns over expenditures (costs) in a transaction or series of transactions; the excess of the selling price of goods over their cost of goods.

Profit system: free enterprise.

Rate of return on the investment: net farm income less unpaid farm labor and operator's labor divided by the total value of the farm investment.

Return: money received from an investment or yield on investment.

Revenue: income that comes back from an investment or the total income produced by a given source.

Roadside market: generally refers to a seasonal farm market located by a road or highway.

Rural life: life in the countryside (non-urban area) or small town in the countryside.

Serf: a peasant who cannot legally leave the land on which he or she works.

Short run: the time horizon of less than one year for grain crops; 1–5 years for fruit crops and where many inputs are considered fixed.

Subsistence farming: providing only the basic needs of living from food, fiber, and forestry products and/or utilizing the production and processing of food, fiber, and lumber to sustain life in rural life.

Technology: the use of science for practical purposes, such as substitution of machinery for labor; computers and programs for accounting, management, and fruit traceability, social media for advertising, genetically improved plants; chemical pest control; and the efficient monitoring and use of lime and fertilizer.

Three 'Rs' of Credit: *Risk* (financial, ability to obtain a loan based on net worth), *Returns* (net farm income), *Repayment* (of loans).

Time value of money: money is worth more today than tomorrow (future date) usually due to uncertainty, alternative uses, and/or inflation.

Traceability: a system or process that traces or tracks the fruit from production and harvest through to the customer.

Urban: in or around a town or city.

Value: a fair return or equivalent in goods, services, or money for something exchanged.

Value added: increasing value to a commodity by completing a process that creates additional value, such as producing jam, pastries, or wine from raspberries and selling the new product at a higher return.

Value chain: steps that a business (farmer) goes through from the beginning step of production through the final step of purchase by the customer, as the end user.

Wages: money paid to workers, usually on an hourly basis.

INDEX

Page numbers in *italics* refer to figures.

water quality 321–322
fertilization 146, 252
  calcium 162–163
  excessive 124, 158, *162*
  magnesium 163
  micronutrients 163–164
  nitrogen 123–124, 155–159,
    164–166
  organic fertilizers 119, 127, 164–166,
    335
  phosphorus 122–123, 159–161
  potassium 122–123, 161
  sulfur 161–162
  in tunnels 127, 334–335
field layouts 115–116, 126–127,
  316–319, *317*
fire blight (anthracnose) 203–204
fish fertilizer 164
flavor 30, 51, 78
floricane-fruiting varieties (biennial)
  248
  fertilization 158
  growth 17–18, 25–26
  leaf sampling 152
  pruning and training 127, 180–182
  season extension 24–25, 252
  temperature requirements 36, 116
  in tunnels 256
floricanes 2, 18, 26–27
  pruning 156, 171, 180
flowers 2, 27–29, *28*
  bud development 22, 25–26
  frost damage 39–42, *40*
  on RCA trellises 197, *198*
foliar nutrient sprays 159
frost damage 38–42, *40*
frozen (IQF) blackberries 289
fruit 1, 29–30, 49–57
  annual/biennial fruiting 17–18
  breeding goals 78–83, 88
  chemistry 50–55, 79
  color 51–52, 53, 79–80, 266, *267*
    reversion 80, 269–270, *269*
  drupelet set 29
  extension of fruiting season 24–25,
    252, 254
  firmness 80, 82–83, 267–268
  flavor 30, 51, 78

gray mold 208–210, 267, *267*
harvesting *see* harvesting
heat/light damage *30*, 43–45, *44*,
  80–81
insect pests 234–236, *235*, *236*, 238,
  238
netting 239
postharvest 78–79, 270–271,
  273–281, 299–300
ripening 29, 30, 81, 266–267, *267*,
  269
shape 83
size 29–30, 81–82, 306
USDA standards 270
fruiting laterals 26–27, *27*
fumigation of soil 120, 125–126, 296
fungal diseases 71, 87, 203–215, 267

generation (G) designation 98–99
genetics 77, 88–89
global production figures 308–313
good agricultural practice (GAP) 273,
  290
graft indexing 94, *94*
grapevine mealybug 218
grass, between-row 129, 133, 143, 223,
  224, *257*
gray mold 208–210, 267, *267*
green June beetle 232
greenhouses 224, 335
growth 17–30
  buds 22, 25–26
  cane architecture 2, 18–19, 83–84
  fruiting 26–30, *27*, *30*, 36
  primocanes *18*, *19*, 21–25, *22*, *23*,
    179, *179*, *180*
  roots 20–21, 132
growth media 105–106

harvesting
  fresh market fruit 268, 270, *271*,
    272–273
  hand *262*, 261–262, 272–273, 304
  mechanical 17, 85, *254*, 258–261,
    270–271, 302, 305
  U-Pick (PYO) 262–263, 304

www.ingramcontent.com/pod-product-compliance
Lightning Source LLC
Chambersburg PA
CBHW042309210326
41598CB00041B/7324